ANALYTICAL CHEMISTRY

Methods and Applications

Research Progress in Chemistry

ANALYTICAL CHEMISTRY
Methods and Applications

Harold H. Trimm, PhD, RSO

*Chairman, Chemistry Department, Broome Community College;
Adjunct Analytical Professor, Binghamton University,
Binghamton, New York, U.S.A.*

Apple Academic Press

TORONTO   NEW JERSEY

Research Progress in Chemistry Series

Analytical Chemistry: Methods and Applications

© Copyright 2011*
Apple Academic Press Inc.

First Published in the Canada, 2011
Apple Academic Press Inc.
3333 Mistwell Crescent
Oakville, ON L6L 0A2
Tel. : (888) 241-2035
Fax: (866) 222-9549
E-mail: info@appleacademicpress.com
www.appleacademicpress.com

First issued in paperback 2021

ISBN 13: 978-1-77463-215-4 (pbk)
ISBN 13: 978-1-926692-58-6 (hbk)

Harold H. Trimm, PhD, RSO

Cover Design: Psqua

Library and Archives Canada Cataloguing in Publication Data
CIP Data on file with the Library and Archives Canada

CONTENTS

ACKNOWLEDGMENTS AND HOW TO CITE

The chapters in this book were previously published in various places and in various formats. By bringing these chapters together in one place, we offer the reader a comprehensive perspective on recent investigations into this important field.

We wish to thank the authors who made their research available for this book, whether by granting permission individually or by releasing their research as open source articles or under a license that permits free use provided that attribution is made. When citing information contained within this book, please do the authors the courtesy of attributing them by name, referring back to theiroriginal articles, using the citations provided at the end of each chapter.

INTRODUCTION

Chemistry is the science that studies atoms and molecules along with their properties. All matter is composed of atoms and molecules, so chemistry is all encompassing and is referred to as the central science because all other scientific fields use its discoveries. Since the science of chemistry is so broad, it is normally broken into fields or branches of specialization. The five main branches of chemistry are analytical, inorganic, organic, physical, and biochemistry. Chemistry is an experimental science that is constantly being advanced by new discoveries. It is the intent of this collection to present the reader with a broad spectrum of articles in the various branches of chemistry that demonstrates key developments in these rapidly changing fields.

Analytical chemistry is the study of which chemicals are present and in what amount. This often involves trying to determine trace amounts of one chemical in a complicated matrix of other chemicals. Analytical chemistry is driven by new and improved instrumentation. Advances in mass spectrometry, chromatography, electrophoresis, electrochemistry, biosensors, and other instruments are allowing analytical chemists to measure smaller and smaller concentrations of chemicals in complex mixtures. This has direct application to the fields of environmental science, medicine, forensics, food safety, and engineering. The determination of STRs in DNA by capillary electrophoresis, fluorescent dyes, and lasers has led to a revolution in criminal investigation and identification in the field of forensics. Present guidelines for the concentration of chemicals in the air we breathe and water we drink are based on the detection limits that analytical chemists can achieve.

Chapters within this book ensure that the analytical chemist can stay current with the latest methods and applications in this important field.

— Harold H. Trimm, PhD, RSO

Micelle Enhanced Fluorimetric and Thin Layer Chromatography Densitometric Methods for the Determination of (±) Citalopram and its S – Enantiomer Escitalopram

Elham A. Taha, Nahla N. Salama and Shudong Wang

ABSTRACT

Two sensitive and validated methods were developed for determination of a racemic mixture citalopram and its enantiomer S-(+) escitalopram. The first method was based on direct measurement of the intrinsic fluorescence of escitalopram using sodium dodecyl sulfate as micelle enhancer. This was further applied to determine escitalopram in spiked human plasma, as well as in the presence of common and co-administerated drugs. The second method was TLC densitometric based on various chiral selectors was investigated. The optimum TLC conditions were found to be sensitive and selective for identification and quantitative determination of enantiomeric purity of escitalopram in drug substance and drug products. The method can be useful to investigate adulteration of pure isomer with the cheap racemic form.

Keywords: citalopram, escitalopram, micelle fluorimetry, spiked plasma, thin layer chromatography densitometry, chiral selectors

Introduction

Citalopram (Fig. 1) a selective serotonine re-uptake inhibitor (SSRI), has been used for the treatment of depression, social anxiety disorder, panic disorder and obsessive-compulsive disorder.[1-3] Citalopram is sold as a racemic mixture, consisting of 50% R-(–)-citalopram and 50% S-(+)-citalopram. As the S-(+) enantiomer has the desired antidepressant effect[4] it is now marketed under the generic name of escitalopram. It has been shown that the R-enantiomer present in citalopram counteracts the activity of escitalopram. Citalopram and escitalopram have demonstrated different pharmacological and clinical effects.[5]

A number of techniques including spectrophotometric,6,7 fluorimetric,8,9 electrochemical,10 chromatography11,12 and capillary electrophoresis13,14 have been developed for the determination of enantiomeric citalopram. Although several chiral methods including LC15–20 and CE21,22 are available for separation of racemic citalopram, there is no report concerning enantiomeric separation of citalopram using thin layer chromatography (TLC).

In this study we develop two simple, economic and validated methods for determination of escitalopram and enantioseparation of its racemic mixture in drug substance and drug products. The fluorimetric method was based on the fluorescence spectral behavior of escitalopram in micellar systems, such as Triton® X-100, Cetylpyridinium bromide; and sodium dodecyl sulfate (SDS). The fluorescence intensity of escitalopram and its racemic mixture citalopram was compared under the same experimental conditions. The method was successfully applied to the analysis of escitalopram in drug substances, drug product as well as spiked human plasma. Furthermore, the method was found to tolerate high concentrations of co-administrated and common drugs without potential interference. In addition to the fluorimetric method, TLC densitometry was proposed for stereoselective separation of (±) citalopram and determination of its enantiomer, escitalopram, using the different chiral selectors namely, brucine sulphate, chondroitin sulphate, heparin sodium and hydroxypropyl-β-cyclodextrin (HP-β-CD). The developed TLC method based on chiral mobile phase additives (CMPAs), tend to be cheap and feasible and offer a potential strategy for simultaneous separation of different chiral drugs on the same plate. The method was validated according to ICH guidelines and can become the method of choice compared to other techniques for fast routine enantiomeric analysis.

Experimental

Instrumentation

Waters-2525 LC system, equipped with a dual wavelength absorbance detector 2487, an auto-sampler injector and Mass Lynx v 4.1, was used. The LC column was C_{18} reverse-phase column (4.6 mm diameter × 100 mm length, 5 μm particles, phenomenex, monolithic).[1] H-NMR spectra were recorded on Bruker Avance-400 spectrometer operating at 400 MHz. FT-IR spectrometer Avatar 360 was used. Spectrofluorimetric measurement was carried out using a Shimadzu spectrofluorimeter Model RF-1501 equipped with xenon lamp and 1-cm glass cells. Excitation and emission wavelengths were set at 242 nm and 306 nm respectively. Pre-coated TLC plates (10 × 10 cm, aluminum plate coated with 0.25 mm silica gel F254) were purchased from Merck Co., Egypt. Samples were applied to the TLC plates with 25 μL Hamilton microsyringe. UV short wavelength lamp (Desaga Germany) and Shimadzu dual wavelength flying spot densitometer, Model CS-9301, PC were used. The experimental conditions of the measurements were as follows: wavelength = 240 nm, photo mode = reflection, scan mode = zigzag, and swing width = 10.

Chemicals and Reagents

All chemicals used were of analytical grade if not stated otherwise. Escitalopram oxalate (Alkan Pharm Co., Egypt) certified to contain 99.60% was used as the reference standard. Cipralex containing 10 mg escitalopram oxalate per tablet (manufactured by Lund beck Co., Denmark, Batch No 2147226, Mfg. D: 2008, Exp. D: 2011) was purchased from the market. Escitalopram oxalate was extracted and purified from cipralex tablets. Citalopram was kindly supplied by Adwia Co., Egypt. Its purity was found to be 99.80% according to official HPLC method.23 Lecital, containing 40 mg citalopram hydrobromide per tablet (manufactured by Joswe Medical Co., Jordon) was purchased from the market. Human plasma was kindly supplied from Vacsera, Egypt. Sodium dodecyl sulfate (BHD, Egypt), brucine sulphate (BHD, Egypt), chondroitin sulfate, (Eva Co., Egypt), heparin sodium and 2-hydroxypropylβ-cyclodextrin (Fluka, Egypt) were purchased. Trifluoroacetic acid (Aldrich, U.K), methanol and acetonitrile (Fisher Scientific, U.K.) were LC grade. Ultra pure water (ELGA, U.K.) was used.

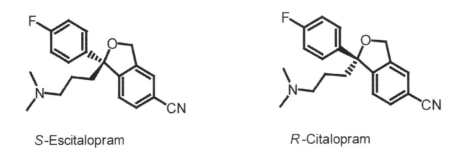

S-Escitalopram   R-Citalopram

Figure 1. Chemical structure of citalopram enantiomers.

Extraction of Escitalopram from Cipralex Tablets

Ten cipralex tablets were finely powdered and transferred to a 100 mL conical flask to which 50 mL methanol was added and stirred for 20 min. The solution was filtered through whatman No. 42 filter paper. The residue was washed several times with small volume of methanol for complete recovery. The combined extract was evaporated and the pure sample was obtained by recryslallization from methanol.

Characterization of Isolated Escitalopram

The weight of escitalopram oxalate obtained by extraction and recrystalization was the same as the labeled value. Characterization of the extracted escitalopram was done using UV, TLC, LC-MS and NMR.

Absorbance spectra were recorded in methanol and TLC separation was carried out using toluene-ethyl acetate-triethylmine (7:3.5:3 v/v/v) as the mobile phase.11 LC-MS was used for establishing the purity of escitalopram using a reverse phase C18 column at flow rate of 1 ml/min and acetonitrile: water:trifluoroacetic acid (60:40:0.01% v/v/v) mobile phase. Further characterization included FT-IR, 1H-NMR in deuterated methanol (CD3OD).

Standard Solutions

Standard stock solutions of (±) citalopram hydrobromide, and S-(+) escitalopram oxalate 0.05 mg mL^{-1} and 4 mg mL^{-1} were prepared by dissolving appropriate amounts of each in water and methanol for fluorimetric and TLC methods respectively. The stock solutions were subsequently used to prepare working standards in methanol. All solutions were stored in refrigerator at 4ºC.

Synthetic Mixtures

For TLC method, synthetic mixtures of escitalopram and citalopram in proportions ranging from 10%–90% were analyzed and the percentage recovery of escitalopram was calculated.

Method Development

Spectrofl Uorimetric Method

1 mL of aqueous stock solution equivalent to (1.25–162.5 μg mL^{-1}) of escitalopram or (1.25–125.0 μg mL^{-1}) citalopram was transferred into a series of 10 mL volumetric fl asks followed by 1 mL SDS (5 mmol aqueous solution). The volume was completed to the mark with methanol. The fluorescence was measured at 306 nm using 242 nm as excitation wavelength. To obtain the standard calibration graph, concentrations were plotted against fluorescence intensity and the linear regression equations were computed.

TLC Method

Chromatograms were developed in clean, dry, paper-lined glass chambers (12 × 24 × 24 cm) pre-equilibrated with developer for 10 minutes. The TLC plates were prepared by running the mobile phase of acetonitrile-water (17:3 v/v) containing 1 mmol chiral selector to the lost front in the usual ascending way and were air-dried. For detection and quantifi cation, 10 μL each of citalopram and escitalopram solutions within the quantification range were applied side-by-side as separate compact spots 20 mm apart and 10 mm from the bottom of the TLC plates using a 25 μL Hamilton micro syringe. The chromatograms were developed up to 8 cm in the usual ascending way using the same mobile phase omitting the chiral selectors, and were then air dried. The plates were visualized at 254 nm or by exposure to I$_2$ vapor and scanned for escitalopram at wavelength 240 nm using the instrumental parameters mentioned above.

For quantitative determination of escitalopram aliquots of standard solution (4 mg mL-1) equivalent to 0.125–4.000 mg were transferred into 10 mL volumetric flasks and made up to volume with methanol. 10 μL of each concentration was applied on the TLC plate, air dried and scanned for escitalopram at 240 nm using the instrumental parameters mentioned above. The average peak areas were calculated and plotted against concentration. The linear relationship was obtained and the regression equation was recorded.

Application to Tablets

An accurately weighed amount of powdered tablets equivalent to 100 mg of escitalopram and citalopram were dissolved in 50 mL methanol. The solutions were stirred with magnetic stirrer for 20 min. Each solution was transferred quantitatively to a 50 mL volumetric flask, diluted to the volume with methanol, and filtered. For fluorimetric analysis, a portion equivalent to 25 mg was evaporated, transferred quantitatively to a 50 mL volumetric flask and made up to volume with water. The procedure was completed as mentioned above.

Application to Spiked Human Plasma

Aliquots equivalent to 0.1–0.4 mg mL^{-1} of escitalopram were sonicated with 1 mL plasma for 5 minutes. Acetonitrile (2 mL) was added and then centrifuged for 30 minutes. One milliliter of supernatant was evaporated and the procedure was completed as described above.

Results and Discussion

In this work a simple method was used for isolation of escitalopram from its drug product rather than the published procedure.[24] The isolated escitalopram was characterized and confirmed by different analytical techniques as mentioned above.

Fluorimetric Method

Escitalopram solution was found to exhibit an intense fluorescence at a wavelength of 306 nm on excitation at 242 nm as shown in Figure 2. Different media such as water, methanol and ethanol were attempted. Maximum fluorescence intensity was obtained upon using methanol as diluting solvent, while water decreases the fl uorescence intensity.

The effect of different surfactants on the fluorescence intensity of escitalopram was studied by adding 1 mL of each surfactant to the aqueous drug solution. CPB and Triton X-100 led to peak broadening and no effect on fluorescence intensity, while SDS caused two fold increasing in the intensity. The fluorescence intensity was stable for at least two hours.

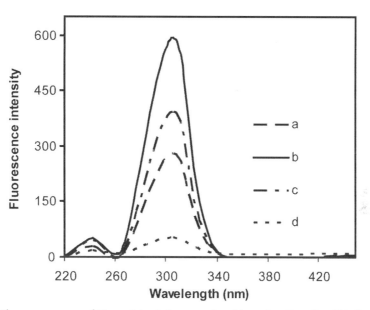

Figure 2. Fluorescence spectra of 10 µg mL^{-1} escitalopram oxalate a) in methanolic medium b) in 5 mM SDS micellar medium c) spiked plasma sample (6.6 µg mL^{-1}) in 5 mM SDS micellar medium d) blank plasma sample.

When compared to its racemic form, escitalopram showed a lower fluorescence intensity. This is concordance with published data giving the molar absorbitivity of escitalopram as 13.630 mol–1cm–1 while that of citalopram is 15.630 mol–1cm–1.8.

The quantum yield was found to be 0.026 for escitalopram and 0.030 for citalopram according to the following equation.[25]

$$Yu = Ys \cdot Fu/Fs \cdot As/Au$$

where Yu and Ys referred to fluorescence quantum yield of escitalopram and quinine sulphate, respectively; Fu and Fs represented the integral fluorescence intensity of escitalopram and quinine sulphate, respectively; Au and As referred to the absorbance of escitalopram and quinine sulphate at the excited wavelength respectively.

The method was validated by testing linearity, specificity, precision and reproducibility as presented in (Table 1).

Table 1. Validation report on the fluorimetric and TLC-densitometric methods for the determination of escitalopram in drug substance.

Parameters	Fluorimetric method	TLC-method
Linearity range	0.125–16.250 µg mL^{-1}	0.50–40.00 µg/spot
Regression equation		
Slope	55.556	264.240
SE of Slope	0.513	2.460
Intercept	25.466	195.730
SE of Intercept	5.377	50.470
Correlation Coefficient (r)	0.9998	0.9997
SE of estimation	8.438	102.360
Accuracy		
Mean[a] ± RSD%	99.71 ± 1.05	100.10 ± 1.63
Precision (Mean ± RSD%)		
Intra-day[b]	99.40 ± 0.41	100.08 ± 1.38
Inter-day[b]	99.54 ± 1.08	99.70 ± 2.09
LOD	0.017	0.014
LOQ	0.056	0.046

[a]n = 6; [b]n = 9.

Calibration plot was found to hold good over a concentration range of 0.125–16.25 µg mL-1 and 0.125–12.50 µg mL-1 for escitalopram and citalopram respectively. The procedure gave good reproducibility when applied to escitalopram drug substance over three concentration levels; 3.30, 6.60 and 13.30 µg mL-1. Whereas the specificity was proved by quantitate the studied drug in its tablet form, confirming non-interference from excipients and additives.

The results were comparable to those given by a reported method8 as revealed by statistical analysis adopting Student's t- and F-tests, where no significant difference was noticed between the two methods as presented in (Table 2). The validity of the procedure was further assured by the recovery of the standard addition. The limit of detection (LOD) and the limit of quantification (LOQ) were found to be 0.017 and 0.056 µgmL-1 respectively.

The high sensitivity attained by the fluorimetric procedure allowed its successful application to the analysis of escitalopram in spiked human plasma. To avoid variation in background fl uorescence, a simple deproteination of plasma samples with acetonitrile was performed followed by centrifugation, the clear supernatant containing escitalopram was analyzed. A calibration graph was obtained by spiking plasma samples with escitalopram in the range 3.30–16.25 µg mL-1. Linear regression analysis of the data gave the equation

$$FI = 37.27\ C + 126$$
$$r = 0.991\ (n = 6)$$

where FI is the fl uorescence intensity, C is the concentration of escitalopram in plasma in μg mL^{-1} and r is correlation coefficient. The limit of detection and quantification in spiked plasma were found to be 0.17 μg mL^{-1} and 0.56 μg mL^{-1}. The average recovery was 98.00% ± 2.80% RSD. The results from analysis of 5 spiked plasma samples are presented in Table 2.

Table 2. Analytical applications of fluorimetric method.

Forms		Fluorimetric method		Reported method[a] (8)
Drug substance				
Mean		99.71		99.28
SD		1.05		0.81
V		1.10		0.66
SE		0.40		0.31
n		7		7
t-test (2.228)[b]		0.84		
F-test (4.53)[b]		1.67		
Cipralex tablet 10 mg/tablet, escitalopram oxalate	Recovery[c] ± RSD%	Added μg mL^{-1}	Found recovery[c] ± RSD%	
		0.250	101.60 ± 0.85	97.30 ± 1.98
	98.90 ± 1.22	5.000	99.00 ± 1.06	
		12.500	100.9 ± 1.57	
Spiked plasma	Recovery[c] ± RSD%			
Drug substance	98.00 ± 2.80			—
Cipralex tablet 10 mg/tablet, escitalopram oxalate	90.00 ± 3.50			

[a]Spectrophotometric method.
[b]Theoretical values, at P = 0.05.
[c]Mean of four experiments.

Table 3. Interference study of different compounds in the determination of 5 μg mL^{-1} of escitalopram by fluorimetric method.

Drugs	Tolerated interference/analyte ratio[a] (w/w)
1.Co-administerated	
Paroxetine HCl	100
Loratadine	100
Propranolol	20
Alprazolam	100
2.Common	
Pseudoephedrine HCl	100
Aspirin	100
Ibuprofen	50

[a]Maximum ratio tested.

The interference due to co-administrated and common drugs was investigated in mixed solutions containing 5 µg mL−1 escitalopram and different concentrations of an interferant. The resulting fluorescence was compared to those obtained for escitalopram only at the same concentration. Tolerance was defined as the amount of interferant that produced an error not exceeding 5% in determination of the analyte. The method was found to be selective enough to tolerate high concentration of co-administerated and common drugs. Table 3, shows the maximum tolerable weight ratio for these drugs.

The fluorimetric method offers simplicity, rapid response and the potential to be efficient for bioavailability assessments and therapeutic drug monitoring of patients treated with citalopram or escitalopram.

TLC-densitometric Method

Compared to other chromatographic techniques, TLC is a simple, economical, rapid and flexible technique allowing sensitive parallel processing of many samples on one plate. For enantiomeric separation, chiral stationary phases and mobile phase additives can be used. Brucine, chondroitin, heparin and HP-β-CD were used as chiral selectors for enantiomeric separation of different pharmaceutical compounds using TLC, LC and CE.[26–28]

The literature reveals that chiral recognition may occur due to formation of inclusion complexes, hydrogen-bonding, π–π interaction, hydrophobic interaction or steric repulsion.[29] For instance, enantioselectivity using brucine arises due to the formation of two diastereomers through simple ionic interactions between racemate and chiral selector, e.g. (+)-citalopram/brucine and (−)-citalopram/ brucine.[30] The enantiomeric resolution by HP-β-CD may involve the inclusion of drug within the CD cavity relative to the comparability of sizes, shapes and hydrophibicities. Whereas steric effect derived from the anion of chondroitin sulphate contributes mainly to the interactions with drug enantiomer,[31] the chiral discriminating capability of heparin is believed to be due to formation of a helical structure in aqueous solution.

In this work, TLC methodology was developed for separation of (±) citalopram and determination of escitalopram using different chiral selectors, the method depending on the difference in Rf values of (R)- and (S)- forms of (±) citalopram. The experimental conditions such as mobile phase composition, chiral selector, pH and temperature were optimized to provide accurate, precise, reproducible and robust separation. Various chiral mobile phase additives including brucine sulphate, chondroitin sulphate, heparin sodium and HP-β-CD were tested. The best resolution was achieved by using 1 mM of brucine sulphate in acetonitrile:water (17:3 v/v) as a mobile phase (Table 4). The order of enantioselectivity was found to be brucine sulphate > HP-β-CD > heparin sodium > chondroitin sulphate as shown in Figure 3. The Rf values were 0.17, 0.22, 0.22, 0.29 for escitalopram and 0.71, 0.70, 0.66, 0.77 for (R)-citalopram for the four selected chiral additives respectively as shown in Figure 4. Due to it's lower health risks, HP-β-CD was chosen over brucine sulphate for the determination of escitalopram. We also investigated the effect of pH and temperature on resolution of racemic citalopram as they have been known to affect chiral recognition.26 The best conditions for discrimination of citalopram enantimers were found at pH 8.0 and 25 ± 2ºC.

Table 4. Effect of mobile phase system on enantiomeric resolution of RS-citalopram and S-escitalopram (20 µg/spot), using brucine (1 mM) as chiral selector.

Mobile phase system CH$_3$CN:CH$_3$OH:H$_2$O	R$_f$ values			R$_f$(R)/R$_f$(S)
	Pure	Racemic mixture		
	S-(+)	R-(−)	S-(+)	
16:1:3	0.22	0.80	0.22	3.64
17:0:3	0.17	0.71	0.17	4.18
20:4:0	0.26	0.75	0.26	2.88

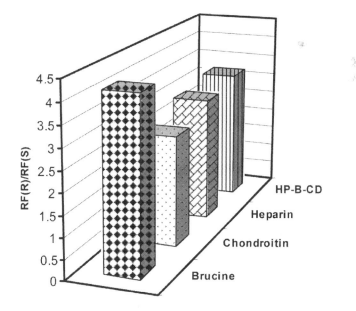

Figure 3. Effect of chiral selectors (1 mM) on enantiomeric resolution of racemic citalopram hydrobromide by silica gel TLC (20 µg/spot).

Method Validation

The method was validated according to ICH regulations by documenting its linearity, accuracy, precision, limit of detection and quantification, specificity and, robustness.[30,33] The good linearity was obtained for seven concentrations in the range of 0.5– 40 µg/spot as shown in Figure 5. The accuracy based on the mean percentage of measured concentrations (n = 6) to the actual concentration is stated in (Table 1). The precision of the method was assessed by determining RSD% values of intra-and inter-day analysis (n = 9) of escitalopram over three days. Two different analysts performed intermediate precision experiments with separate mobile phase systems according to the proposed procedure. The RSD% values of the intermediate precision are less than 2% for drug substance and drug product. The LOD and LOQ were found to be 0.014 and 0.076 µg/spot respectively (Table 1). The specificity of the method was assessed by analyzing synthetic mixtures of escitalopram and citalopram in different proportions as shown in (Table 5). The conditions for this method were modifi ed slightly with respect to mobile phase ratio, pH and temperature, the results indicating its ability to remain unaffected by small changes in the method's parameters, thus the method is considered robust.

The standard addition recoveries were carried out by adding a known amount of escitalopram to the powdered tablets at three different levels (5, 10 and 20 µg) with each level in triplicates (n = 3). The recovery percentage was evaluated by the ratio of the amount found to added. The average recovery was calculated and presented in (Table 6).

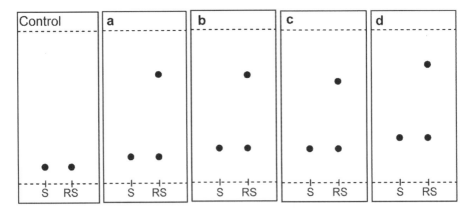

Figure 4. Thin layer chromatogram showing resolution of racemic citalopram hydrobromide, 20 µg/spot using different chiral selectors 1 mM, a) brucine sulphate, b) chondroitin sulfate c) heparin sodium, d) HP-β-CD; acetonitrile:water (17:3 v/v); solvent front 8 cm, 25 ± 2 oC, compared with control without chiral selector.

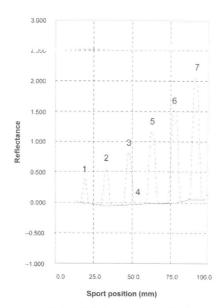

Figure 5. Densitometric scanning profile for TLC-chromatogram of different concentrations of escitalopram oxalate, (0.5–40 μg/spot) at 240 nm.

Conclusion

The present work makes use of micelle enhanced intrinsic fluorescence of escitalopram for its determination in drug substance, commercial showing satisfactory data for all validation tablets and spiked human plasma. It was found parameters tested. Both methods offer simplicity, to be selective and tolerate high concentrations rapid response and economy. of other co-administrated and common drugs. The TLC method developed was effective for enantioseparation and determination of enantiomers of citalopram. A comparative study using different chiral selectors was described with the methods being completely validated, showing satisfactory data for all validation parameters tested. Both methods offer simplicity, rapid response and economy.

Table 5. Determination of escitalopram in presence of racemic citalopram in synthetic mixtures by TLC Method.

Sample	Escitalopram: citalopram (μg: μg)	Recovery[a] (%)	RSD (%)
1	36:4	98.00	1.48
2	16:24	99.13	0.23
3	32:8	98.33	1.66
4	4:36	98.44	1.98

[a]All measurements were made in triplicates, 25 °C ± 2 °C.

Table 6. Analysis results for determination of escitalopram in cipralex tablets and application of standard addition technique by TLC method.

Sample	Recovery[a] ± RSD%	Standard addition	
		Escitalopram authentic added µg/spot	Found recovery[a] ± RSD%
1	100.52 ± 0.97	5	102.00 ± 2.09
2	101.19 ± 1.29	10	101.87 ± 1.99
3	98.45 ± 0.59	20	98.71 ± 1.07

[a] $n = 3$.

Acknowledgements

E. Taha and N. Salama would like to thank the NODCAR and School of Pharmacy, University of Nottingham, U.K., for the visiting scholarship awards.

Disclosure

The authors report no conflicts of interest.

References

1. Brunton L, Blumenthal D, Buxton I, Parker K. Goodman and Gilman Manual of Pharmacology and Therapeutics. 2007.
2. In: http://en.wikipedia.org/wiki/Citalopram.
3. Sweetman SC. (Ed.), Martindale – The complete drug reference, Pharmaceutical Press, London, U.K. 2005.
4. Sanchez C, Bogeso KP, Ebert B, Reines EH, Braestrup C. Psychopharmacology (Berl). 2004;174:163–176.
5. Sanchez C. Clin Pharmacol Toxicol. 2006;99:91–95.
6. Raza A. Chem Pharm Bull (Tokyo). 2006;54:432–434.
7. Pillai S, Singhvi I. Indian J Pharm Sci. 2006;68:682–684.
8. Serebruany V., Malinin A., Dragan V., Atar D., van Zyl L., Dragan A. Clin Chem Lab Med. 2007;45:513–520.
9. El-Sherbiny DT. J AOAC Int. 2006;89:1288–1295.
10. Nouws H., Delerue-Matos C., Barros A. Anal Lett. 2006;39:1907–1915.
11. Nilesh D., Santosh G., Shweta S., Kailash B. Chromatographia. 2008;67:487–490.
12. Greiner C., Hiemke C., Bader W., Haen E. Biomed Life Sci. 2007;848:391–394.

13. Andersen S., Halvorsen TG., Pedersen-Bjergaard S., Rasmussen KE., Tanum L., Refsum H. J Pharm Biomed Anal. 2003;33:263–273.

14. Mandrioli R., Fanali S., Pucci V., Raggi MA. Electrophoresis. 2003;24:2608–2616.

15. Buzinkaiova T., Polonsky J. Electrophoresis. 2000;21:2839–2841.

16. Haupt D. J Chromatogr B Biomed Appl. 1996;685:299–305.

17. Yang XM., Liu X., Yan YC., Xu JP. Di Yi Jun Yi Da Xue Xue Bao. 2004;24: 716–717.

18. El-Gindy A., Emara S., Mesbah MK., Hadad GM. J AOAC Int. 2006;89: 65–70.

19. Rao RN., Meena S., Nagaraju D., Rao AR. Biomed Chromatogr. 2005;19: 362–368.

20. Singh SS., Shah H., Gupta S., et al. J.Chromatogr. B Analyt Technol. Biomed Life Sci. 2004;811:209–215.

21. Berzas Nevado JJ., Guiberteau Cabanillas C., Villasenor Llerena MJ., Rodriguez Robledo V. J Chromatogr A. 2005;1072:249–257.

22. Berzas-Nevado JJ., Villasenor-Llerena MJ., Guiberteau-Cabanillas C., Rodriguez-Robledo V. Electrophoresis. 2006;27:905–917.

23. The United Stated Pharmacopoeia, The National Formulary USP 31, United States Pharmacopoeial Convection Inc. 2008. p. 1778.

24. Michael R. http://employees.csbsju.edu/mross/research/research.html. In, 2005.

25. Tang B., Wang X., Jia B., et al. Anal Lett. 2003;36(14):2985–2997.

26. Aboul-Enein HY., El-Awady MI., Heard CM. J Pharm Biomed Anal. 2003;32:1055–1059.

27. Guo Z., Wang H., Zhang Y. J Pharm Biomed Anal. 2006;41:310–314.

28. Nishi H., Kuwahara Y. J Biochem Biophys Methods. 2001;48:89–102.

29. Gubitz G., Schmid MG. Biopharm Drug Dispos. 2001;22:291–336.

30. Bhushan R., Gupta D. J Biomed Chromatogr. 2004;18:838–840.

31. Du Y., Di B., Chen J., Zheng Z. Se Pu. 2004;22:382–385.

32. ICH Q2A. validation of analytical methods: definitions and terminology, In IFPMA (ed), International Conference on Harmonisation, Geneva. 1994.

33. ICH Q2B. validation of analytical procedure: methodology, In IFPMA (ed), International Conferences on Harmonisation, Geneva. 1996.

CITATION

Originally published under the Creative Commons Attribution License or equivalent. Taha EA, Salama NN, Wang S. Micelle Enhanced Fluorimetric and Thin Layer Chromatography Densitometric Methods for the Determination of (±) Citalopram and its S – Enantiomer Escitalopram. Anal Chem Insights. 2009; 4: 1–9. http://www.ncbi. nlm.nih.gov/pmc/articles/PMC2716673/

A Multidisciplinary Investigation to Determine the Structure and Source of Dimeric Impurities in AMG 517 Drug Substance

Maria Victoria Silva Elipe, Zhixin Jessica Tan,
Michael Ronk and Tracy Bostick

ABSTRACT

In the initial scale-up batches of the experimental drug substance AMG 517, a pair of unexpected impurities was observed by HPLC. Analysis of data from initial LC-MS experiments indicated the presence of two dimer-like molecules. One impurity had an additional sulfur atom incorporated into its structure relative to the other impurity. Isolation of the impurities was performed, and further structural elucidation experiments were conducted with high-resolution LC-MS and 2D NMR. The dimeric structures were confirmed, with one of the impurities having an unexpected C-S-C linkage. Based on the synthetic route of AMG 517, it was unlikely that these impurities were generated during the last two steps of the process. Stress studies on the enriched impurities were carried out to further confirm the existence of the C-S-C linkage in the benzothiazole portion of AMG 517. Further investigation revealed that these two dimeric impurities originated from existing impurities in the AMG 517 starting material, N-acetyl benzothiazole. The characterization of these two dimeric impurities allowed for better quality control of new batches of the N-acetyl benzothiazole starting material. As a result, subsequent batches of AMG 517 contained no reportable levels of these two impurities.

Introduction

In the early stages of new drug development, understanding the impurity profiles of the drug substance is critical when interpreting the data from toxicology and clinical studies. There is a body of regulatory requirements with regard to identification and control of impurities. A commonly used framework used in the pharmaceutical industry is Q3A(R2), the International Conference on Harmonization (ICH) guidance for controlling impurities in new drug substance [1]. Although this guidance is intended only for products approaching application for final market registration, many companies consider similar elements when evaluating impurities in new chemical entities during the clinical phases of development.

Impurities in drug substances are classified into several categories in the ICH guideline Q3A(R2): organic impurities, inorganic impurities, and residual solvents. The organic impurities are of major concern for a new drug substance produced by chemical synthesis because the potential toxicity of most of these impurities is unknown. These impurities can originate from starting materials, by-products, intermediates, degradation products, reagents, ligands, and catalysts [1]. Knowledge of impurity structures can provide important insight into the chemical reactions responsible for forming these impurities as well as understanding potential degradation pathways [2]. Such information is essential in establishing critical control points in the drug substance synthetic process and eventually ensuring its overall quality and safety.

HPLC with UV detection is the most common analytical methodology used in the pharmaceutical industry to monitor organic impurities in new drug substances [2, 3]. These HPLC-UV methods are frequently used to track impurity profiles across various batches of drug substance which are often produced by different synthetic routes and at different scales. This is especially important in the earlier phases of clinical development when, due to resources and time constraints, the synthetic process is dynamic and not completely characterized, and the source/quality of starting materials has not been thoroughly evaluated [4]. When a new impurity is detected above a particular threshold (e.g., > 0.10% according to ICH Q3A(R2) for commercial products), structural elucidation of that impurity is typically initiated. LC-MS systems are widely available these days and are routinely used in initial impurity identification efforts during early drug development phases [5]. The sensitivity of LC-MS allows for the analysis of the impurities without isolation, which is often time consuming. Coupled with knowledge of the sample's history (e.g., synthetic scheme, purification process, storage conditions, stress conditions, etc.), it is often possible to propose the chemical structure of the impurity solely based on LC-MS data [6, 7]. However, the LC-MS data alone may not provide sufficient information to derive a chemical structure. In such cases, NMR spectroscopy (1D and/or 2D) is often employed to gather further structural information for impurity identification [8, 9]. Although online LC-NMR has gained some popularity in recent years [10, 11], isolation or enrichment of impurity component for offline NMR studies is still one of the most common approaches [12, 13]. Frequently, publications detailing the identification of pharmaceutical impurities will focus on the application of a selected technique and will document the proposed formation reaction for the impurity. Rarely does the publication involve multiple analytical disciplines used to both identify the impurity and to trace back to its ultimate source through a complex synthetic scheme [14].

Preparation for the first kilogram-scale production of one of Amgen's investigational anti-inflammatory drugs, AMG 517, provides a case in which a multidisciplinary investigation involving HPLC-UV, LC-MS, NMR, preparative HPLC, and forced degradation was required for unequivocal impurity identification. Two unexpected late eluting impurities were detected by an HPLC-UV method during release testing of this first scale-up batch of AMG 517 (see Figure 1). This first kilogram-scale batch of AMG 517 was manufactured with a process that was not well characterized (see Figure 2), using starting materials from outside vendors with which we had very little prior experience. Such situation is not uncommon in early clinical drug development. As the new batch was slated for use in first-in-human clinical trials, characterization of these impurities was required to enable process development which would lead to better process control. As a result of LC-MS and NMR analyses, the structures of these impurities were proposed as a simple dimer of AMG 517 and a thioether-linked dimer. A typical impurity investigation may end here with proposal of impurity structures. However, the formation of these impurities could not be explained by the synthesis scheme shown in Figure 2. A forced degradation study of the dimeric impurities provided a degree of certainty to the proposed structure for the thioether impurity. The desire to understand the origin of these impurities in the drug sustance led to investigation of starting materials using HPLC-UV and LC-MS. Information compiled from these studies allowed us to work back through the synthetic scheme for AMG 517 to determine the source of the dimeric impurities. Knowing the origin of these impurities ultimately allowed for better quality control of the AMG 517 drug substance.

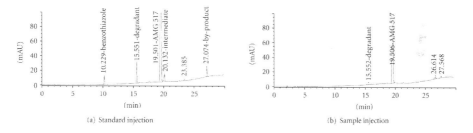

Figure 1. (a) HPLC-UV chromatogram of a standard mixture and (b) a representative AMG 517 sample containing the unknown impurities. Chromatographic conditions are in the experimental section and Table 1.

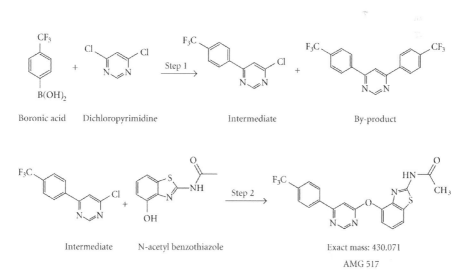

Figure 2. Synthetic pathway of AMG 517 during the early stages of clinical development.

Experimental

Materials and Reagents

HPLC grade acetonitrile (ACN, Burdick and Jackson, Muskegon, Mich, USA), trifluoroacetic acid (TFA, J. T. Baker, Phillipsburg, NJ, and Pierce, Rockford, Ill, USA), and purified water from a Milli-Q unit (Millipore, Molsheim, France) were used in the preparation of various mobile phases and diluents in chromatographic analysis. Dimethyl-d6 sulfoxide (DMSO-d6) "100%" (D, 99.96%), used for NMR analysis, was from Cambridge Isotope Laboratories (Andover, Mass, USA).

Samples of AMG 517 drug substance, N-(4-hydroxy- benzo[d]thiazol-2-yl) acetamide (N-acetyl benzothiazole), and the enriched impurity fraction were provided by the Chemical Process Research and Development Department of Amgen inc., (Thousand Oaks, Calif, USA).

HPLC

Analytical-scale chromatographic analyses were performed on an Agilent (Wilmington, Del, USA) 1100 series HPLC system. Mobile phase A was 0.1% TFA in water; mobile phase B was 0.1% TFA in ACN. A Phenomenex (Torrance, Calif, USA) Luna C18(2) HPLC column (5 μm, 150×4.6 mm, at 30°C) was used for the separation and quantitation of the AMG 517 impurities. Two different gradients with different flow rates were employed for the separation of AMG 517 and N-acetyl benzothiazole (see Table 1). A UV detection wavelength of 254 nm and an injection volume of 30 μL were used in the analysis of both compounds.

Table 1. Gradient conditions used for the HPLC-UV and LC-MS analyses of AMG 517 and N-acetyl benzothiazole.

Compound	AMG 517 (HPLC-UV and LC-MS)		N-acetyl Benzothiazole (HPLC-UV)		(LC-MS)	
	Time (min)	%B	Time (min)	%B	Time (min)	%B
	0	5	0	5	0	5
Gradient program	15	65	10	30	10	30
	20	70	15	50	15	50
	27	98	20	75	20	75
	30	98			25	95
Flow rate	1.0 mL/min		1.5 mL/min		1.0 mL/min	
Sample diluent	50% ACN/50% water		10% ACN/90% water			

LC-MS

LC-MS experiments with accurate mass determination via high resolution mass spectrometry were performed using an Agilent 1100 HPLC (configured with a diode array UV detector) interfaced with a Waters (Milford, Mass, USA) Micromass Q-Tof Ultima API quadrupole time-of-flight mass spectrometer. The mass spectrometer was configured with a lockspray electrospray ionization (ESI) source to allow for the introduction of an internal mass calibration solution, which provides for a 5 ppm mass error specification when used in conjunction with tune settings producing ~20,000 mass resolution on the instrument.

LC-MS analyses of the enriched impurities, and of their hydrolysates, were accomplished using a Phenomenex Luna C18(2) HPLC column (3 μ, 100 Å, 2.0×150 mm) and mobile phase consisting of 0.1% aqueous TFA (mobile phase A) and 0.1% TFA in ACN (mobile phase B). A flow rate of 0.2 mL/minute was used, and a column temperature of 30°C was maintained throughout each HPLC run. Gradient conditions listed in Table 1 for the HPLC-UV analysis of AMG 517 were also used for the LC-MS analysis of AMG 517 and its impurities.

LC-MS analysis of the AMG 517 starting material, N-acetyl benzothiazole, was accomplished using a Phenomenex Luna C18(2) HPLC column (3 μ, 100 Å, 4.6×150 mm). The same mobile phase system described above was used at a flow rate of 1.0 mL/minute. Column temperature was also maintained at 30°C. The gradient conditions used for the LC-MS analysis of N-acetyl benzothiazole are listed in Table 1.

NMR

Spectra were acquired at 25°C and 27°C on Bruker DPX 400 and Bruker AVANCE 600 NMR instruments (Bruker BioSpin Corporation, Billerica, Mass, USA) equipped with 5 mm and 2.5 mm multinuclear inverse z-gradient probes, respectively. 1H NMR experiments were carried out at 400.13 and 600.13 MHz, respectively, and 13C NMR experiments were carried out at 100.61 and 150.90 MHz, respectively. The data processing was performed on the spectrometers. Chemical shifts are reported in the δ scale (ppm) by assigning the residual solvent peak at 2.50 and 39.51 ppm to DMSO for 1H and 13C, respectively. The 1D 1H and 13C NMR spectra were determined using a 30° flip angle with 1 second and 2 seconds equilibrium delays, respectively. The 90° pulses used were 7.7 and 4.5 microseconds for 1H, and 22.0 and 12.50 microseconds for 13C in experiments carried out on the 400 and 600 MHz spectrometers, respectively. The 1H, 1H-2D correlation spectroscopy (COSY) spectra were acquired into 2K data points in the f2dimension with 128 increments in the f1 dimension, using a spectral width of 4789.3 Hz on the 400 MHz spectrometer and 7788.2 Hz on the 600 MHz instrument. The nuclear Overhauser effect spectroscopy (NOESY) experiments were determined with an 800 milliseconds mixing time, and with the same spectral width for f2 dimensions as COSY experiments, but with 256 increments in f1 dimension. The delays between successive pulses were 1.5 and 2 seconds for 2D COSY and NOESY, respectively. Both the 1H, 13C-2D heteronuclear single-quantum correlation (HSQC) and 1H, 13C-2D heteronuclear multiple bond correlation (HMBC) spectra were determined using gradient pulses for coherence selection. The 1H, 13C-2D heteronuclear multiple-quantum correlation (HMQC) and the HSQC spectra were determined with decoupling during acquisition. The 2D HMQC and 2D HMBC experimental data were acquired on the 400 MHz spectrometer with spectral widths of 4789.3 Hz for H1 and 20123.9 Hz for 13C, into 1K data points in the f2 dimension with 128 increments in the f1 dimension. The 2D HSQC and 2D HMBC experimental data carried out on the 600 MHz spectrometer were acquired with spectral widths of 6009.6 and 7788.2 Hz for H1 for HSQC and HMBC, respectively, and 27162.5 Hz for 13C dimension. The data were acquired into 1K and 4K data points in the f2 dimension for HSQC and HMBC, respectively, and with 256 and 128 increments in the f1 dimension for HSQC and HMBC, respectively. Delays corresponding to one bond 13C–1H coupling (ca. 145 Hz) for the low-pass filter and to two-to-three bond 13C–1H long range coupling (7.7 Hz) were used for the HMBC experiments. All 2D NMR data were processed using sine and qsine weighting window functions with some line broadening.

Results and Discussion

Impurity Profiles in Kilogram-Scale Batches of AMG 517 Drug Substance

A stability indicating HPLC-UV method was developed to separate and quantify AMG 517 along with its potential impurities and possible degradants. Figure 1(a) represents a typical separation of a standard mixture of AMG 517 in the presence of its known impurities and degradants. This method was used to analyze the first six drug substance batches of AMG 517 during release and stability testing of these lots. A pair of unexpected late-eluting unknown impurities was observed in all six batches of AMG 517. The area percent levels of impurity Unknown 1 ranged from 0.15% to 0.44%, while Unknown 2 ranged from 0.06% to 0.21%. A representative chromatogram of an AMG 517 drug substance lot containing these impurities is shown in Figure 1(b). These two impurities were not detected at a reportable level in the previous small-scale batches of AMG 517. Since these two unknown impurities eluted near the retention time of the by-product in step 1 of the AMG 517 synthetic reaction. It was concluded that these new impurities were highly hydrophobic and may have structural features similar to the by-product (see Figure 2).

Preliminary low-resolution LC-MS analysis on the drug substance provided molecular mass and tandem mass spectrometry (MS/MS) fragment ion information for these two impurities (data not shown). The observed mass for the protonated Unknown 1 and Unknown 2 was 859 Da and 891 Da, respectively. Since the exact mass for AMG 517 is 430.0711 Da, an observed mass of 859 Da for Unknown 1 suggested that it could be some sort of dimeric structure related to AMG 517. MS/MS data also suggested dimeric structures for both impurities. Fragment ions that corresponded to the neutral loss of multiple acetyl and hydrofluoric functional groups were observed in MS/MS experiments performed on the protonated ions of both Unknown 1 and Unknown 2. The mass difference between the two unknowns was 32 Da, which could be attributed to either one additional sulfur atom or two additional oxygen atoms in Unknown 2 relative to Unknown 1. However, the preliminary LC-MS analysis alone could not conclusively identify the structures of these impurities due to the possible existence of multiple isomeric structures consistent with the mass data. To aid in the structural elucidation efforts, an enriched fraction of these two impurities was isolated via preparative-scale HPLC. The isolated fraction contained about 35% of Unknown 1 and 62% of Unknown 2 based on UV detection at 254 nm. LC-MS and NMR experiments were performed to characterize this enriched fraction.

Accurate Mass Determination for Unknowns 1 and 2 in the Enriched Fraction

An accurate mass of 859.1342 Da was determined for Unknown 1 in the enriched fraction. Elemental composition analysis was performed for this protonated mass. Instrument performance, the synthetic pathway for AMG 517, and information gained from the preliminary LC-MS analysis of the impurities were taken into account in setting parameters for this analysis. Based upon the performance of the mass spectrometer, the error between the observed and calculated masses was limited to 5 ppm or less. MS/MS analysis indicated the presence of two trifluoromethyl groups, so the number of atoms of F required was set to six. MS/MS analysis also indicated the presence of two acetyl groups, so the minimum number of atoms of O required was set to two. Consideration of the synthetic pathway for AMG 517 suggested that a molecule containing less than four atoms of N was unlikely. All elemental composition analyses performed as part of this investigation utilized a similar strategy to logically identify the most likely elemental formula for an observed mass.

The elemental composition analysis for the observed mass of Unknown 1 determined that the elemental formula $C_{40}H_{24}N_8O_4F_6S_2$ was the best fit for the impurity. This elemental composition was consistent with a dimer of AMG 517 minus two hydrogens (elemental formula $C_{20}H_{13}N_4O_2F_3S$). The mass error between the observed mass for Unknown 1 and the calculated mass for a dimer of AMG 517 was 0.3 ppm.

An accurate mass of 891.1058 Da was determined for Unknown 2 in the enriched fraction. Elemental composition analysis using this protonated mass determined that the elemental formula $C_{40}H_{24}N_8O_4F_6S_3$ was the best fit for the impurity. This elemental composition was consistent with a dimer of AMG 517 with the addition of a sulfur atom [dimer+S]. The mass error between the observed mass for Unknown 2 and the calculated mass for [dimer+S] was 0.8 ppm.

Another possible elemental formula for Unknown 2 is $C_{40}H_{24}N_8O_6F_6S_2$ which corresponded to an AMG 517 dimer with two additional oxygen atoms [dimer+2O]. The mass error between the observed mass for Unknown 2 and the calculated mass for the [dimer+2O] was 20.7 ppm. Based on the accurate mass data, it was concluded that [dimer+S] was a more likely structure for Unknown 2.

The MS data was consistent with dimeric structures for both Unknown 1 and Unknown 2 but provided no definitive structural linkage information. The structure of AMG 517 itself and the synthetic scheme shown in Figure 2 did not provide any obvious possible point of linkage. Therefore, NMR analyses were performed on the enriched fraction to help elucidating the structures of these impurities.

NMR

AMG 517 and its enriched impurity fraction containing Unknowns 1 and 2 were first analyzed by ^1H and ^{13}C NMR to further investigate the connectivity. Proton assignments were made based on chemical shifts, proton-proton coupling constants, and COSY and NOESY spectra (see Tables 2 and 3). Carbon assignments were based on chemical shifts, carbon-fluorine coupling constants, and HMQC, HSQC, and HMBC spectra (see Tables 2 and 3). All assignments referring to the structures of AMG 517 and impurities are depicted in these two tables.

Table 2. 1H and 13C chemical shifts (δ/ppm) of AMG 517 standard in DMSO-d6 (400 MHz).

Position	^{1}H (δ/ppm, J/Hz)[a]	^{13}C (δ/ppm)[a]
1		130.9 (q, J_{C-F} = 31.9 Hz)
2, 6	7.92 (d, 2H, J = 8.2 Hz)[b]	125.9 (q, J_{C-F} = 3.7 Hz)
3, 5	8.44 (d, 2H, J = 8.2 Hz)	128.0
4		139.7
7		163.4
9	8.79 (s, 1H)	158.6[e]
11		170.3[d]
12	7.97 (s, 1H)	104.2
13		124.0 (q, J_{C-F} = 272.2 Hz)
15		143.5
16	7.35 (m, 1H)[c]	119.1
17	7.39 (t, 1H, J = 7.7 Hz)[c]	124.2
18	7.93 (m, 1H)[b]	119.6
19		133.6
20		141.4
22		158.4[e]
24	12.42 (s, 1H)	
25		169.5[d]
26	2.13 (s, 3H)	22.6

[a]Signal splitting patterns: s = singlet, d = doublet, t = triplet, q = quartet, m = multiplet; [b], [c]overlapping; [d], [e]interchangeable assignment.

Table 3. ^{1}H and ^{13}C chemical shifts (δ/ppm) of the enriched impurity fraction in DMSO-d6.

Position	^{1}H (δ/ppm, J/Hz)[a]	^{13}C (δ/ppm)[a]
1		130.9 (q, J_{C-F} = 31.9 Hz)
2, 6	7.92 (d, 2H, J = 8.2 Hz)[b]	125.9 (q, J_{C-F} = 3.7 Hz)
3, 5	8.44 (d, 2H, J = 8.2 Hz)	128.0
4		139.7
7		163.4
9	8.79 (s, 1H)	158.6[e]
11		170.3[d]
12	7.97 (s, 1H)	104.2
13		124.0 (q, J_{C-F} = 272.2 Hz)
15		143.5
16	7.35 (m, 1H)[c]	119.1
17	7.39 (t, 1H, J = 7.7 Hz)[c]	124.2
18	7.93 (m, 1H)[b]	119.6
19		133.6
20		141.4
22		158.4[e]
24	12.42 (s, 1H)	
25		169.5[d]
26	2.13 (s, 3H)	22.6

[a]Signal splitting patterns: s = singlet, d = doublet, t = triplet, q = quartet, m = multiplet; [b], [c]overlapping; [d], [e]interchangeable assignment.

NMR analysis was also conducted on AMG 517 for comparison (see Table 2). The 1H NMR spectrum showed the presence of all the protons of the molecule including the exchangeable NH proton. The 1H NMR spectrum showed the presence of three aromatic systems, an AA'BB' spin system (δ 7.92 and 8.44 ppm) for a p-disubstituted benzene ring, two singlets (δ 7.97 and 8.79 ppm) for another aromatic ring, and an ABX spin system (δ 7.35, 7.39, and 7.93 ppm) for a 1,2,3-trisubstituted benzene ring. The downfield chemical shift of the singlet at 8.79 ppm together with the singlet at 7.97 ppm suggested a 4,6-disubstituted pyrimidine as one of the aromatic rings in the molecule. The C13 NMR spectrum showed the presence of all the carbons of the molecule. Three of these carbons were coupled to 19F; C-13 as a quartet through one C–F bond (δ 124.0, 1J [13C, 19F] = 272.2 Hz), C-1 as a quartet through one C–C and one C–F bonds (δ 130.9, 2J [13C, 19F] = 31.9 Hz), and C-2, 6 as a quartet through two C–C and one C–F bonds (δ 125.9, 3J [13C, 19F] = 3.7 Hz) (see Table 2).

The 1H NMR spectrum of the enriched fraction containing the impurities indicated that the sample was a mixture of two components structurally related to AMG 517, present at a ratio of 1:1.94 based on the areas of their related aromatic signals. Based on the HPLC-UV data from the enriched fraction, the major component present corresponded to Unknown 2, and the minor component to Unknown 1. The 1H NMR spectrum of the impurities contained signals corresponding to the same substitution patterns observed for AMG 517 (see Figure 3). 1H NMR and 1H, 1H-2D NOESY spectra indicated the presence of a p-disubstituted benzene ring, a 4,6-disubstituted pyrimidine, a 2,4,7-trisubstituted benzothiazole ring, and an N-acetyl group. The only difference between AMG 517 and these two related compounds is the substitution pattern of the benzothiazole. The 1H NMR spectrum showed more distinct chemical shift differences for the protons H-16 and H-17 from these two AMG 517-related compounds (see Figure 3 and Table 4). The signals from Unknown 1 were shifted downfield compared to Unknown 2. The elemental molecular formulae for Unknowns 1 and 2 were indicative of dimer structures. Only one set of resonances was observed for each of the two unknowns. This indicated that the unknowns were symmetrical dimers. The monomers were connected through carbon C-18 based on the presence of an AB system, their chemical shifts, and the coupling constants for the benzothiazole ring. The 1H, C13-2D HSQC spectrum supported the 1H NMR data showing only two aromatic C–H (C-16 and 17) on the benzothiazole ring of the impurities. The absence of a C–H signal for C-18, as was observed in AMG 517, was noted in the NMR spectra in both of the unknowns (see Figure 4). C13 NMR, 1H, C13-2D HSQC, and 1H, C13-2D HMBC spectra of the impurities showed more distinct chemical shift differences for the carbons C-16, C-17, C-18, and C-19. This indicated that the difference between these two impurities was in the linkage through C-18, either directly or through a heteroatom (see Table 4). The possibility of having a sulfur atom connecting the two AMG 517 monomers for Unknown 2 was considered very plausible based on the MS data and the chemical shift data (see Table 4).

Table 4. Partial ^{1}H and ^{13}C chemical shifts (δ /ppm) of the benzothiazole ring for AMG 517 and the enriched impurity fraction in DMSO-d6.

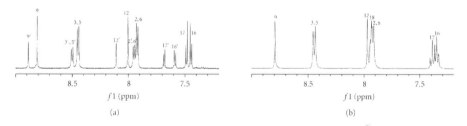

AMG 517 R_1= H
Unknown 2 R_1 = S-AMG 517 monomer
Unknown 1 R_1 = AMG 517 monomer

	^{1}H (δ/ppm, J/Hz)			^{13}C (δ/ppm)		
Position	AMG 517[a]	Unknown 2[b]	Unknown 1[b]	AMG 517[a]	Unknown 2[b]	Unknown 1[b]
16	7.35 (m, 1H)[c]	7.45 (d, 1H, J = 8.2 Hz)	7.59 (d, 1H, J = 8.0 Hz)	119.1	120.4	119.9
17	7.39 (t, 1H, J = 7.7 Hz)[c]	7.49 (d, 1H, J = 8.2 Hz)	7.68 (d, 1H, J = 8.0 Hz)	124.2	128.1	123.9
18	7.93 (m, 1H)			119.6	122.6	131.2
19				133.6	136.6	132.8

[a]Data from 400 MHz NMR instrument; [b]data from 600 MHz instrument; [c]overlapping signals.

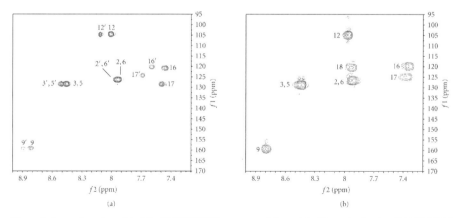

Figure 3. Aromatic region of the ^{1}H NMR spectra of the enriched impurity fraction ((a) 600 MHz) and AMG 517 ((b) 400 MHz) in DMSO-d6. In (a), numbers designated as prime (e.g., 3′) represent signals of Unknown 1, with all others representing signals of Unknown 2.

Figure 4. Aromatic region of the ^{1}H, ^{13}C-2D HSQC spectrum of the enriched impurity fraction ((a) 600 MHz) and the H1, ^{13}C-2D HMQC spectrum of AMG 517 ((b) 400 MHz) in DMSO-d6. In (a), numbers designated as prime (e.g., 3′) represent signals of Unknown 1, with all others representing signals of Unknown 2.

LC-MS Analysis of the Hydrolysate of the Enriched Impurities

The MS data for the two impurities strongly supported a thioether-linked dimer of AMG 517 as the structure for Unknown 2. The ^{1}H and ^{13}C NMR data provided indirect evidence of such thioether linkage but could not afford direct measurement of the heteroatom. However, the formation of this impurity in the synthesis of AMG 517 (see Figure 2) did not seem as plausible as the oxidation of a heteroatom from a reaction mechanistic standpoint. There was a significant difference between the calculated mass values for the two potential structures for Unknown 2, however, the relatively high mass of the impurity resulted in a large number of potential elemental formulae. To simplify the elemental composition analysis, a chemical degradation experiment was performed. The enriched fraction was treated with 0.5 equivalent of aqueous HCl in DMSO-d6 and heated overnight at 70°C. This experiment furnished low-molecular-weight fragments of the impurity that could not be generated via MS/MS. These low-mass fragments resulted in a small number of potential elemental formulae for each observed mass.

Multiple hydrolysis fragments were observed in LC-MS after forced degradation of the enriched fraction with hydrochloric acid (see Figure 5). Accurate mass data collected in the LC-MS analysis of the acid treated enriched impurity fraction was used to identify peaks corresponding to the expected hydrolysis fragments (see Figure 5). The scheme in Figure 6 shows the expected acid hydrolysis fragments from Unknown 2, with Unknown 2 and its fragments presented using both the thioether and bis-sulfoxide structures being considered for the impurity. A number of deacetylation products were also observed. This LC-MS analysis demonstrated that all of the expected fragments for Unknown 2 (and some for Unknown 1) were formed during the forced degradation.

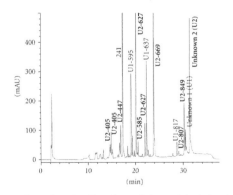

Figure 5. UV chromatogram from the LC-MS analysis of the acid hydrolyzed impurities. Labeled peaks correspond to hydrolysis fragments of Unknown 1 (U1) and Unknown 2 (U2).

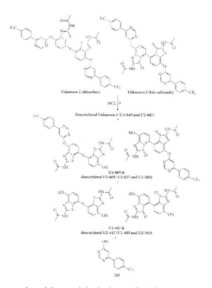

Figure 6. Potential fragments produced by acid hydrolysis of Unknown 2. Structures on the left represent fragments expected to be generated from the thioether, those on the right from the bis-sulfoxide. Fragment 241 would be common to both structures.

Table 5 shows the accurate mass assignments for the acid hydrolysis products of Unknown 2 as well as the calculated exact mass for each product that was expected to arise from both the proposed thioether and bis-sulfoxide structures. The mass error (observed mass versus calculated mass of the hydrolysis fragments) is shown for each proposed structure for Unknown 2. The mass error range for hydrolysis products arising from the thioether structure was 0 to 3.5 ppm; the mass error range for the corresponding bis-sulfoxide was 17.5 to 40.7 ppm. Thus, the accurate mass data collected for the acid hydrolysis fragments allowed for elimination of the bis-sulfoxide as a potential structure for Unknown 2.

Table 5. Mass error analysis of the observed accurate mass for fragments generated by the acid hydrolysis of the enriched impurity fraction. The analysis is conducted for both the thioether and bis-sulfoxide structures proposed for Unknown 2. (ND: not detected; N/A: not applicable.)

	Obs. mass $(M + H)^+$ (Da)	Calc. mass thioether $(M + H)^+$ (Da)	Calc. mass bis-sulfoxide $(M + H)^+$ (Da)	Mass error thioether (ppm)	Mass error bis-sulfoxide (ppm)
Unknown 2	891.1081	891.1060	891.1237	2.4	17.5
Mono-deacetyl	849.0967	849.0954	849.1132	1.5	19.4
Bis-deacetyl	807.0874	807.0848	807.1026	3.2	18.8
U2-669	669.0674	669.0655	669.0832	2.8	23.6
Mono-deacetyl	627.0552	627.0549	627.0727	0.5	27.9
	627.0558			1.4	27.0
Bis-deacetyl	585.0458	585.0443	585.0621	2.6	27.9
U2-447	447.0250	447.0250	447.0428	0.0	39.8
Mono-deacetyl	405.0158	405.0144	405.0322	3.5	40.5
	405.0157			3.2	40.7
Bis-deacetyl	ND	363.0039	363.0216	N/A	N/A
241	241.0579	241.0583	241.0583	1.7	1.7

These mass error results strongly supported an elemental formula of $C_{40}H_{24}F_6N_8O_4S_3$ (thioether) for Unknown 2, and essentially ruled out an elemental formula of $C_{40}H_{24}F_6N_8O_6S_2$ (bis-sulfoxide) for the impurity.

LC-MS Analysis of N-Acetyl Benzothiazole Starting Material

Although the MS and NMR data provided great confidence in the proposed dimeric structures for these two late-eluting impurities, the chemical reactions described in Figure 2 were not likely to generate such impurities. Since the dimeric linkages are in the benzothiazole portion of AMG 517, it was possible that these two impurities were originated from existing impurities in the AMG 517 starting material, N-acetyl benzothiazole, which was prepared via multistep synthesis from 2-methoxybenzenamine by a contract manufacturer. To determine if N-acetyl benzothiazole was a potential source for generating Unknowns 1 and 2, additional experiments were performed to evaluate the impurity profiles of N-acetyl benzothiazole.

A different HPLC method was developed for the analysis of N-acetyl benzothiazole (see Table 1). Although the supplier's Certificate of Analysis indicated an HPLC purity of >99% area for various batches of N-acetyl benzothiazole, retrospective analysis by Amgen's HPLC method resulted in purities ranging from 96.2 to 98.2% area. LC-MS analysis was performed on lot A, the starting material used in the production of the six kilogram-scale AMG 517 batches. Analysis of this lot indicated that there were multiple impurities, some of which had the potential to generate Unknowns 1 and 2 (see Figure 7). These impurities were designated by their nominal mass values as determined by the LC-MS analysis (e.g., MS447 corresponds to a compound with an observed mass of 447 Da). Table 6 provides a summary of the proposed structures for the observed impurities of N-acetyl benzothiazole.

An accurate mass of 447.0247 Da was determined for the protonated ion of impurity MS447 in N-acetyl benzothiazole. Elemental composition analysis using the observed mass determined that the elemental formula $C_{18}H_{14}N_4O_4S_3$ was the best fit for this impurity. This elemental formula, along with MS/MS analysis of MS447 (see Figure 8), supported a thioether linked dimer of benzothiazole as the structure for MS447 (see Table 6). This symmetrical thioether compound could participate in the same reaction as AMG 517 step 2 to generate Unknown 2 (see Figure 9).

Table 6. Summary of proposed structures of impurities observed in the LC-MS analysis of N-acetyl benzothiazole.

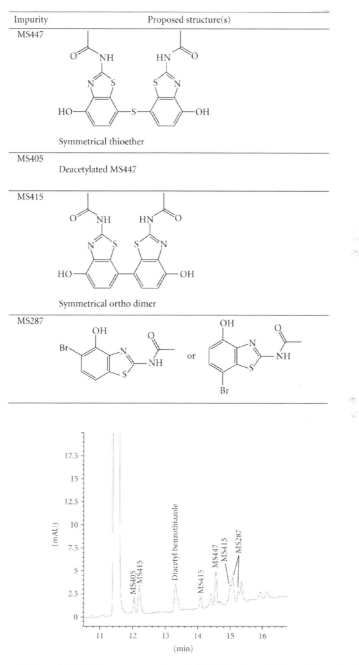

Impurity	Proposed structure(s)
MS447	Symmetrical thioether
MS405	Deacetylated MS447
MS415	Symmetrical ortho dimer
MS287	

Figure 7. Expanded view of the UV chromatogram from the LC-MS analysis of N-acetyl benzothiazole lot A (ca. 0.1 mg/mL in 10% ACN, 3 μg injected on column).

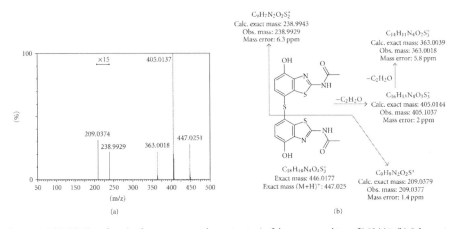

Figure 8. (a) MS/MS analysis (with accurate mass determination) of the protonated ion of MS447. (b) Schematic of the MS/MS fragmentation interpretation of MS447.

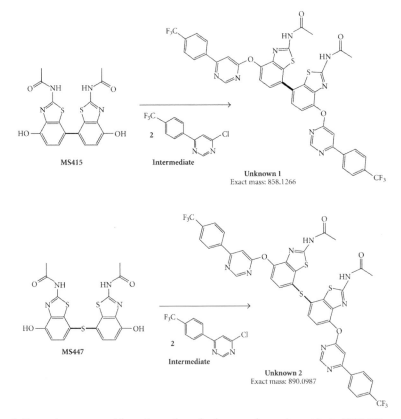

Figure 9. Proposed structures and formation pathway for the two unknown impurities in AMG 517.

An accurate mass of 405.0144 Da was determined for the protonated ion of impurity MS405. Elemental composition analysis using the observed mass determined that the elemental formula $C_{16}H_{12}N_4O_3S_3$ was the best fit for the impurity. MS405 was proposed to be the mono-deacetylated form of MS447.

Accurate mass determination for each of the three peaks designated MS415 led to the assignment of exact mass values that were in close agreement with each other (415.0522 Da, 415.0540 Da, and 415.0536 Da, in order of elution), and elemental composition analysis using these observed mass values points to the same elemental composition ($C_{18}H_{14}N_4O_4S_2$) as the most likely formula for each. These three MS415 impurities in N-acetyl benzothiazole could be positional isomers to each other. One of these isomers, a symmetrical ortho dimer (see Table 6), was a plausible precursor for the proposed structure of Unknown 1 (see Figure 9).

Accurate mass determination for each of the two peaks designated MS287 led to the assignment of exact mass values that are in close agreement with each other (286.9486 Da and 286.9475 Da, in order of elution), and elemental compositio analysis using these observed mass values points to the same elemental composition ($C_9H_7N_2O_2SBr$) as the most likely formula for each. Structures consistent with these elemental formulae are shown in Table 6. The presence of these molecules in the benzothiazole synthetic process could lead to the formation of impurities MS415 and MS447.

Conclusion

An extensive investigation successfully utilized multiple analytical disciplines to elucidate structures for two complex impurities in AMG 517 drug substance and to trace the source of the impurities to a starting material used in the manufacture of AMG 517.

The structures of two unknown impurities in AMG 517 drug substance were identified through extensive HPLC, LC-MS, high resolution MS, MS/MS, and 1D and 2D NMR studies. The existence of an unexpected C-S-C linkage in one of the impurities was confirmed. Further investigation revealed that these impurities originated from existing impurities in the N-acetyl benzothiazole starting material used in AMG 517 synthesis. This information was shared with the supplier of this starting material, and the process for N-acetyl benzothiazole preparation was re-evaluated. Better synthetic process controls and tighter specifications were established resulting in higher quality N-acetyl benzothiazole batches. These two dimeric impurities were not observed in subsequent larger-scale AMG 517 production runs.

Acknowledgements

The authors would like to thank Lauren Krance and Carlos Orihuela for their contributions to this investigation.

References

1. "Q3A(R2): Impurities in New Drug Substances," ICH Harmonized Tripartite Guideline.

2. F. Qiu and D. L. Norwood, "Identification of pharmaceutical impurities," Journal of Liquid Chromatography and Related Technologies, vol. 30, no. 5–7, pp. 877–935, 2007.

3. R. Nageswara Rao and V. Nagaraju, "An overview of the recent trends in development of HPLC methods for determination of impurities in drugs," Journal of Pharmaceutical and Biomedical Analysis, vol. 33, no. 3, pp. 335–377, 2003.

4. A. Abdel-Magid and S. Caron, Fundamentals of Early Clinical Drug Development: From Synthesis Design to Formulation, John Wiley & Sons, New York, NY, USA, 2006.

5. M. S. Lee, LC/MS Applications in Drug Development, John Wiley & Sons, New York, NY, USA, 2002.

6. J. Ermer, "The use of hyphenated LC-MS technique for characterisation of impurity profiles during drug development," Journal of Pharmaceutical and Biomedical Analysis, vol. 18, no. 4-5, pp. 707–714, 1998.

7. A. Kocijan, R. Grahek, and L. Zupančič-Kralj, "Identification of an impurity in pravastatin by application of collision-activated decomposition mass spectra," Acta Chimica Slovenica, vol. 53, no. 4, pp. 464–468, 2006.

8. N. Lindegårdh, F. Giorgi, B. Galletti, et al., "Identification of an isomer impurity in piperaquine drug substance," Journal of Chromatography A, vol. 1135, no. 2, pp. 166–169, 2006.

9. Ch. Bharathi, Ch. S. Prasad, D. V. Bharathi, et al., "Structural identification and characterization of impurities in ceftizoxime sodium," Journal of Pharmaceutical and Biomedical Analysis, vol. 43, no. 2, pp. 733–740, 2007.

10. B. C. M. Potts, K. F. Albizati, M. O'Neil Johnson, and J. P. James, "Application of LC-NMR to the identification of bulk drug impurities in GART inhibitor AG2034," Magnetic Resonance in Chemistry, vol. 37, no. 6, pp. 393–400, 1999.

11. G. J. Sharman and I. C. Jones, "Critical investigation of coupled liquid chromatography-NMR spectroscopy in pharmaceutical impurity identification," Magnetic Resonance in Chemistry, vol. 41, no. 6, pp. 448–454, 2003.

12. M. V. Silva Elipe, "Advantages and disadvantages of nuclear magnetic resonance spectroscopy as a hyphenated technique," Analytica Chimica Acta, vol. 497, no. 1-2, pp. 1–25, 2003.

13. C. Szántay, Jr., Z. Béni, G. Balogh, and T. Gáti, "The changing role of NMR spectroscopy in off-line impurity identification: a conceptual view," Trends in Analytical Chemistry, vol. 25, no. 8, pp. 806–820, 2006.

14. K. M. Alsante, P. Boutros, M. A. Couturier, et al., "Pharmaceutical impurity identification: a case study using a multidisciplinary approach," Journal of Pharmaceutical Sciences, vol. 93, no. 9, pp. 2296–2309, 2004.

CITATION

Originally published under the Creative Commons Attribution License or equivalent. Elipe MVS, Tan ZJ, Ronk M, Bostick T. A Multidisciplinary Investigation to Determine the Structure and Source of Dimeric Impurities in AMG 517 Drug Substance. International Journal of Analytical Chemistry, Volume 2009 (2009), Article ID 768743. http://dx.doi.org/10.1155/2009/768743.

Selective Spectrophotometric and Spectrofluorometric Methods for the Determination of Amantadine Hydrochloride in Capsules and Plasma via Derivatization with 1,2-Naphthoquinone-4-sulphonate

Ashraf M. Mahmoud, Nasr Y. Khalil, Ibrahim A. Darwish
and Tarek Aboul-Fadl

ABSTRACT

New selective and sensitive spectrophotometric and spectrofluorometric methods have been developed and validated for the determination of amantadine *hydrochloride (AMD) in capsules and plasma. The methods were based on the condensation of AMD with 1,2-naphthoquinone-4-sulphonate (NQS) in an alkaline medium to form an orange-colored product. The spectrophotometric method involved the measurement of the colored product at 460 nm. The spectrofluorometric method involved the reduction of the product with potassium borohydride, and the subsequent measurement of the formed fluorescent reduced AMD-NQS product at 382 nm after excitation at 293 nm. The variables that affected the reaction were carefully studied and optimized. Under the optimum conditions, linear relationships with good correlation coefficients (0.9972–0.9974) and low LOD (1.39 and 0.013 μg mL^{-1}) were obtained in the ranges of 5–80 and 0.05–10 μg mL^{-1} for the spectrophotometric and spectrofluorometric methods, respectively. The precisions of the methods were satisfactory; RSD ≤ 2.04%. Both methods were successfully applied to the determination of AMD in capsules. As its higher sensitivity, the spectrofluorometric method was applied to the determination of AMD in plasma; the recovery was 96.3–101.2 ± 0.57–4.2%. The results obtained by the proposed methods were comparable with those obtained by the official method.*

Introduction

Amantadine hydrochloride (AMD), Scheme 1, is an antiviral agent used against infection with influenza type A virus and to ameliorate symptoms when administered during the early stages of infection as well as in the management of herpes zoster [1]. It has mild anti-Parkinsonism activity and thus it has been used in the management of Parkinsonism, mainly in the early disease stage and when the symptoms are mild. AMD is usually given by mouth as the hydrochloride salt [2].

Scheme 1

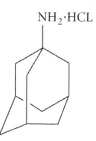

NH$_2$·HCL

Amantadine hydrochloride (AMD)

Spectrophotometry and spectrofluorometry are considered as the most convenient analytical techniques in pharmaceutical analysis because of their inherent simplicity and availability in most quality control and clinical laboratories [3–9]. However, AMD does not possess any chromophore or fluorophore in its molecule, which are the essential requirements for the direct analysis by either spectrophotometric or spectrofluorometric techniques. Therefore, derivatization of AMD was necessary for its determination by either of the two techniques. For spectrophotometric determination of AMD, it has been derivatized with different reagents. The involved derivatization reactions that have been published prior to 1983 have been reviewed by Kirschbaum [10]. The derivatizing reagents used thereafter included iodine [11], acetaldehyde/chloranil [11], α,α-diphenyl-β-picrylhydrazyl [12], bromocresol green [13], tetracyanoethylene [14], iron(III) [15], and cyclodextrin [16]. Few spectrofluorometric methods have been reported for the analysis of AMD. These methods were based on its oxidation with cerium(IV) [7] or its derivatization with 2,3-diphenylquinolizinium bromide [17], fluorescamine [18], and 9-isothiocynatoacridine [19]. As well, many derivatization techniques coupled with chromatography have been established for the determination of AM D in the dosage forms and biological matrices: TLC [20], HPLC [21–23], GC [24], and capillary electrophoresis [25]. 1,2-Naphthoquinone-4-sulphonate (NQS) has been used for the determination of many compounds [26–28]. The reaction between NQS and AMD has not been investigated yet. Therefore, the present study was devoted to explore NQS as a derivatizing reagent in the development of selective and sensitive spectrophotometric and spectrofluorometric methods for the determination of AMD in capsules and plasma.

Experimental

Apparatus

UV-1601 PC (Shimadzu, Kyoto, Japan) ultraviolet-visible spectrophotometer with matched 1 cm quartz cells was used for all spectrophotometric measurements. Spectrofluorimeter, Kontron SFM 25 equipped with a 150 W xenon high-pressure lamp was used for measuring the fluorescence intensity. MLW type thermostatically controlled water bath (Memmert GmbH, Co. Schwa bach, Germany). Biofuge Pico centrifuge (Heraeus Instruments, Germany).

Chemicals and Materials

Amantadine hydrochloride (AMD; Sigma-Aldrich Chemie GmbH, Steinheim, Germany) was obtained and used as received, its purity was 100.02±1.25%.

1,2-naphthoquinone-4-sulphonate; (NQS; El-Nasr Pharmaceutical Chemical Co., Abo-Zaabal, Egypt). Potassium borohydride (Sigma-Aldrich Chemie GmbH, Steinheim, Germany). Adamine capsules (Kameda Co. for Pharmaceutical Industries & Diagnostic Reagents, Cairo, Egypt) are labeled to contain 100 mg of AMD per capsule. Human plasma samples were collected from normal healthy volunteer at King Khaled University Hospital (Riyadh, Kingdom of Saudi Arabia), and they were stored at –20°C until analysis. All solvents and materials used throughout this study were of analytical grade. Double distilled water was obtained through WSC-85 water purification system (Hamilton Laboratory Glass Ltd., Ky, USA), and used throughout the work.

Preparation of Standard and Sample Solutions

Amantadine Hydrochloride (AMD) Standard Solutions

An accurately weighed amount (100 mg) of AMD was quantitatively transferred into a 50 mL calibrated flask, dissolved in 30 mL distilled water, completed to volume with the same solvent to obtain a stock solution of 2 mg mL–1. This stock solution was further diluted with water to obtain working solutions in the ranges of 50–800 and 0.5–100 μg mL–1 for the spectrophotometric and spectrofluorometric methods, respectively.

1,2-Naphthoquinone-4-Sulphonate (NQS) Derivatizing Reagent

Accurately weighed amount of NQS (150 mg) was transferred into a 25 mL calibrated flask, dissolved in 5 mL distilled water, completed to volume with water to obtain a solution of 0.6% (w/v). The solution was freshly prepared and protected from light during use.

Capsules Sample Solution

Twenty capsules were carefully evacuated; their contents were weighed and finely powdered. An accurately weighed quantity of the capsule contents equivalent to 100 mg of AMD was transferred into a 100 mL calibrated flask, and dissolved in about 40 mL of distilled water. The contents of the flask were swirled, sonicated for 5 minutes, and then completed to volume with water. The contents were mixed well and filtered rejecting the first portion of the filtrate. The prepared solution was diluted quantitatively with distilled water to obtain a suitable concentration for the analysis.

Spiked Plasma Samples

Aliquots of 1.0 mL of plasma were spiked with different concentration levels of AMD. The spiked plasma samples were treated with 0.1 mL of 70% perchloric acid and vortexed for 1 minute. The samples were centrifuged for 20 minutes at 13000 rpm. The supernatants were transferred into test tubes and neutralized with 1 M NaOH solution.

General Recommended Procedure

Spectrophotometric Method

One milliliter of the standard or sample solution (50–800 µg mL^{-1}) was transferred into a test tube. One milliliter of 0.01 M NaOH and 1 ml of NQS reagent (0.6%, w/v) were added. The contents of the tubes were heated in a water bath at 80 ± 5°C for 45 minutes and then cooled in ice water for 2 minutes. The contents of the test tubes were transferred quantitatively into a separating funnel containing anhydrous sodium sulphate and extracted with two portions (5 mL) of chloroform. The combined chloroformic extracts were transferred into 10 mL calibrated flasks and the solutions were completed to volume with chloroform if necessary. The absorbances of the resulting solutions were measured at 460 nm against reagent blanks treated similarly.

Spectrofluorometric Method

One milliliter of the standard or sample solution (0.5–100 µg mL^{-1}) was transferred into a test tube. One milliliter of 0.01 M NaOH and 1 mL of NQS reagent (0.6%, w/v) were added. The contents of the test tubes were heated in a water bath at 80 ± 5°C for 45 minutes and then cooled in ice water for 2 minutes. The contents of the test tubes were transferred quantitatively into a separating funnel and extracted with two portions (5 mL) of chloroform. The combined chloroformic extracts were evaporated under stream of air and the residues were reconstituted in 2 mL of methanol and quantitatively transferred into 10 mL calibrated flask. A one milliliter of KBH4 solution (0.03%, w/v in methanol) was added and the reaction was allowed to proceed for 5 minutes at room temperature (25 ± 5°C). The solution was diluted to volume with 0.025 M ethanolic HCl and the fluorescence intensity of the resulting solution was measured at 382 nm after excitation at 293 nm against reagent blanks treated similarly.

Determination of the Molar Ratio of the Reaction

The Job's method of continuous variation [29] was employed. Master equimolar (2.5×10^{-2} M) aqueous solutions of AMD and NQS were prepared. Series of 5 mL portions of the master solutions of AMD and NQS were made up comprising different complementary proportions (0:10, 1:9, ..., 9:1, 10:0, inclusive) in test tubes. One milliliter of 0.01 M NaOH was added to each tube, and the tubes were further manipulated as described under the general recommended procedure for the spectrophotometric method.

Results and Discussion

Strategy for Assays Development, Involved Reaction, and Spectral Characteristics

Because of the absence of any chromophoric group in the AMD molecule, it has no absorption in the ultraviolet-visible region above 200 nm, and it has no native fluorescence as well. Therefore, direct spectrophotometric and fluorimetric determination of AMD were not possible. Therefore, derivatization of AMD was attempted in the present study for the development of both spectrophotometric and spectrofluorometric methods for its determination. NQS has been used as chromogenic and fluorogenic reagent for primary and secondary amines [26–28, 30], however, its reaction with AMD has not been investigated yet. Therefore, the present study was devoted to explore NQS as a derivatizing reagent in the development of spectrophotometric and spectrofluorometric methods for the determination of AMD in capsules and plasma. Our preliminary experiments in investigating the reaction between AMD and NQS revealed that NQS-AMD product is orange colored exhibiting a maximum absorption at 460 nm and it is insoluble in water, but soluble in organic solvents. Since the present work was directed to involve plasma samples, the interference of endogenous amines was a major concern. It is well known that NQS reacts with the endogenous amines (e.g., amino acids) and yields water-soluble products [30]. For this reason, an extraction step was necessary for the development of selective methods for the determination of AMD in the presence of the endogenous amines. As well, the reduced AMD-NQS derivative was found to be fluorescent and exhibited one emission maximum at 382 nm and three excitation maxima at 293, 325, and 344 nm. The highest fluorescence intensity was obtained after excitation 293 nm, thus the excitation in the present study was performed at this wavelength. Scheme 2 shows the reaction pathway between AMD and NQS, and Figure 1 shows the absorption, excitation, and emission spectra of the reaction product. The following sections describe the optimization of the assay variables and validation for the performance of both spectrophotometric and spectrofluorometric methods.

Scheme 2. Scheme for the reaction pathway of amantadine hydrochloride (AMD) with 1,2-naphthoquinone-4-sulphonate (NQS).

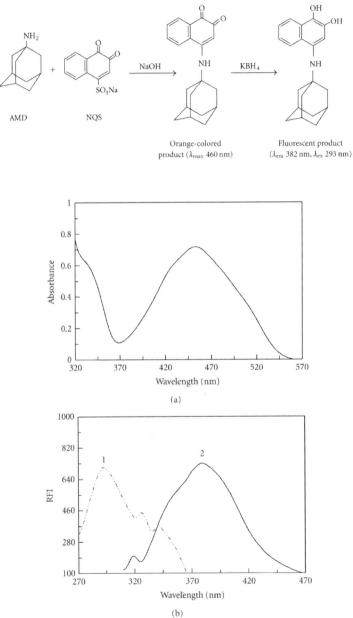

Figure 1. Absorption, excitation, and emission spectra on the reaction products of AMD with NQS (0.06%, w/v). (a) Absorption spectrum of the reaction product of AMD (65 μg mL^{-1}) with NQS after extraction with chloroform. (b) (1) Excitation and (2) emission spectra of the reduced reaction product. RFI is the relative fluorescence intensity.

Method Development

Optimization of Derivatization Reaction and Spectrophotometric Procedure

The factors affecting the derivatization reaction (the concentrations of NQS and NaOH, reaction time, temperature, diluting solvent, and the extracting solvent) were investigated by altering each variable in a turn while keeping the others constant. The studying of NQS concentrations revealed that the reaction was dependent on NQS reagent (Figure 2). The highest absorption intensity was attained when the concentration of NQS was 0.06–0.075% (w/v) in the final solution; further experiments were carried out at NQS concentration of 0.06% (w/v). The results of investigating the effect of NaOH concentration on the reaction revealed that the optimum NaOH concentration was 0.01 M (Figure 2). The effect of temperature on the derivatization reaction was studied by carrying out the reaction at different temperatures (25–100°C) and the maximum readings were obtained at 70–100°C (Figure 3). For more precise readings, further experiments were carried out at 80 ± 5°C. The effect of heating time on the formation of the reaction product was investigated by carrying out the reaction at different times. The maximum absorbance intensity was attained after 40 minutes, and longer reaction time did not affect the absorbance intensity (Figure 3). For more precise results, further experiments were carried out at 45 minutes.

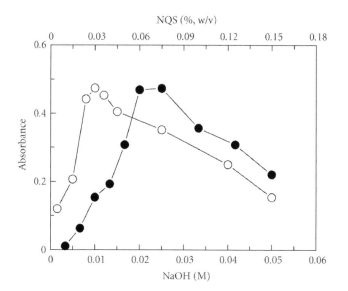

Figure 2. Effect of NaOH (□) and NQS(•) concentrations on the reaction of AMD (45 μg mL-1) with NQS.

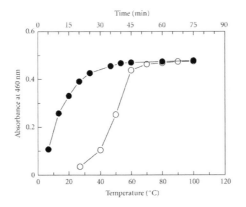

Figure 3. Effect of heating temperature (□) and time (•) on the reaction of AMD (45 μg mL–1) with NQS.

It was found that the colored AMD-NQS product is insoluble in the aqueous reaction medium. For measurements, the reaction product might be either dissolved in a miscible organic solvent of lower polarity than water or extracted with an immiscible extractive solvent. Different solvents were tested for dilution; methanol, ethanol, acetonitrile, dimethylsulphoxide, isopropanol, 1,4-dioxane, and acetone. The highest readings were obtained when dioxane was used for dilution (data not shown). As well, different nonmiscible solvents were tested for the extraction of the AMD-NQS product: carbon tetrachloride, chloroform, dichloromethane, ethyl acetate, toluene, and benzene. The highest readings were obtained when chloroform was used for extraction (Table 1). The results revealed that the extractive procedure is more sensitive (1.5 fold) than the nonextractive procedure. This was attributed to the effective decrease in the blank readings and consequently enhanced the sensitivity of the assay.

Table 1. Effect of diluting and extracting solvents on the intensity of the reaction product of AMD with NQS (0.06%, w/v). Values for all solvents are mean of three determinations; the RSDs for the readings were <3.

Spectrophotometric method		Spectrofluorometric method	
Extracting solvent	Absorbance	Diluting solvent	Fluorescence intensity
Carbon tetrachloride	0.487	Methanol	65.66
Chloroform	0.718	Ethanol	78.38
Dichloromethane	0.616	Isopropanol	63.02
Ethyl-acetate	0.362	Acetone	2.00
Toluene	0.267	Acetonitrile	16.18
Benzene	0.239	Dimethylformamide	3.94
		1,4-Dioxane	76.11

Optimization of Spectrofluorometric Procedure

For developing the spectrofluorometric procedure, a reduction step for AMD-NQS product was necessary. However, the reduced NQS reagent itself is also fluorescent and it had the same excitation and emission maxima of the AMD-NQS product, therefore, a selective extraction step for the AMD-NQS product from the remaining NQS reagent was necessary before carrying out the reduction step. Furthermore, the extraction step is essential to provide the required selectivity for analyzing the plasma samples as the NQS products with endogenous amines were water soluble. Based on the reported efficiency [30], potassium borohydride as a reducing reagent was selected for NQS-derivatives. In order to investigate the effect of potassium borohydride concentration on the reduction, the reaction was performed using varying concentrations (0.0005–0.01%, w/v). The highest fluorescence intensity was obtained when the concentration was 0.003% in the final solution (1 mL of 0.03%, w/v). Concentrations more than 0.003% did not affect the fluorescence intensity (Figure 4). The effect of pH on the fluorescence intensity was also studied and the results showed that the highest fluorescence intensity was obtained at pH 2.0 (Figure 5). This pH could be attained by diluting the reaction mixture with 0.025 M ethanolic HCl solution.

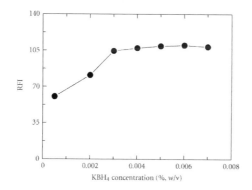

Figure 4. Effect of KBH4 concentration on the fluorescence intensity of the reaction product of AMD (1 µg mL^{-1}) with NQS (0.06%, w/v).

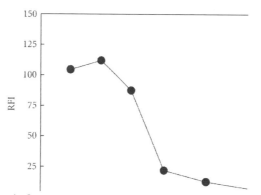

Figure 5. Effect of pH on the fluorescence intensity of the reduced reaction product of AMD-NQS. RFI is the relative fluorescence intensity.

Stoichiometry of Derivatization Reaction

Under the optimum conditions, the stoichiometry of the reaction between AMD and NQS was investigated by Job's method [29] and was found to be 1:1 because AMD molecule contains only one center (primary amino group) available for this condensation reaction. Based on this ratio, the reaction pathway was postulated to be proceeded as shown in Scheme 2.

Method Validation

Linearity, Limits of Detection and Quantitation

In the proposed methods, linear plots (n=6) with good correlation coefficients (0.9974 and 0.9972) were obtained in the concentration ranges of 5–80 μg mL^{-1} for and 0.05–10 μg mL^{-1} for the spectrophotometric and the spectrofluorometric methods, respectively (Table 2). The limits of detection (LOD) and quantitation (LOQ) were determined [31] using the formula LOD or LOQ = κSDa/b, where κ= 3.3 for LOD and 10 for LOQ, SDa is the standard deviation of the intercept, and b is the slope. The LOD values were 1.39 and 0.013 μg mL–1 for the spectrophotometric and spectrofluorometric methods, respectively (Table 2).

Table 2. Quantitative parameters and statistical data for determination of amantadine hydrochloride by the proposed spectrophotometric and spectrofluorometric methods.

Parameter	Spectrophotometric method	Spectrofluorometric method
Range (μg mL^{-1})	5.00–80.0	0.05–10.0
Intercept (a)	0.0759 ± 0.0041	−2.457 ± 0.332
Slope (b)	0.00974 ± 0.00035	81.969 ± 2.154
Correlation coefficient (r)	0.9974	0.9972
$\varepsilon \times 10^3$ (L mol^{-1} cm^{-1})	2.058	—
LOD (μg mL^{-1})	1.39	0.013
LOQ (μg mL^{-1})	4.21	0.041

Precision and Accuracy

The precision of the proposed methods was determined by replicate analysis of five separate sample solutions at three concentration levels of AMD. The relative standard deviations (RSDs) were 0.83–0.96 and 0.46–1.01% for the spectrophotometric and spectrofluorometric methods, respectively (Table 3), indicating the good reproducibility of the proposed methods. Furthermore, the inter- and intra-assay precisions of the proposed spectrofluorometric method were determined from the recovery studies of spiked human plasma samples. The RSD values of the recovery were 0.57–2.04 and 0.72–4.20% for the intra- and inter-assay determinations, respectively (Table 4). The accuracy of the proposed methods was evaluated by standard addition method. The obtained recovery values were 98.8–100.2±1.04–1.54% indicating the high accuracy of the proposed methods. Moreover, the accuracy of the proposed spectrofluorometric method was evaluated by the recovery studies of spiked human plasma samples. The obtained recovery values were 96.3–101.2±0.57–4.2% (Table 4). These recovery results of the spiked human plasma indicate the suitability of the proposed spectrofluorometric method for the analysis of AMD in human plasma.

Table 3. The precision of the proposed methods at three concentration levels of AMD.

Method	Nominal concentration (μg mL^{-1})	RSD%[a]
	8.0	0.96
Spectrophotometric	40.0	0.83
	60.0	0.87
	0.1	1.01
Spectrofluorometric	4.0	0.58
	8.0	0.46

[a] Values are mean of five determinations.

Table 4. Recovery studies for the proposed spectrofluorometric method for the determination of AMD in spiked human plasma.

Spiked concentration (μg mL^{-1})	Recovery (% ± RSD)[a]	
	Intra-assay	Inter-assay
0.05	98.2 ± 2.04	96.3 ± 4.20
0.10	99.3 ± 1.01	101.1 ± 2.0
0.20	100.5 ± 0.99	99.1 ± 1.50
0.40	100.9 ± 0.74	98.5 ± 1.52
0.80	100.1 ± 0.75	101.2 ± 1.44
1.60	98.6 ± 0.82	100.3 ± 1.12
3.20	98.9 ± 0.57	99.7 ± 0.72

[a] Values are mean of three and five determinations for intra- and inter-assay, respectively.

Interference Studies

The results of the interferences study showed that no interferences were found from any of the excipients studied; lactose, sucrose, starch, talc, gum acacia, glucose, and magnesium stearate; the recovery of AMD was 98.15–100.72%. This indicated the absence of interferences from these excipients. Moreover, the interferences from the amino acids with the assay procedures were also studied using glycine as an example for the amino acids. The results of this study revealed that the amino acids could interfere with the spectrophotometric procedures. However, there is no any interference coming from the amino acids after an extraction step for the derivatized AMD product because the derivatized amino acids products are water soluble. Therefore, the extraction step increased both the sensitivity and selectivity by removing the interferences caused by both amino acids and proteins in the plasma samples.

Robustness and Ruggedness

Robustness was examined by evaluating the influence of small variation of method variables, including concentration of analytical reagents and reaction time on the performance of the proposed methods. In these experiments, one parameter was changed whereas the others were kept unchanged, and the recovery percentage was calculated each time. It was found that small variation of method variables did not significantly affect the procedures. This provided an indication for the reliability of the proposed method during its routine appli cation for the analysis of AMD. Ruggedness was also tested by applying the proposed methods to the assay of AMD using the same operational conditions but using two different instruments at two different laboratories and different elapsed time. Results obtained from lab-to-lab and day-to-day variations were found to be reproducible, the full range of recovery values was 98.4–101.3% and the RSD was 0.73 and 1.06% for the spectrophotometric and spectrofluorometric methods, respectively.

Application of the Proposed Method to Analysis of AMD in Capsules

It is evident from the above-mentioned results that the proposed methods gave satisfactory results with AMD in bulk. Thus, its capsules were subjected to the analysis of their contents from the active ingredient by the proposed methods and the official (nonaqueous titration) method [32]. The capsule contents, as percentage, were 98.70 ± 1.79 and $98.91 \pm 1.93\%$ for the spectrophotometric and spectrofluorometric methods, respectively (Table 5). These results were compared with those obtained from the official method by statistical analysis with respect to the accuracy (t-test) and precision (F-test). No significant differences were found between the calculated and theoretical values of t- and F-tests at 95% confidence limit proving similar accuracy and precision in the analysis of AMD in its dosage form.

Table 5. Analysis of AMD in capsules by the proposed and official methods.

Method	Recovery (% ± SD)[a]	t-value[b]	F-value[b]
Spectrophotometric	98.70 ± 1.79	0.42	1.64
Spectrofluorometric	98.91 ± 1.91	0.19	1.85
Official HPLC[c]	99.13 ± 1.41	—	—

[a] Values are mean of five determinations.
[b] Theoretical values for t and F at 95% confidence limit and $n = 5$ were 2.31 and 6.39, respectively.
[c] Reference [32].

Conclusions

The present study described the use of NQS reagent for the development of selective, sensitive, and accurate spectrophotometric and spectrofluorometric methods for the determination of AMD in bulk, capsules, and plasma. The described methods are superior to the previously reported spectrophotometric or spectrofluorometric methods for analysis of AMD in terms of their selectivity and sensitivity. The linear ranges of the proposed spectrophotometric and spectrofluorometric methods were 5–80 and 0.05–10 µg/mL, respectively, which are much less than some reported methods 100–1300, 15–90, 25–75, 94–940 µg/mL [11, 14, 15] which were based on either the nonselective oxidation or charge-transfer complex formation of AMD base. Although the sensitivity of the proposed spectrofluorimetric method is comparable to that described by Darwish et al. [7], which was based on the nonselective oxidation of AMD with cerric sulphate, however, our proposed methods are more selective. Also, The proposed methods involved spectrophotometric and spectrofluorometric measurements with comparable analytical performance devoid from any potential interference. This gives the advantage of flexibility in performing the analysis on any available instrument. Furthermore, all the analytical reagents are inexpensive, have excellent shelf life, and are available in any analytical laboratory. Therefore, these methods can be recommended for the routine analysis of AMD in quality control and clinical laboratories.

References

1. I. T. Prud'homme, O. Zoueva, and J. M. Weber, "Amantadine susceptibility in influenza A virus isolates: determination methods and lack of resistance in a Canadian sample, 1991–94," Clinical and Diagnostic Virology, vol. 8, no. 1, pp. 41–51, 1997.

2. Martindale, The Complete Drug Reference, Pharmaceutical Press, London, UK, 33rd edition, 2002.

3. I. A. Darwish, I. H. Refaat, H. F. Askal, and M. A. Marzouq, "Generic nonextractive spectrophotometric method for determination of 4-quinolone antibiotics by formation of ion-pair complexes with β-naphthol," Journal of AOAC International, vol. 89, no. 2, pp. 334–340, 2006.

4. H. F. Askal, I. H. Refaat, I. A. Darwish, and M. A. Marzouq, "A selective spectrophotometric method for determination of rosoxacin antibiotic using sodium nitroprusside as a chromogenic reagent," Spectrochimica Acta Part A, vol. 69, no. 4, pp. 1287–1291, 2008.

5. I. A. Darwish, A. S. Khedr, H. F. Askal, and R. M. Mohamed, "Application of inorganic oxidants to the spectrophotometric determination of ribavirin in bulk and capsules," Journal of AOAC International, vol. 89, no. 2, pp. 341–351, 2006.

6. I. A. Darwish, S. A. Hussein, A. M. Mahmoud, and A. I. Hassan, "A sensitive spectrophotometric method for the determination of H 2-receptor antagonists by means of N-bromosuccinimide and p-aminophenol," Acta Pharmaceutica, vol. 58, no. 1, pp. 87–97, 2008.

7. I. A. Darwish, A. S. Khedr, H. F. Askal, and R. M. Mahmoud, "Simple fluorimetric method for determination of certain antiviral drugs via their oxidation with cerium (IV)," Farmaco, vol. 60, no. 6-7, pp. 555–562, 2005.

8. I. A. Darwish, H. M. Abdel-Wadood, and N. Abdel-Latif, "Validated spectrophotometric and fluorimetric methods for analysis of clozapine in tablets and urine," Annali di Chimica, vol. 95, no. 5, pp. 345–356, 2005.

9. H. M. Abdel-Wadood, N. A. Mohamed, and A. M. Mahmoud, "Validated spectrofluorometric methods for determination of amlodipine besylate in tablets," Spectrochimica Acta Part A, vol. 70, no. 3, pp. 564–570, 2008.

10. J. Kirschbaum, Analytical Profile of Drug Substances, vol. 12, Academic Press, New York, NY, USA, 1983.

11. I. A. Darwish, A. S. Khedr, H. F. Askal, and R. M. Mahmoud, "Simple and sensitive spectrophotometric methods for determination of amantadine hydrochloride," Journal of Applied Spectroscopy, vol. 73, no. 6, pp. 792–797, 2006.

12. S. Salman and N. Bayrakdur, Eczaclik Bull., vol. 25, no. 2, pp. 30–33, 1983.

13. M. Sultan, "Spectrophotometric determination of amantadine in dosage forms," Current Topics in Analytical Chemistry, vol. 4, pp. 103–109, 2004.

14. M. S. Rizk, S. S. Toubar, M. A. Sultan, and S. H. Assaad, "Ultraviolet spectrophotometric determination of primary amine-containing drugs via their

charge-transfer complexes with tetracyanoethylene," Microchimica Acta, vol. 143, no. 4, pp. 281–285, 2003.

15. A. A. Mustafa, S. A. Abdel-Fattah, S. S. Toubar, and M. A. Sultan, "Spectro-photometric determination of acyclovir and amantadine hydrochloride through metals complexation," Journal of Analytical Chemistry, vol. 59, no. 1, pp. 33–38, 2004.

16. T. Kuwabara, H. Nakajima, M. Nanasawa, and A. Ueno, "Color change indicators for molecules using methyl red-modified cyclodextrins," Analytical Chemistry, vol. 71, no. 14, pp. 2844–2849, 1999.

17. M. A. Martin and B. Del Castillo, "2,3-diphenylquinolizinium bromide as a fluorescent derivatization reagent for amines," Analytica Chimica Acta, vol. 245, no. 2, pp. 217–223, 1991.

18. J. A. F. De Silva and N. Stronjny, "Spectrofluorometric determination of pharmaceuticals containing aromatic or aliphatic primary amino groups as their fluorescamine (Fluram) derivatives," Analytical Chemistry, vol. 47, no. 4, pp. 714–718, 1975.

19. J. E. Sinsheimer, D. D. Hong, J. T. Stewart, M. L. Fink, and J. H. Burckhalter, "Fluorescent analysis of primary aliphatic amines by reaction with 9-isothiocyanatoacridine," Journal of Pharmaceutical Sciences, vol. 60, no. 1, pp. 141–143, 1971.

20. H. F. Askal, A. S. Khedr, I. A. Darwish, and R. M. Mahmoud, "Quantitative thin-layer chromatographic method for determination of amantadine hydrochloride," International Journal of Biomedical Science, vol. 4, no. 2, pp. 155–160, 2008.

21. T.-H. Duh, H.-L. Wu, C.-W. Pan, and H.-S. Kou, "Fluorimetric liquid chromatographic analysis of amantadine in urine and pharmaceutical formulation," Journal of Chromatography A, vol. 1088, no. 1-2, pp. 175–181, 2005.

22. Y. Higashi and Y. Fujii, "Simultaneous determination of the binding of amantadine and its analogues to synthetic melanin by liquid chromatography after precolumn derivatization with dansyl chloride," Journal of Chromatographic Science, vol. 43, no. 4, pp. 213–217, 2005.

23. Y. Higashi, S. Nakamura, H. Matsumura, and Y. Fujii, "Simultaneous liquid chromatographic assay of amantadine and its four related compounds in phosphate-buffered saline using 4-fluoro-7-nitro-2,1,3-benzoxadiazole as a fluorescent derivatization reagent," Biomedical Chromatography, vol. 20, no. 5, pp. 423–428, 2006.

24. H. J. Leis, G. Fauler, and W. Windischhofer, "Quantitative analysis of memantine in human plasma by gas chromatography/negative ion chemical ionization/

mass spectrometry," Journal of Mass Spectrometry, vol. 37, no. 5, pp. 477–480, 2002.

25. N. Reichová, J. Pazourek, P. Poláková, and J. Havel, "Electrophoretic behavior of adamantane derivatives possessing antiviral activity and their determination by capillary zone electrophoresis with indirect detection," Electrophoresis, vol. 23, no. 2, pp. 259–262, 2002.

26. L. Gallo-Martinez, A. Sevillano-Cabeza, P. Campíns-Falcó, and F. Bosch-Reig, "A new derivatization procedure for the determination of cephalexin with 1,2-naphthoquinone 4-sulphonate in pharmaceutical and urine samples using solid-phase extraction cartridges and UV-visible detection," Analytica Chimica Acta, vol. 370, no. 2-3, pp. 115–123, 1998.

27. H. Y. Wang, L. X. Xu, Y. Xiao, and J. Han, "Spectrophotometric determination of dapsone in pharmaceutical products using sodium 1,2-naphthoquinone-4-sulfonic as the chromogenic reagent," Spectrochimica Acta Part A, vol. 60, no. 12, pp. 2933–2939, 2004.

28. I. A. Darwish, "Kinetic spectrophotometric methods for determination of trimetazidine dihydrochloride," Analytica Chimica Acta, vol. 551, no. 1-2, pp. 222–231, 2005.

29. P. Job, in Advanced Physicochemical Experiments, Ann. Chem. 16 (1936), p. 54, Oliner and Boyd, Edinburgh, UK, 2nd edition, 1964.

30. M. Pesez and J. Bartos, Colorimetric and Fluorimetric Analysis of Organic Compounds and Drugs, Marcel Dekker, New York, NY, USA, 1974.

31. ICH guideline and Q2(R1), "Validation of Analytical Procedures: Text and Methodology," London, 2005.

32. The United States Pharmacopeia 25, The National Formulary 20, US Pharmacopeial Convention Inc, Rockville MD, 101–103, 2256–2259, 2002.

CITATION

Originally published under the Creative Commons Attribution License or equivalent. Mahmoud AM, Khalil NY, Darwish IA, Aboul-Fadl T. Selective Spectrophotometric and Spectrofluorometric Methods for the Determination of Amantadine Hydrochloride in Capsules and Plasma via Derivatization with 1,2-Naphthoquinone-4-sulphonate. Int J Anal Chem. 2009; 2009:810104. doi:10.1155/2009/810104.

Analytical Applications of Reactions of Iron(III) and Hexacyanoferrate(III) with 2,10-Disubstituted Phenothiazines

Helena Puzanowska-Tarasiewicz, Joanna Karpińska
and Ludmiła Kuźmicka

ABSTRACT

The presented review is devoted to analytical applications of reactions of Fe(III) and $K_3[Fe(N)_6]$ with 2,10-disubstituted phenothiazines (PT). It was found that iron(III) and hexacyanoferrate(III) ions in acidic media easily oxidized PT with the formation of colored oxidation products. This property has been exploited for spectrophotometric determination of iron(III) ions and phenothiazines. Some flow-injection procedures of the determination of PT *based on the oxidation reaction by means of the above-mentioned oxidants have been proposed. In the presented review, the application of 2,10-disubstituted phenothiazines as indicators in complexometric titration of iron(III) as well as procedures of PT determination based on generation of ternary compound in the system Fe(III)-SCN⁻ - PT was also described.*

Introduction

Phenothiazines constitute one of the largest chemical classes of organic compounds in official compendia. Over 4 thousands compounds have been synthesized and about 100 have been used in clinical practice [1]. 2,10-Disubstituted phenothiazines are very important drugs which are widely used in psychiatric treatment as tranquillizers. Invention and introduction of phenothiazine derivatives into treatment of mental disease has changed the modern psychiatry. This fact has improved the life style of patients and allowed quick development of ambulatory system of treatment for such sickness. The common use of phenothiazines has generated the need for fast and reliable methods for quality control of phenothiazine pharmaceuticals and monitoring them in clinical samples. Over fifty years of the medical use of phenothiazines have resulted in countless number of analytical procedures devoted to resolve this problem [2, 3].

Phenothiazine derivatives are interesting from analytical point of view due to their characteristic structure—the presence of chemically active sulfur and nitrogen atoms in positions 5 and 10 and substituents in position 2 and alkylamine side chain at N10 atom. Phenothiazine and its derivatives are characterized by low ionization potentials [1, 4]. They are easily oxidized by different chemical, electrochemical, photochemical, and enzymatic agents with the formation of colored oxidation product—intermediate cation radical [1]. Colors of formed intermediate depend on a presence and a structure of substituents in positions 2 and 10 (Table 1).

The run of reactions with the oxidants (e.g., Fe(III), [Fe(CN)6]3–, Cr2O72–, IO3–, IO4–, BrO3–, H2O2, chloramine T) has been studied and employed for determination of used phenothiazine or oxidant. The stability of oxidation products depends on acidity, concentration of oxidizing agents, time, temperature, and the presence of some salts [5]. Recently the redox properties of 2,10-disubstituted phenothiazines radicals have been studied by Madej and Wardman [6]. They have established the reduction potentials of phenothiazine radicals and equilibrium constants using a pulse radiolysis and a cyclic voltammetry. The further oxidation leads to a generation of colorless sulphoxide. A lot of published works have been based on the oxidation behavior of 2,10-disubstituted phenothiazines [1, 7, 8].

The oxidation involves a series of one-electron steps providing free radicals and cations [1]. The distribution of π-electrons in the 2,10-disubstituted phenothiazines, according to theoretical considerations, may lead to the formation of some resonance forms of free cation radical [9].

Table 1. The structures of 2,10-disubstituted phenothiazines studied with Fe(III) and K3[Fe(CN)6] and colors of their cation radicals.

No.	Trivial name	$-R_1$	$-R_2$	Color of oxidation products (radicals)
1	Promazine–HCl	$-(CH_2)_3-N\begin{smallmatrix}CH_3\\CH_3\end{smallmatrix}$	–H	Orange
2	Chlorpromazine–HCl	$-(CH_2)_3-N\begin{smallmatrix}CH_3\\CH_3\end{smallmatrix}$	–Cl	Red
3	Levomepromazine–$C_4H_4O_4$	$-CH_2-CH-CH_2-N\begin{smallmatrix}CH_3\\CH_3\end{smallmatrix}$ (CH_3)	–OCH$_3$	Violet
4	Methopromazine–$C_4H_4O_4$	$-(CH_2)_3-N\begin{smallmatrix}CH_3\\CH_3\end{smallmatrix}$	–OCH$_3$	Violet
5	Perazine	$-(CH_2)_3-N\bigcirc N-CH_3$	–H	Orange
6	Trifluperazine–2HCl	$-(CH_2)_3-N\bigcirc N-CH_3$	–CF$_3$	Orange
7	Propericiazine	$-(CH_2)_3-N\bigcirc N-OH$	–CN	Red
8	Perphenazine	$-(CH_2)_3-N\bigcirc N-(CH_2)_2OH$	–Cl	Red
9	Fluphenazine–2 HCl	$-(CH_2)_3-N\bigcirc N-(CH_2)_2OH$	–CF$_3$	Orange
10	Thioridazine–HCl	$-(CH_2)_2\bigcirc$ (CH_3)	–SCH$_3$	Blue

Phenothiazines have exhibited complexing properties due to the presence of the condensed three-ring aromatic system and amine nitrogen atom in a side chain in position 10. They have reacted with some metal ions or thiocyanate complexes of metals forming colored, hard soluble in water but easy soluble in organic solvents compounds [1, 10]. Some organic substances (e.g., picric, flavianic acid, alizarin S, brilliant blue, pyrocatechol violet) have formed with 2,10-disubstituted phenothiazines colored ion-association compounds sparingly soluble in water, but quantitatively extracted into organic phase [10, 11]. The conducted spectroscopic studies have confirmed ion-association nature of these conjunctions [10, 11]. Phenothiazines also have created charge-transfer complexes with nitroso-R-salt [12] and chloranilic acid [13].

In our earlier works, it has been found that 2,10-disubstituted phenothiazines are useful as redox indicators [8, 19] and spectrophotometric reagents [10, 11,

20]. Some elements have a catalytic influence on a run of oxidation of phenothiazines [21, 22]. The catalytic effect of presence of iodide, nitrite, vanadium, and iron ions on the reactions of phenothiazines with KBrO3, H2O2 has been described by Mohamed [23] in his dissertation.

After careful analysis of articles concerned with phenothiazines determination, it could be stated that iron ions and its anionic complexes are reagents most often used for this purpose. The mild oxidation potential of Fe(III)/Fe(II) couple and stability of its complexes make them very convenient reagents for phenothiazine derivatives assay. The oxidation reaction conducted in the system Fe(III)-NO3--CIO4- (Forrest's reagent) is still used for quick examination of presence of phenothiazines in studied sample [24]. Taking the above-mentioned facts into account, we have decided to get together the most important information focused on analytical applications of the reactions of 2,10-disubstituted phenothiazines with Fe(III) and [Fe(CN)6]3- ions.

Reactions of Phenothiazines with Iron(III) and Hexacyanoferrate(III) Ions

As mentioned above, the most important property of phenothiazines is their susceptibility to oxidation by many oxidizing agents, for example, Fe(III), $[Fe(CN)_6]^{3-}$ with the formation of colored oxidation products (free radicals) in acidic media [8, 21]. The stability of the oxidation products depends on the nature of the substituents at positions 2 and 10 (Table 1) [25]. The stability of 57 radicals has been studied by Levy et al. [26] using ESR method and the Hammett metasubstituent constant. The kinetics [27] of oxidation reactions of eight phenothiazines with $[Fe(H_2O)]^{3+}$ and $[Fe(CN)_6]^{3-}$ and an influence of micellar system [28] on the run of studied processes have been investigated by Pelizzetti and Mentasti [27] and Pelizzetti et al. [28].

Basavaiah and Swamy [29] have studied oxidation reaction of five phenothiazine derivatives with hexacyanoferrate(III). The reduced hexacyanoferrate(II) reacted further with ferrin forming ferroin. The measurements of absorbance of final product allowed to determine studied compounds in the concentration range 1–12 μg/mL with the molar absorption coefficient ranged from 2.08×104 for chlorpromazine to 3.49 × 104 for prochlorpromazine.

In our previous works [14, 15], we have described the optimal conditions for the formation of colored products of 2,10-disubstituted phenothiazines with FeCl3 and K3[Fe(CN)6] in acidic media. The absorption spectra of these products in aqueous solutions have been recorded. The spectra of the nonoxidized and colored oxidation product, for example, promazine hydrochloride (PM) are given in Figure 1. We did not obtain the spectrum of sulphoxides using FeCl3 [14] and K3[Fe(CN)6] [15] as oxidants.

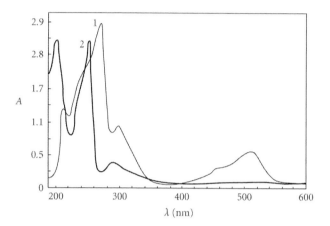

Figure 1. UV-VIS spectra of 1-colored product of reaction PM-Fe(III) (C_{PM}=3 × 10^{-4} M, $C_{Fe(III)}$=3 × 10^{-4} M; 2-nonoxidized form of PM (C_{PM}=4 × 10^{-4} M).

The investigations in the UV-region have testified that the reaction of PT with FeCl3 or K3[Fe(CN)6] proceeds only to the first steep, for example, promazine (Scheme 1).

Scheme 1: The oxidation of PM by means of FeCl3.

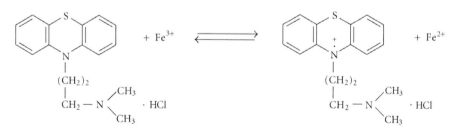

The presence of iron(II) after the oxidation of phenothiazines studied has been confirmed by means of 2,2′-bipirydyl [15, 17]. A reversible nature of oxidation process of phenothiazines has been reentered by the electrochemical method—cyclic voltammetry [14]. Figure 2 shows a cyclic voltammogramm for promazine. As can be seen, promazine exhibits an oxidative peak at 0.55 V. Acorresponding reductive peak (0.47 V) appears when the polarization of electrode is reversed. This suggests that the reaction can be regarded as reversible (Scheme 1).

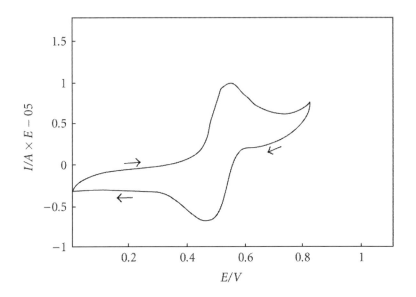

Figure 2. Cyclic voltammogramm of promazine (PM) at Pt electrode CPM=5 × 10⁻⁵ M, in a mixture of 5 × 10⁻¹ M K₂SO₄ and 5 × 10⁻¹ M KHSO₄ (pH=1.3), scan rate 50 mV/s; in potential window 0.0–1.05 V.

From already used various oxidants, iron(III) ion has been chosen as a mild oxidation agent for spectrophotometric determination of 2,10-disubstituted phenothiazines. The formal redox potential of the Fe(III)-Fe(II) couple ($E°$=0.77 V, v ME) does not allow further oxidation of the colored free radicals to the uncolored sulphoxides [9]. The formed oxidation products are stable. These properties have been exploited in chemical analysis [1, 8, 19, 20]. The oxidation properties of iron(III)-phenothiazines system have been employed for the spectrophotometric determination of iron(III) ions [16] and diethazine [18].

Application of The Reactions in Fia Systems

Some flow injection procedures have been described for the determination of PT. They have been based on the oxidation reaction of 2,10-disubstituted phenothiazines with iron(III) or hexacyanoferrate(III) (Table 2).

Phenothiazines solutions have been injected into a stream of distilled water, which has been merged with the stream of iron(III) chloride in hydrochloric acid [30] or with the stream of iron(III) perchlorate in perchlorate acid medium [31]. One of the used FIA manifolds is presented in Figure 3.

Table 2. Determination of some PF with Fe(III) and $K_3[Fe(CN)_6]$ using flow injection methods.

Oxidant	Compound	Determination range (μg mL^{-1})	Pharmaceutical formulation	Ref.
FeCl$_3$	Promazine	10–130		
	Thioridazine	10–130	Promazin (injection)	
				[14]
	Promazine	6–128		
	Chlorpromazine	6–124	Largactil (tablets)	
	Levomepromazine	7–133		
	Promethazine	6–117		[15]
Fe(ClO$_4$)$_3$	Fluphenazine	5–312	Phenergan (tablets)	
	Thioridazine	11–230		
	Thioproperazine	12–248	Majeptil (tablets)	[15]
	Trifluoperazine	12–230		
	Promethazine	0,5–8,0	Syrup, cream and tablets	[16]
	Trifluoperazine	0,5–10		
	Promazine	2,5–25	Promazine (injection)	[17]
	Chlorpromazine			
K$_3$[Fe(CN)$_6$]	Fluphenazine			
	Thioridazine	1–2	Urine	[18]
	Promethazine			
	Methotrimeprazine			

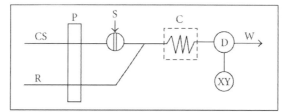

Figure 3. Schematic diagram of the FIA manifold used for the assay of promazine; R: iron(III) in hydrochloric acid solution; CS: water; P: peristaltic pump; S: samples injector; C: reactor; D: spectrophotometer adjusted to the corresponding wavelength of the oxidized form of phenothiazines; X: recorder; W: waste.

Another method for determination of promazine [32] has been proposed by Kojło et al. This assay has been based on promazine oxidation by K3[Fe(CN)6] previously retained on anion exchange column. The oxidation has been carried out at room temperature in aqueous acidic medium.

The flow injection (FIA) methods are preferable to other conventional methods because they are fast (from 50 to 200 samples studied per h) and precise (RSD values ranging from 0.6 to 2.5%). Additional advantage of flow methods is a possibility of combination with preconcentration of assayed phenothiazine on sorption column mounted on line [33]. Another applied approach is the use a detection cell filled with suspension of an appropriate sorbent (so-called a solid spectrophotometry mode) [34].

Application of Phenothiazines As Visual Indicators in Complexometric Titration

2,10-Disubstituted phenothiazines are useful redox indicators. The radical cations, which are stable enough under acidic conditions, exhibit quite intense color [1, 4, 6]. This property allows employing phenothiazines as redox indicators in many redoxometric determinations. The values of reduction potentials of some PT established by Madej and Wardman [6] and Gowda and Ahmed [35] are given in Table 3.

Table 3. Reduction potentials of PT.

Phenothiazines	mV	Phenothiazines	mV
Chlorpromazine*	860	Propericiazine*	966
Promethazine*	925	Trifluoperazine	880
Diethazine	845	Prochlorperazine	799
Thioridazine*	789	Butaperazine	865

*Determined by Madej and Wardman at pH ~ 5–7 [6]; others values established by Kojło et al. in 0,5 M H_2SO_4 [32].

Phenothiazines have been used as indicators for complexometric determination of iron(III) with dissodium versenate [36]. They form with Fe(III) ions colored oxidation products (red, orange, or blue). The addition of dissodium versenate to the titrated solution containing iron(III) solution and phenothiazines as indicator has caused a change of the test solution in end point of titration as shown in Table 4.

Table 4

Colorless → red	Colorless → orange	Colorless → blue
Chlorpromazine	Propericiazine	Thioridazine
Diethazine	Trifluoperazine	
Promethazine	Prochlorperazine	
	Butaperazine	

The usefulness of PT (chlorpromazine, promazine, perphenazine, methopromazine) as redox indicators in chromatometric determination of K4[Fe(CN)6] has been described by Puzanowska-Tarasiewicz et al. [38].

Phenothiazines indicators are superior to conventional indicators (e.g., ferroin, variamin blue). They give sharper end point and act over a wider range of acidity than other conventional indicators.

Application of Complexation Reactions in Assay of Phenothiazines

As it was mentioned in introduction section, phenothiazines show the ability to create stable compounds in reaction with anionic complexes of metal ions, some organic anions or with π-electron acceptors.

According to Ozutsumi et al. [39], the formation of the thiocyanate-iron(III) complexes in aqueous solution and the development of the red color are related to [Fe(SCN)]2+, [Fe(SCN)2]+, [Fe(SCN)3], [Fe(SCN)4]–, [Fe(SCN)5]2–, [Fe(SCN)6]3–. Tarasiewicz [37] has found that one of these complexes reacts with 2,10-disubstituted phenothiazines forming red-brown compounds. The optimal conditions for the formation of the compounds have been established and the composition determined. The absorption spectra have been recorded in UV-VIS and IR regions [37]. On the basis of obtained data, the following reaction course in PT-Fe(III)-SCN– system has been suggested:

$$Fe(III) + m\ SCN^- \rightarrow [Fe(SCN)_m]^{n-},$$

$$PT + H+ \rightarrow (PT\bullet H)+, (1)$$

$$(PT\bullet H)^+ + [Fe(SCN)_m]^{n-} \rightarrow (PT\bullet H)_n[Fe(SCN)_m],$$

where n=2, m=5.

The spectral properties of these compounds suggest that the formation of color compound occurred due to interaction between opposite charged ions (large phenothiazines cation and an anionic thiocyanate complex of a metal). The obtained color precipitate is quantitatively extracted with chloroform or dissolved in acetone with formation colored and stable solution. This property has been the basis of sensitive extractive-spectrophotometric or spectrophotometric method of determination of phenothiazines [37] (Table 5).

Table 5. Extractive-spectrophotometric and spectrophotometric methods of determination of some of 2,10-disubstituted phenothiazines.

Organic phase	Phenothiazines	Determination range ($\mu g\ mL^{-1}$)	Ref.
Chloroform	Chlorpromazine	120–300	[37]
	Levomepromazine	140–400	
	Promethazine	160–550	
Acetone	Chlorpromazine	20–150	[37]
	Levomepromazine	20–300	
	Promethazine	60–400	

Chlorpromazine and some phenothiazines react with ferro- and ferricyanate ions [40] and nitroso-ferricyanate [41] to form ion association-compounds sparingly soluble in water. The composition of these compounds has been established and physicochemical properties have been investigated.

Valero [42] has stated, using UV-VIS spectrophotometry, that iron(III) forms ternary complexes with pyrocatechol violet (PCV) and chlorpromazine (CPZ) which compositions have been established as: Fe:PCV:CPZ = 1:2:3 and Fe:PCV:CPZ = 1:3:4. The last complex can be used for spectrophotometric determination up to 1.6 ppm of iron(III).

Other Applications

Some ions of d-electron elements exhibit a catalytic effect on the oxidation of 2,10-disubstituted phenothiazines [21]. Fukasawa et al. [22] have described a spectrophotometric determination of trace amounts of iron by its catalytic effect on the thioridazine-H2O2 reaction. It was found that others d-electron ions of metals have the catalytic effect on phenothiazine reactions with H_2O_2 [21].

It is known that stability of color cation radical depends mainly on oxidation agent used. In the case of strong oxidant, the color of radical disappears quickly due to the second step of reaction which leads to the formation of a colorless sulphoxide. This effect resulted in decrease of sensitivity of assay and reproducibility. In purpose to improve these analytical properties indirect methods of phenothiazines determination have been proposed. One of them has been described by Basavaiah and Swamy [43]. They have applied potassium dichromate and iron-thiocyanate for spectrophotometric investigations of phenothiazines (chlorpromazine, promethazine, triflupromazine, trifluoperazine, fluphenazine, prochlorperazine). They have used a combination of dichromate and iron(III)-thiocyanate system for the determination of phenothiazines (Scheme 2).

Potassium dichromate as a strong oxidizing agent (couple Cr2O72-/Cr3+ E0=1.33 V, versus standard hydrogen electrode) oxidizes 2,10-disubstituted phenothiazines via colored radical cation to a colorless sulphoxide [43]. The excess of used oxidant has been further reduced by iron(II) ions. Next, the thiocyanate ions have been used for quantification of produced iron(III) ions. It has been stated that the absorbance of iron(III) thiocyanate solution is proportional to amount of the determined phenothiazines. The sensitivity of the proposed method has been the best at the molar ratio K2Cr2O7:PT equal to 1:6 at room temperature [43].

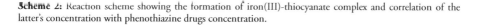

Scheme 2: Reaction scheme showing the formation of iron(III)-thiocyanate complex and correlation of the latter's concentration with phenothiazine drugs concentration.

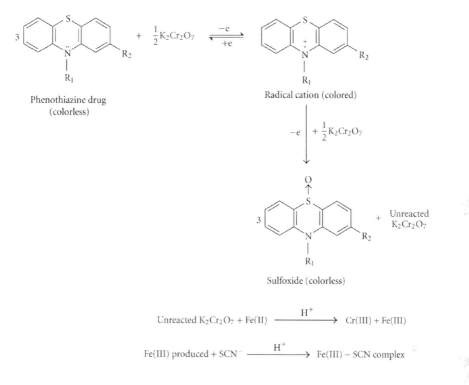

Conclusions

Based on information gathered in the presented review, it can be concluded that iron(III) ion and its anionic complexes are valuable reagents useful in an analysis of phenothiazines (PT). The mild oxidation potential of iron(III) and $K_3[Fe(CN)_6]$ allows quantification of phenothiazines in batch and flow systems. The proposed methods are characterized by simplicity, sensitivity, and good precision. The determination of PT by flow injection (FIA) methods is preferable to other conventional methods because they are fast (from 50 to 200 samples studied per hour) and precise (RSD values ranging from 0.6 to 2.5%).

The ability to create ion-pair compounds can be employed for selective and sensitive determination of iron ions(III) and phenothiazines as well.

References

1. R. R. Gupta, Ed., Phenothiazines and 1, 4-Benzothiazines: Chemical and Biomedical Aspects, R. R. Gupta, Ed., vol. 4 of Bioactive Molecules, Elsevier, Amsterdam, The Netherlands, 1988.

2. H. Puzanowska-Tarasiewicz and J. Karpińska, "Determination of phenothiazines in drugs," Pharmazie, vol. 47, no. 12, pp. 887–892, 1992.

3. J. Karpińska, B. Starczewska, and H. Puzanowska-Tarasiewicz, "Analytical properties of 2- and 10-disubstituted phenothiazine derivatives," Analytical Sciences, vol. 12, no. 2, pp. 161–170, 1996.

4. L. E. Lyons and J. C. Mackie, "Electron-donating properties of central sympathetic suppressants," Nature, vol. 197, no. 4867, p. 589, 1963.

5. I. Jelínek, I. Němcová, and P. Rychlovský, "Effects of salts on the stability of the cationic radical of phenothiazine derivatives," Talanta, vol. 38, no. 11, pp. 1309–1313, 1991.

6. E. Madej and P. Wardman, "Pulse radiolysis and cyclic voltammetry studies of redox properties of phenothiazine radicals," Radiation Physics and Chemistry, vol. 75, no. 9, pp. 990–1000, 2006.

7. H. Puzanowska-Tarasiewicz, M. Tarasiewicz, J. Karpińska, A. Kojło, E. Wołyniec, and E. Kleszczewska, "Analytical application of reactions of 2-and 10-disubstituted phenothiazines with some oxidizing agents," Chemia Analityczna, vol. 43, no. 2, pp. 159–178, 1998.

8. H. Puzanowska-Tarasiewicz, L. Kuźmicka, J. Karpińska, and K. Mielech-Łukasiewicz, "Efficient oxidizing agents for determination of 2,10-disubstituted phenothiazines," Analytical Sciences, vol. 21, no. 10, pp. 1149–1153, 2005.

9. E. Udsin, H. Eckert, and I. S. Forrest, Eds., Phenothiazines and Structurally Related Drugs: Basic and Clinical Studies, E. Udsin, H. Eckert, and I. S. Forrest, Eds., vol. 7 of Developments in Neuroscience, Elsevier, Amsterdam, The Netherlands, 1980.

10. M. Tarasiewicz, E. Wołyniec, and H. Puzanowska-Tarasiewicz, "Analytical application of the reactions of 2,10-disubstituted phenothiazines with organic substances," Pharmazie, vol. 53, no. 3, pp. 151–155, 1998.

11. W. Misiuk, H. Puzanowska-Tarasiewicz, L. Kuźmicka, and K. Mielech, "Application of the reaction of promazine hydrochloride with chromium(VI) in volumetric and spectrophotometric analysis," Journal of Trace and Microprobe Techniques, vol. 20, no. 3, pp. 305–316, 2002.

12. M. Jayarama, M. V. D'Souza, H. S. Yathirajan, and Rangaswamy, "Interaction of phenothiazines with nitroso-R salt and extractive spectrophotometric determination of phenothiazine drugs," Talanta, vol. 33, no. 4, pp. 352–354, 1986.

13. K. Basavaiah, "Determination of some psychotropic phenothiazine drugs by charge-transfer complexation reaction with chloranilic acid," Farmaco, vol. 59, no. 4, pp. 315–321, 2004.

14. W. Misiuk, L. Kuźmicka, K. Mielech, and H. Puzanowska-Tarasiewicz, "Examination of iron (III) and hexacyanoferrate (III) ions as reagents for the spectrophotometric determination of promazine and perazine," Acta Poloniae Pharmaceutica, vol. 58, no. 6, pp. 421–426, 2001.

15. H. Puzanowska-Tarasiewicz and J. Karpińska, "Analytical studies and application of reaction of promazine and thioridazine hydrochlorides with some oxidants," Acta Poloniae Pharmaceutica, vol. 60, no. 6, pp. 409–415, 2003.

16. M. Tarasiewicz, "Phenothiazine derivatives as new reagents in chemical analysis. II. colorimetric determination of iron(III)Phenothiazine derivatives as new reagents in chemical analysis. II. colorimetric determination of iron(III)," Chemia Analityczna, vol. 16, pp. 1179–1187, 1971.

17. J. Karpińska and H. Puzanowska-Tarasiewicz, "Application of the coupled redox and complexation reactions to flow injection spectrophotometric determination of promazine," Analytical Letters, vol. 30, no. 13, pp. 2365–2375, 1997.

18. H. Puzanowska-Tarasiewicz, M. Tarasiewicz, and Cz. Matel, "Use of ferric chloride for diethazine spectrophotometric determination," Farmacja Polska, vol. 36, no. 8, pp. 475–478, 1980.

19. A. Kojło, J. Karpińska, L. Kuźmicka, W. Misiuk, H. Puzanowska-Tarasiewicz, and M. Tarasiewicz, "Analytical study of the reaction of phenothiazines with some oxidants, metal ions, and organic substances (review article)," Journal of Trace and Microprobe Techniques, vol. 19, no. 1, pp. 45–70, 2001.

20. M. Tarasiewicz, H. Puzanowska-Tarasiewicz, W. Misiuk, A. Kojło, A. Grudniewska, and B. Starczewska, "Analytical applications of the reactions of 2-and 10-disubstituted phenothiazines with some metal ions," Chemia Analityczna, vol. 44, no. 2, pp. 137–155, 1999.

21. J. Karpińska, A. Kojło, W. Misiuk, B. Starczewska, and H. Puzanowska-Tarasiewicz, "Application of phenothiazine derivatives as reagents in kinetic-catalytic determination of some d-electron elements," Journal of Trace and Microprobe Techniques, vol. 18, no. 3, pp. 369–379, 2000.

22. T. Fukasawa, J. Karpińska, and H. Puzanowska-Tarasiewicz, "Thioridazine hydrochloride as a new reagent for high sensitive analysis of trace iron," Journal of Trace and Microprobe Techniques, vol. 13, no. 4, pp. 421–429, 1995.

23. A. A. Mohamed, "Catalytic determination of some phenothiazines," Qualifying thesis, Ain Shams University, Cairo, Egypt, 1995.

24. J. Kubalski and H. Tobolska-Rydz, The Addictive Agents, PZWL, Warsaw, Poland, 1984.

25. A. K. Davies, E. J. Land, S. Navaratnam, B. J. Parsons, and G. O. Phillips, "Pulse radiolysis study of chlorpromazine and promazine free radicals in aqueous solution," Journal of the Chemical Society, Faraday Transactions 1, vol. 75, pp. 22–35, 1979.

26. L. Levy, T. N. Tozer, L. Dallas Tuck, and D. B. Loveland, "Stability of some phenothiazine free radicals," Journal of Medicinal Chemistry, vol. 15, no. 9, pp. 898–905, 1972.

27. E. Pelizzetti and E. Mentasti, "Cation radicals of phenothiazines. Electron transfer with aquoiron(II) and -(III) and hexacyanoferrate(II) and -(III)," Inorganic Chemistry, vol. 18, no. 3, pp. 583–588, 1979.

28. E. Pelizzetti, E. Fisicaro, C. Minero, A. Sassi, and H. Hidaka, "Electron-transfer equilibria and kinetics of N-alkylphenothiazines in micellar systems," The Journal of Physical Chemistry, vol. 95, no. 2, pp. 761–766, 1991.

29. K. Basavaiah and J. M. Swamy, "Spectrophotometric determination of some phenothiazines using hexacyanoferrate(III) and ferriin," Chemia Analityczna, vol. 47, no. 1, pp. 139–146, 2002.

30. J. Karpińska, A. Kojło, A. Grudniewska, and H. Puzanowska-Tarasiewicz, "An improved flow injection method for the assay of phenothiazine neuroleptics in pharmaceutical preparations using Fe(III) ions," Pharmazie, vol. 51, no. 12, pp. 950–954, 1996.

31. M. A. Koupparis and A. Barcuchová, "Automated flow injection spectrophotometric determination of some phenothiazines using iron perchlorate: applications in drug assays, content uniformity and dissolution studies," Analyst, vol. 111, no. 3, pp. 313–318, 1986.

32. A. Kojło, H. Puzanowska-Tarasiewicz, and J. Martinez Calatayud, "Immobilization of hexacyanoferrate (III) for a flow injection-spectrophotometric determination of promazine," Analytical Letters, vol. 26, no. 3, pp. 593–604, 1993.

33. C. C. Nascentes, S. Cárdenas, M. Gallego, and M. Valcárcel, "Continuous photometric method for the screening of human urines for phenothiazines," Analytica Chimica Acta, vol. 462, no. 2, pp. 275–281, 2002.

34. M. J. Ruedas Rama, A. Ruiz Medina, and A. Molina Díaz, "Bead injection spectroscopy-flow injection analysis (BIS-FIA): an interesting tool applicable to pharmaceutical analysis: determination of promethazine and trifluoperazine,"

Journal of Pharmaceutical and Biomedical Analysis, vol. 35, no. 5, pp. 1027–1034, 2004.

35. H. S. Gowda and S. A. Ahmed, "N-substituted phenothiazines as redox indicators in bromatometry," Talanta, vol. 26, no. 3, pp. 233–235, 1979.

36. H. Basińska, H. Puzanowska-Tarasiewicz, and M. Tarasiewicz, "Phenothiazine derivatives as new reagents in chemical analysis," Chemia Analityczna, vol. 14, p. 883, 1969.

37. M. Tarasiewicz, "Determination of phenothiazine derivatives. V. The application of iron-thiocyanate complexes for colorimetric determination of chloropromazine, levomepromazine and promethazine," Acta Poloniae Pharmaceutica, vol. 29, no. 6, pp. 578–584, 1972.

38. H. Puzanowska-Tarasiewicz, Cz. Wyszyńska, and M. Tarasiewicz, "Phenothiazine derivatives as new redoks indicators in chemical analysis. IV. Chromatometric determination of hexacyanoferrate(II)," Chemia Analityczna, vol. 29, p. 105, 1984.

39. K. Ozutsumi, M. Kurihara, and T. Kawashima, "Structure of iron(III) ion and its complexation with thiocyanate ion in N,N-dimethylformamide," Talanta, vol. 40, no. 5, pp. 599–607, 1993.

40. M. Tarasiewicz, "Studies of the compounds of phenothiazine derivatives with acidocomplexes of metals. I. The properties of the compounds of chlorpromazine with hexacyanoferrite and hexacyanoferrate," Roczniki Chemii Annales Societatis Chimicae Polonorum, vol. 46, p. 2175, 1972.

41. H. Puzanowska-Tarasiewicz and J. Karpińska, "Use of sodium nitroso-ferricyanide for spectrophotometric determination of promazine hydrochloride," Farmacja Polska, vol. 44, no. 8, pp. 458–461, 1988.

42. J. Valero, "Formacion de complejos ternaries del Fe(III) y del Cu(II) con violeta de pirocatecol y clorpromacina," Química analítica, vol. 4, pp. 66–71, 1985.

43. K. Basavaiah and J. M. Swamy, "Application of potassium dichromate and iron-thiocyanate in the spectrophotometric investigations of phenothiazines," Il Farmaco, vol. 56, no. 8, pp. 579–585, 2001.

CITATION

Originally published under the Creative Commons Attribution License or equivalent. Puzanowska-Tarasiewicz H, Karpińska J, Kuźmicka L. Analytical Applications of Reactions of Iron(III) and Hexacyanoferrate(III) with 2,10-Disubstituted Phenothiazines. Int J Anal Chem. 2009;2009:302696. doi:10.1155/2009/302696.

Quantitative Mass Spectrometric Analysis of Ropivacaine and Bupivacaine in Authentic, Pharmaceutical and Spiked Human Plasma without Chromatographic Separation

Nahla N. Salama and Shudong Wang

ABSTRACT

The present study employs time of flight mass spectrometry for quantitative analysis of the local anesthetic drugs ropivacaine and bupivacaine in authentic, pharmaceutical and spiked human plasma as well as in the presence of *their impurities 2,6-dimethylaniline and alkaline degradation product. The method is based on time of flight electron spray ionization mass spectrometry technique without preliminary chromatographic separation and makes use of bupivacaine as internal standard for ropivacaine, which is used as internal standard for bupivacaine. A linear relationship between drug concentrations and the peak intensity ratio of ions of the analyzed substances is established. The method is linear from 23.8 to 2380.0 ng mL⁻¹ for both drugs. The correlation coefficient was ≥ 0.996 in authentic and spiked human plasma. The average percentage recoveries in the ranges of 95.39%–102.75% was obtained. The method is accurate (% RE < 5%) and reproducible with intra- and inter-assay precision (RSD% < 8.0%). The quantification limit is 23.8 ng mL-1 for both drugs. The method is not only highly sensitive and selective, but also simple and effective for determination or identification of both drugs in authentic and biological fluids. The method can be applied in purity testing, quality control and stability monitoring for the studied drugs.*

Keywords: ropivacaine, bupivacaine, 2,6-dimethylaniline, time of flight mass spectrometry, pharmaceutical formulations, spiked plasma

Introduction

Mass spectrometry (MS) is one of the most powerful analytical techniques, particularly for pharmaceutical analysis, where good selectivity and high sensitivity are often needed. In the pharmaceutical industry measurements of drugs and their metabolites in plasma are essential for drug discovery and development. The more accurate and rapid these measurements, the more quickly a drug can progress towards regulatory approval. Time-of-flight mass spectrometer (TOF-MS) delivers high sensitivity, resolution, and exact mass measurements .A variety of ion source and software options makes MS a versatile choice for a range of analytical challenges.[1-5]

Bupivacaine (Bup),amemberofthepipecoloxylidide group (Fig. 1), is the most commonly used local anesthetic. Commercial Bup is the optically inactive racemic (RS) mixture of R-and S-Bup. Several recent studies have demonstrated that systemic exposure to excessive quantities of Bup result in cardio toxicity due to its high affinity for, and dwell time at, voltage-gated sodium channels. A promising alternative to Bup is ropivacaine. Ropivacaine (Rop) is a structural derivative of Bup that differs only by the replacement of the butyl group on the piperidine nitrogen atom of Bup with a propyl group (Fig. 1). The minor structural modification leads to a reduced hydrophobicity and the decreased ability to diffuse into the heart and brain. As a result, Rop has lower systemic toxicity than Bup. In addition, Rop is manufactured as a pure S-enantiomer, further lowering the cardiotoxic potential. Both drugs act by blocking the conduction of impulses in target nerve structures, primarily located within the subarachinoid space.6–8 Several methods using GC with9,10 or without MS,11 high performance liquid chromatography (HPLC) with UV,12–14 mass spectrometery (MS)15–17 or amperometric detection,18 and capillary electrophoresis19 have been developed to analyze the drugs and their impurities in pharmaceutical formulations and/or biological fluids.

The aim of this study is to develop rapid, accurate and simple method for determination of both drugs in authentic, pharmaceutical, spiked human plasma as well as in presence of their impurities without chromatographic separation. The described method was not investigated previously.

In this work we describe how a simple TOF ES-MS analytical method can be used to determine Bup and Rop in authentic and pharmaceutical formulations as well as in spiked human plasma. The technique has many advantages; no need for method development, a short analytical time (1.5 min), and a minimal amount of solvent being required, coupled with high sensitivity, selectivity and exact mass measurements.

Experimental

Materials and Reagents

Ropivacaine was kindly supplied by AstraZeneca Co., UK, certified to contain 99.00%, CAS No. 132112-35-7. NaropinTM vial containing 7.5 mg mL^{-1} ropivacaine hydrochloride (AstraZeneca Co., UK) was purchased from local market. Bupivacaine was kindly supplied by Al-Debeiky Pharma Co., Egypt, its purity was found to be 99.60% according to BP 2008. Bucain vial containing 5.0 mg mL^{-1} bupivacaine hydrochloride (DeltaSelect, GmbH, Germany) was purchased from the market. Human plasma was kindly donated from volunteers. The following reagents and solvents were purchased and used without further purification: 2,6-dimethylaniline (99.00% Aldich UK, CAS No. 87-62-7), methanol, chloroform (HPLC grade, Fisher Scientific, UK), acetonitrile (LC-MS grade, Reidel-dehaen, UK), ultra pure water (ELGA, UK), formic acid (Sigma-Aldrich, UK), sodium hydroxide (BDH, UK) and hydrochloric acid (Certified Fisher Scientific, UK).

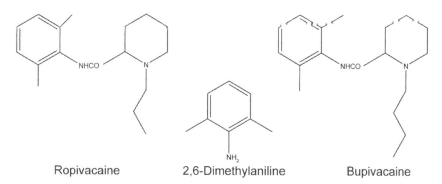

Figure 1. Chemical structures of ropivacaine, bupivacaine and their impurity 2,6-dimethylaniline.

TLC-Separation

Pre-coated TLC plates (10 × 10 cm, aluminium plates coated with 0.25 mm silica gel F254) were purchased from Merck UK. UV-Radiation (Cole-Parmer Instrument, France) detective wavelength was 254 nm.

TOF-ES MS Measurements

The TOF-ES-MS measurements were performed using WATERS—2795 (Waters, UK) equipped with an autosampler injector (10 μL) and Mass Lynx v 4.1. The system was operating in the following regime:electrosprayvoltage,3 kv;capillarytemperature, 150°C; sample solution flow rate, 0.1 mL/min. All analysis was performed in the positive ion detection mode. All samples were dissolved in a 50% solution of acetonitrile in water containing 0.1% formic acid.

Preparation of Alkaline Degradation Products

The degradation products were laboratory prepared by heating 10 mg of bupivacaine hydrochloride in 1 M NaOH (10 mL) for 4 hours on hot plate at temperature 100°C. The solution was neutralized with 1M HCl then the solution was evaporated and diluted to 10 mL with methanol. Complete degradation was monitored by TLC. The TLC separation was carried out by using chloroform-methanol (9:1 v/v) as the mobile phase.[20] The R_f value is 0.71% ± 0.002% for both drugs. The major alkaline degradation product, namely 2,6-dimethylaniline, which has an R_f value of 0.81% ± 0.001% as identified by comparison with the reference standard.

Standard Solutions and Calibration Curves

Stock solutions of ropivacaine and bupivacaine were prepared in methanol at a concentration of 1 mg mL $^{-1}$, and stored at 4°C. These were further diluted with 50% aqueous acetonitrile containing 0.1% formic acid to give the appropriate working solutions. Working solutions of each drug were prepared to yield final concentrations of 23.8, 59.5, 119.0, 238.0, 595.0, 1190.0 and 2380.0 ng mL $^{-1}$ by further dilution with the same solvent. Bupivacaine (2380.0 ng mL^{-1}) was used as IS for ropivacaine. Ropivacaine (2380.0 ng mL^{-1}), was used as IS for bupivacaine.

Preparation of Spiked Human Plasma

Aliquots equivalent to 71.4 – 7140.0 ng mL^{-1} of each drug and 7140.0 ng mL^{-1} of the internal standard in 1 mLplasma were sonicated for 5 minutes. Acetonitrile (2 mL) was added and then centrifuged at 7000 rpm for 30 minutes. One milliliter of the supernatant was evaporated.

Laboratory Prepared Solutions

Mixturesofropivacaineandbupivacainewereprepared by mixing different concentrations of each drug with its impurity (2,6-dimethylaniline) and alkaline degradation product, where the ratio 0.1%–10.0% of the mixtures were obtained.

Procedure

Ten μL each of the above solutions was injected in the TOF-ES-MS under conditions mentioned above. The characteristic m/z ions used for identification and determination of Rop, Bup and 2,6-DMA were m/z = 275, m/z = 289 and m/z = 122 [M + H]$^{+}$, respectively.

Analysis of Pharmaceutical Preparations

Milliliters equivalent to 50 mg from the corresponding drug vial were transferred quantitatively into 50 mL volumetric flasks and made up to the volume with methanol. The procedure was completed as mentioned above.

Calculations

The calibration curves were calculated by unweighted least-squares linear regression analysis of the concentrations of the analyte versus the peak intensity ratio of ions of analyzed substance of ropivacaine (m/z 275) to that of the IS (m/z 289). As for bupivacaine (m/z 289) to that of IS (m/z 275) was used. Concentrations of unknown samples were determined by applying the linear regression equation of the standard curve to the unknown sample's peak intensity ratio.

Method Validation

Limit of Quantification

The limit of quantification of the two drugs was defined as the lowest concentration of the calibration curve.

Precision and Accuracy

Precision and accuracy were assessed by assaying freshly prepared solutions of the two drugs in triplicate at three concentration levels; 59.5, 1190.0 and 2380.0 ng mL^{-1}. Precision is reported as relative standard deviation (RSD%) of the estimated concentrations and accuracy (% Relative error) expressed as [measured-nominal/nominal $\times$ 100].

Selectivity and Specificity

Specificity is the ability of the method to measure the analyte response in the presence of impurity and or degradation products. For specificity determination, synthetic mixtures of different percentages of 2,6-dimethylaniline and degradation products of each drug were added to each pure drug sample. The recovery percent was calculated.

Results and Discussion

The work includes (1) mass spectrometric identification and determinations of ropivacaine and bupivacaine; (2) generation of the standard calibration curves; (3) identification and determination of drug substances in spiked human plasma; (4) determination of both drugs in presence of alkaline degradation and/or impurity(2,6-DMA); and (5) quantitative analysis of the individual ropivacaine, bupivacaine in their pharmaceutical preparations.

The mass spectra of ropivacaine and bupivacaine and their internal standards are shown in Figure 2. Under the conditions of TOF ES-MS in positive mode, the spectra displays intense peaks of [M + H]+ with ions of the highest mass to charge, e.g. m/z = 275.2154 for ropivacaine and 289.2226 for bupivacaine, respectively. Linearity range was found to hold good over a concentration range of 23.8 – 2380.0 ng mL-1 for both drugs. The results of regression data are presented in Table 1.

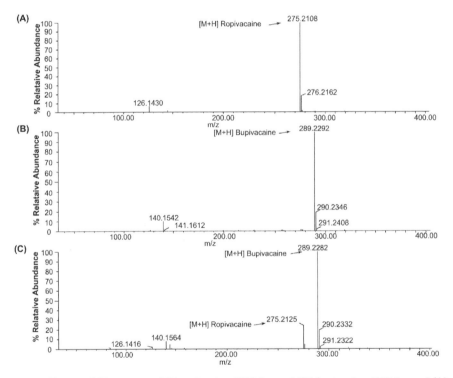

Figure 2. The typical Mass spectra of (A) ropivacaine, 2380.0 ng mL⁻¹(B) bupivacaine, 2380.0 ng mL⁻¹(c) ropivacaine (1190.0 ng mL⁻¹, analyte) and bupivacaine (2380.0 ng mL⁻¹, internal standard) in 50% aqueous solution of acetonitrile containing 0.1% formic acid.

Analysis of studied drugs in spiked human plasma are shown in Figure 3. The linearity was in the range of 59.5–2380.0 ng mL-1 for Rop while it is within 23.8–2380.0 ng mL-1 for Bup in 1 mL plasma sample, which is the anticipated concentration range in clinical investigation of drug pharmacokinetics.

The maximum plasma level of Rop and Bup after different rout of administrations were found to be more than 100 ng mL-1 21–23 which could be assessed by the proposed method. The high sensitivity of the proposed method allowed the determination of both drugs in spiked human plasma. Linear regression analysis of the data gives the equations, A = 0.02707 C + 0.0082, r = 0.996, for Rop and A = 0.5441 C – 0.0137, r = 0.999, for Bup. Where A is the peak intensity ratio for m/z = 275/289 for Rop and 289/275 for Bup, C is the concentration in ng mL-1 and r is correlation coefficient.

The data stated in Table 1, indicate that the method is efficient for determination of the studied drugs in biological fluids as there is no significant differences between the results for determinations of both drugs in authentic and spiked plasma samples.

Validation Data

Linearity and Limit of Quantification

Calibration curves for ropivacaine and bupivacaine exhibited good linearity over the concentration range studied (23.8–2380.0 ng mL^{-1}) for both drugs in authentic and spiked human plasma as stated in Table 1. From the results, it is clear that there is no interference from the plasma matrix demonstrating the efficiency for determination of the drugs in biological fluids by TOF ES-MS. The limit of quantification (LOQ) was chosen as the lowest calibration standard concentration (23.8 ng mL^{-1}) for the studied drugs in authentic and spiked human plasma.

Precision and Accuracy

Table 2 summarizes mean values, precision, and accuracy of intra-and inter-assay analysis.Precision and accuracy were within the ranges acceptable for analytical and bio-analytical purposes. Intra-day precision ranged from 0.60 to 3.61% for ropivacaine while 2.16 to 3.33% for bupivacaine in drug substances. While in spiked human plasma ranged from 1.07 to 7.98% for ropivacaine and from 0.95 to 5.13 for bupivacaine (Table 3). Inter-day precision did not exceed 8.0% over the three level concentrations for three days in drug substances and spiked human plasma. The accuracy of the technique was considered satisfactory, since between-day variation over the concentration range studied was found to be less than 5%.

Table 1. Linearity, recovery and LOQ of TOF-ES MS assay for ropivacaine and bupivacaine in authentic and spiked human plasma.

Parameters	Ropivacaine		Bupivacaine	
	Authentic	Spiked human plasma	Authentic	Spiked human plasma
Linearity ng mL^{-1}	23.8–2380.0	59.5–2380.0	23.8–2380.0	23.8–2380.0
Regression equation				
Slope (b)[a]	0.3645	0.02707	0.6227	0.5441
SE of slope	0.007813	0.011644	0.016894	0.013801
Intercept (a)[a]	0.0133	0.0082	0.0266	−0.0137
SE of intercept	0.009564	0.01303	0.018896	0.016894
Correlation coefficient (r)	0.999	0.996	0.998	0.999
SE of estimation	0.014768	0.023327	0.034049	0.026085
Recovery Mean[b] ± RSD%	98.83 ± 3.03	95.39 ± 3.64	99.61 ± 3.20	99.96 ± 2.88
LOQ ng mL^{-1}	23.8	59.9	23.8	23.8

[a]Regression equation, A = a + bc, where A is the peak intensity ratio for m/z = 275.0 /289.0 for Rop, and A is the peak intensity ratio for m/z = 289.0/275.0 for Bup, C is the concentration.
[b]n = 6.

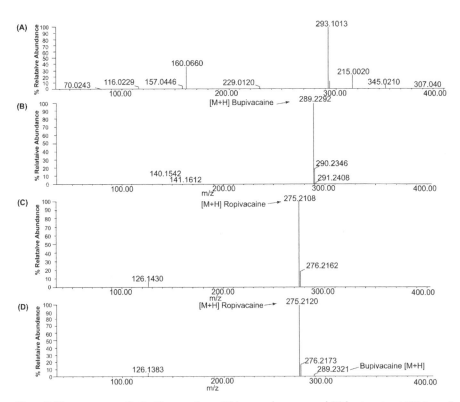

Figure 3. The mass spectra of spiked human plasma (A) human plasma, control (B) bupivacaine, 1190.0 ng mL^{-1} (c) ropivacaine, 1190.0 ng mL^{-1}, (D) mixture of bupivacaine, 23.8 ng mL^{-1} (analyte) and ropivacaine, 2380.0 ng mL^{-1} (Is) in 50% aqueous solution of acetonitrile containing 0.1% formic acid.

Table 2. Intra and inter-day precision and accuracy of TOF-ES MS assay for ropivacaine and bupivacaine in authentic samples.

Drug substances	Conc. ng mL^{-1}	Precision[a] RSD%		Accuracy[a] RE%	
		Inter	Intra	Inter	Intra
Ropivacaine	59.5	3.61	3.57	−4.67	4.59
	1190.0	1.72	2.02	−0.58	−2.19
	2380.0	0.60	1.18	−2.34	2.00
Bupivacaine	59.5	3.33	5.16	2.86	3.12
	1190.0	2.16	2.01	1.05	2.30
	2380.0	2.61	0.96	1.50	3.12

[a]n = 3.

Table 3. Intra and inter-day precision and accuracy of TOF-ES MS assay for ropivacaine and bupivacaine in spiked human plasma.

Drug substances	Conc. ng mL^{-1}	Precision[a] RSD%		Accuracy[a] RE%	
		Intra	Inter	Intra	Inter
Ropivacaine	59.5	7.98	6.50	−3.67	−2.00
	1190.0	3.88	5.75	1.90	3.04
	2380.0	1.07	1.40	0.85	2.50
Bupivacaine	59.5	5.13	4.85	3.66	−1.62
	1190.0	1.95	2.74	−1.07	0.22
	2380.0	0.95	2.99	0.33	0.10

[a]n = 3.

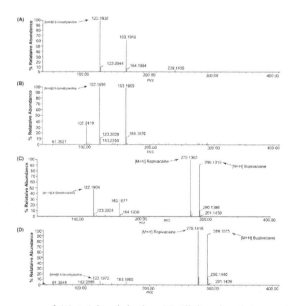

Figure 4. The mass spectra of (A) 2,6-dimethylaniline (B) Alkaline degradation products (c) mixture of ropivacaine, 2380.0 ng mL^{-1} (internal standard), bupivacaine 1190.0 ng mL^{-1} (analyte) and their impurity 2,6-dimethylaniline (D) mixture of both drugs with alkaline degradation products, in 50% aqueous solution of acetonitrile containing 0.1% formic acid.

Selectivity and Specificity

Ropivacaine and bupivacaine are amides expected to alkaline degradation through cleavage of amide linkage with production of 2,6-dimethylaniline. Solutions of alkaline degradations of each were tested by TLC against pure sample of 2,6-dimethylaniline. The same spots of each have the same R_f (0.81) of the pure compound. For further confirmation, TOF-ES-MS was carried for each of both compounds. The product obtained from alkaline degradation has m/z = 122.1935 corresponds to protonated 2,6-dimethylaniline.

The mass spectrometric determinations of Rop and Bup in the presence of their alkaline degradation products are shown in Figure 4. Synthetic compound 2,6-dimethylaniline was used as control, and in the spectrum (A) the ion of the mass to charge (m/z) 122.1935 corresponding to 2,6-dimethylaniline was identified. The highest ion peak was m/z = 163.1910 (122 + 41) which might be resulted in acetonitrile solvent interference in the system. The same peaks appeared in the spectrum (B) and 2,6-dimethylaniline was the major degradation product of bupivacaine. In addition, a relative low of molecular ion peak at m/z = 102.2419 was observed which may be assigned as a m-xylene ion. The spectra (C) and (D) display the intensive ion peaks with m/z = 122.1935 [M + H]+ for 2,6-dimethylaniline clearly indicating 2,6-dimethylaniline to be the major alkaline degradation product of Rop and Bup. The specificity was also assessed by analyzing synthetic mixtures of each drug with its alkaline degradation product in concentration ranging from 0.1 to 10.0%. The results reveal the high selectivity and sensitivity of the method which can determine the impurity in concentration down to 0.1% present in both drugs (Table 4).

Table 4. Specificity of TOF ES-MS assay for ropivacaine and bupivacaine in authentic samples.

Degradation products and/or 2,6-DMA%	Recovery[a] % ± RSD%	
	Ropivacaine	Bupivacaine
0.1	99.79 ± 2.47	100.04 ± 2.77
1.0	101.05 ± 2.16	100.02 ± 1.04
10.0	98.48 ± 1.29	98.18 ± 1.67

[a]Mean of four different experiments.

Table 5. results for the determination of ropivacaine and bupivacaine in pharmaceutical formulations by the proposed TOF-ES MS method.

Preparations	TOF ES-MS	
	Mean recovery[a] %	RSD%
Naropin vial, 7.5 mg mL^{-1} ropivacaine hydrochloride	102.75	1.75
Bucain vial, 5.0 mg mL^{-1} bupivacaine hydrochloride	100.33	2.51

[a]Mean of five experiments.

Analysis of Pharmaceutical Preparations

The method was applied to determine ropivacaine and bupivacaine in Naropin and Bucaine vial respectively. The % RSD was less than 3.0%, indicating the precision of the method, the results are presented in Table 5.

Conclusion

In this manuscript, we described a newly developed TOF-MS based method for quantitative determination of ropivacaine andbupivacaine in authentic, pharmaceutical dosage forms and the spiked human plasma without chromatographic separation. The strategy of this approach consists indirect multi-ion detection of analytes with reference to internal standards with close structures to the analyte. The method can also be used to identify the degradation products in minute amounts in presence of the corresponding ropivacaine or bupivacaine. The method could be routinely used for quantitative drug analysis in pharmaceutical formulations and biological media as well as for assessing drug purity and stability.

Acknowledgements

N. Salama would like to thank the NODCAR, Egypt and School of Pharmacy, The University of Nottingham, UK, for the visiting scholarship awards.

Disclosure

The authors report no conflicts of interest.

References

1. Karatasso YO., Logunova IV., Sergeeva MG., Nikolaev EN., Varfolomeev SD., Chistyakov VV. Pharmaceutical Chemistry Journal. 2007;41:45.

2. Jayasimhulu K., Hunt SM., Kaneshiro ES., Watanabe Y., Giner JL. Am J Mass Spectrom. 2006;18:394.

3. Han X., Yang K., Yang J., et al. Mass Spectrom. 2006;17:264.

4. Delong CJ., Baker PR., Samuel M., Cui Z., Thomas MJ. Lipid Res J. 2001;42:1959.

5. Koivusalo M., Haimi P., Heikinheimo L., Kostainen R., Somerharju P. Lipid Res J. 2001;42:663.

6. The United States Pharmacopoeia. The National Formulary USP 30 United States Pharmacopoeial Convection Inc. 2007. p. 1562.

7. Sweetman SC., Martindale. The Extra Pharmacopoeia, 34 ed, Pharmaceutical Press, London 2005. p. 1371, 1372, 1377.

8. Brunton LL., Lazo JS., Parker KL. Goodman and Gilman's. The Pharmacological Basis of Therapeutics, 11th. Ed. Mc Grow—Hill, New York, U.S.A. 2006. p. 377.

9. Tahraoui A., Watson DG., Skellern GG., Hudson SA., Petrie P., Faccenda K. J Pharm Biomed Anal. 1996;15:251.

10. Watanabe T., Namera A., Yashiki M., Iwasaki Y., Kojima T. J Chromatogr B. 1998;709:225

11. Baniceru M., Croitoru O., Popescu SM. J Pharm Biomed Anal. 2004;593:35.

12. Gross AS., Nicolay A., Eschalier A. J Chromatogr B. 1999;728:107.

13. Einosuke T., Takako N., Shinichi I., Katsuya H. J Chromatogr B. 2006;834:213.

14. Rifai N., Hsin O., Hope T., Skamoto M. Thera Drug Monitor. 2001;23:182.

15. Kawano S., Murakita H., Yamamoto E., Asakawa N. J Chromatogr B Anal Tech Biomed and Life Sci. 2003;792:49.

16. Koehler A., Oertel R., Kirch W. J Chromatogr B. 2005;1088:126.

17. Abdel-Rehim M., Bielenstein M, Askemark Y. Anal Chim Acta. 2003;49:253.

18. Zbigniew F., Emil B., Agata P., Malgorzata WG. J Pharm Biomed Anal. 2005;37:913.

19. Krisko RM., Schieferecke MA., Williams TD. Lunte CE. Electrophoresis. 2003;24:2340.

20. Florey K. Analytical Profiles of Drug Substances and Excipients; Academic Press: Inc., 1993;19:8.

21. Paut O., Schreiber E., Lacroix F., Meyrieux V., Simon N., Lavrut T., Camboulives J., Bruguerolle B. Br J Anaesth. 2004;92:416.

22. Concepcion M., Richard Arthur G., Susan M., Steele Angela M., Bader BG. Covino Anesth Analg. 1990;70:80.

23. Covino BG., Feldman HS., Arthur GR. Anesth Analg. 1988;67:1053.

CITATION

Kinetic Spectrophotometric Determination of Certain Cephalosporins in Pharmaceutical Formulations

Mahmoud A. Omar, Osama H. Abdelmageed
and Tamer Z. Attia

ABSTRACT

A simple, reliable, and sensitive kinetic spectrophotometric method was developed for determination of eight cephalosporin antibiotics, namely, Cefotaxime sodium, Cephapirin sodium, Cephradine dihydrate, Cephalexin monohydrate, Ceftazidime pentahydrate, Cefazoline sodium, Ceftriaxone sodium, and Cefuroxime sodium. The method depends on oxidation of each of studied drugs with alkaline potassium permanganate. The reaction is followed spectrophotometrically by measuring the rate of change of absorbance at 610 nm. The initial rate and fixed time (at 3 minutes) methods are utilized for construction of calibration graphs to determine the concentration of the studied drugs. The calibration graphs are linear in the concentration ranges 5–15 µg mL–1 and 5–25 µg mL–1 using the initial rate and fixed time methods, respectively. The results are validated statistically and checked through recovery studies. The method has been successfully applied for the determination of the studied cephalosporins in commercial dosage forms. Statistical comparisons of the results with the reference methods show the excellent agreement and indicate no significant difference in accuracy and precision.

Introduction

Cephalosporins consist of a fused β-lactam-Δ3-dihydrothiazine two-ring system, known as 7-aminocephalosporanic acids (7-ACAs) and vary in their side chain substituents at C3 (R2) and C7 (acylamido, R1). The chemical structure of the studied cephalosporins in this work is shown in Table 1. They are used for treatment of infection caused by both gram-negative and gram-positive bacteria [1, 2]. A wide variety of analytical methods have been reported for determination of cephalosporins in pure form, in pharmaceutical preparations, and in biological fluids. These methods include spectrophotometry [2–5], atomic absorption spectrophotometry [6], fluorometry [7–12], liquid chromatography [13–20], Micellar electrokinetic capillary chromatography [21, 22], chemiluminescence [23–28], potentiometric [29, 30], and polarographic [31–34] methods. Kinetic spectrophotometric methods became of great interest in chemical and pharmaceutical analyses [35]. The literature is still poor in analytical procedure based on kinetics, especially for determination of drug in commercial dosage forms. We aimed to improve on the current methods by employing the kinetic colorimetric oxidation of cephalosporins to increase selectivity, avoid interference of colored and/or turbidity background of samples and consequently determination of low concentration of the cited drugs as possible.

Experimental

Apparatus

Spectronic Genesys 2PC. Ultraviolet/Visible spectrophotometer (Milton Roy Co, USA) with matched 1 cm quartz cell was used for all measurements connected to IBM computer loaded with winspec application software.

Table 1. Structural formula of the studied cephalosporins.

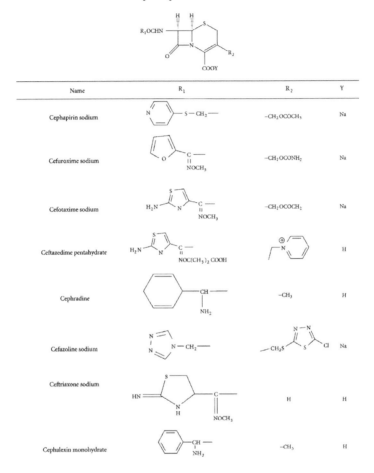

Name	R$_1$	R$_2$	Y
Cephapirin sodium		$-CH_2OCOCH_3$	Na
Cefuroxime sodium		$-CH_2OCONH_2$	Na
Cefotaxime sodium		$-CH_2OCOCH_3$	Na
Ceftazedime pentahydrate			H
Cephradine		$-CH_3$	H
Cefazoline sodium		$-CH_2S$	Na
Ceftriaxone sodium		H	H
Cephalexin monohydrate		$-CH_3$	H

Materials and Reagents

All the materials were of analytical reagent grade, and the solutions were prepared with double-distilled water. Samples of cephalosporin were generously supplied by their respective manufacturers and were used without further purification.

(i) Cephalexin monohydrate, Ceftazedime pentahydrate, and Cefuroxime sodium (Galaxowelcome Egypt, S.A.E, El Salam City, Cairo, Egypt).

(ii) Cephapirin sodium, Cefazoline sodium (Bristol Myers- Squib Pharmaceutical Co., Cairo, Egypt).

(iii) Cefotaxime sodium, cephradine, and Ceftriaxone sodium (EIPICO, Tenth of Ramadan City, Cairo, Egypt).

The purity of authentic samples was checked by UV assay methods and was not less than 99.26±0.72.

(i) Potassium permanganate (Merck, Germany); 6×10-3 Molar solution was prepared by dissolving 100 mg in 100 mL of double-distilled water followed by boiling and filtration through sintered glass.

Potassium permanganate solution should be freshly prepared and its molarity was checked titrimetrically.

(i) Sodium hydroxide (El Nasr chemical co., Abuo Zabbel, Egypt); 0.75 M prepared by dissolving 3 g in 100 mL of double-distilled water.

(ii) Methanol (Merck, Darmstadt, Germany).

Pharmaceutical Formulations

The following available commercial preparations were analyzed.

Ceporex tablets and vials, Fortam vials and Zinnat vials (Galoxowelcome Egypt, S.A.E, El Salam City, Cairo, Egypt), labeled to contain 250 mg cephalexin monohydrate per tablet and 500 mg cephalexin sodium equivalent to 500 mg cephalexin monohydrate per vial, 250 mg ceftazidime pentahydrate per vial, and 250 mg cefuroxime sodium per vial, respectively. Cefatrexyl vials, Totacef vials and Velosef tablets and suspensions (Bristol Myers-Squibb Pharmaceutical Co., Cairo, Egypt), labeled to contain 500 mg cephapirin sodium per vial, 500 mg cefazoline sodium per vial, and 500 mg cephradine per tablet and suspensions, respectively. Cefotax vials (EIPICO. Tenth of Ramadan City, Cairo, Egypt), labeled to contain 250 mg cefotaxime sodium per vial. Ceftriaxone vials (Novartis Pharma S.A.E, Cairo, Egypt), labeled to contain 500 mg ceftriaxone sodium per vial.

Preparation of Standard Solution

Stock solution containing 1 mg mL^{-1} of each cephalosporins was prepared in double-distilled water, working standard solutions containing 100 µg mL^{-1} were prepared by suitable dilution of the stock solutions with double-distilled water.

Recommended Procedure for Cephalosporins Determination

Initial Rate Method

Aliquots of 50–150 µg mL^{-1} of studied cephalosporins test solutions were pipetted into a series of 10 mL volumetric flask. 1.2 mL of sodium hydroxide solution (0.75 M) was added followed by 3.0 mL of potassium permanganate solution

(6×10^{-3} M) to each flask and then diluted to the volume with double-distilled water at $30 \pm 1°C$. The content of mixture of each flask was mixed well and the increase in absorbance at 610 nm was recorded as a function of time for 15 minutes against reagent blank treated similarly. The initial rate of the reaction (v) at different concentrations was obtained from the slope of the tangent to absorbance time curves. The calibration graphs were constructed by plotting the logarithm of the initial rate of the reaction (log v) versus logarithm of molar concentration of the studied drugs (log C).

Fixed Time Method

In this method, the absorbance of each sample solution at preselected fixed time (3 minutes) was accurately measured and plotted against the final concentration of the drug.

Determination of the Studied Drugs in Pharmaceutical Formulations

Procedure for Tablets

An accurately weighed amount equivalent to 100.0 mg of each drug from composite of 20 powdered tablets was transferred into a 100 mL volumetric flask. Dissolved in about 20 mL methanol, swirled, and sonicated for 10 minutes, the resultant mixture was filtered into round bottom flask for removal of any pharmaceutically "inert" ingredients (cellulose, disaccharides) that could be subject to oxidation by permanganate. The residue was washed thoroughly with about 5 mL methanol and the combined filtrate as well as washing solutions was subjected to evaporation under vacuum till dryness. The residue lifted was dissolved in about 20 mL distilled water and filtered into 100 mL volumetric flask. The filter paper was washed thoroughly with double distilled water, and then the combined filtrate as well as washing solutions was mixed well and completed to volume with the same solvent to obtain solution of 1.0 mg mL^{-1}. The final solution was diluted quantitatively with the same solvent to obtain working standard solution of 100.0 µg mL^{-1}, then the general procedure was followed.

Procedure for Capsules and Suspension

The contents of 20 capsules were evacuated and well mixed. Then an accurately weighed amount equivalent to 100.0 mg evacuated capsules or dry powder suspension of each drug was transferred into a 100 mL beaker, and then the procedure was continued as described under tablets.

Procedure for Vials

A 100.0 mg quantity of each vial was transferred into a 100 mL volumetric flask, dissolved, and completed to the mark with double-distilled water to obtain solution of 1.0 mg mL^{-1}. Further dilutions with double-distilled water were made to obtain sample solutions (100.0 μg mL^{-1}), then the general procedure was followed.

Results and Discussion

Potassium permanganate as strong oxidizing agent has been used in oxidimetric analytical method for determination of many compounds [36–39]. During the course of the reaction, the valence of manganese changes. The heptavalent manganese ion changes to the green color (Mn VI), while in neutral and acidic medium, the permanganate is further reduced to colorless (Mn II). The behavior of permanganate was the basis for its uses in development of spectrophotometric method. The absorption spectrum of aqueous potassium permanganate solution in alkaline medium exhibited an absorption band at 530 nm. The additions of any of the studied drugs to this solution produce a new characteristic band at 610 nm (Figure 1). This band is due to formation of manganate ion, which resulted from the oxidation of cephalosporin by potassium permanganate in alkaline medium. The intensity of the color increases with time; therefore a kinetically based method was developed for determination of cephalosporins in their pharmaceutical dosage formulations. The different variables that affect the formation of manganate ion were studied and optimized.

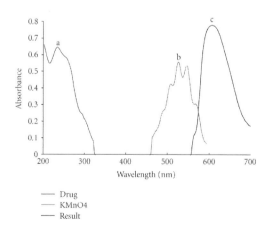

Figure 1. Absorption spectra of (a) alkaline potassium permanganate (6×10^{-3} M) (b) cefotaxime (15 μg mL^{-1}) and (c) the reaction product.

Effect of Potassium Permanganate Concentration

The absorbance increases substantially with increasing the concentration of potassium permanganate (Figure 2). Maximum absorbance was obtained when 2.5 mL of 6×10^{-3} M of potassium permanganate was used. Thus, the adoption of 3 mL of potassium permanganate in the final solution proved to be adequate for the maximum concentration of cephalosporin used in determination process (the concentration of the final assay was 1.8×10^{-3} M).

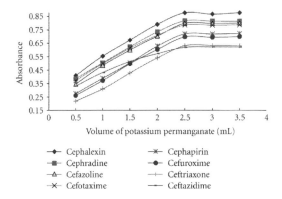

Figure 2. Effect of potassium permanganate (3×10^{-6}) on the reaction between the investigated cephalosporins $(15 \ \mu g \ mL^{-1})$ and alkaline potassium permanganate.

Effect of Sodium Hydroxide Concentration

Maximum absorption was obtained when 1 mL of 0.75 M NaOH was used. Over this volume no change in absorbance could be detected. So 1.2 mL of 0.75 M of NaOH was used as an optimum value (Figure 3).

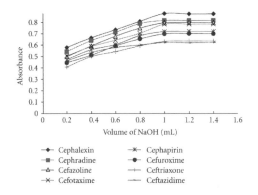

Figure 3. Effect of sodium hydroxide concentration (0.75 M) on the reaction between the investigated cephalosporins $(15 \ \mu g \ mL^{-1})$ and alkaline potassium permanganate.

Effect of Temperature

At room temperature the reaction rate increases substantially with time, although heating the solution was found to increase the rate of the reaction however but MnO_2 was precipitated, therefore room temperature was selected as the optimum temperature.

Stoichiometry and Reaction Mechanism

The stoichiometric ratio between potassium permanganate and each of investigated cephalosporins was determined by Job's method [40, 41] and was found to be 1:1 (Figure 4). Cephalosporins were found to be susceptible for oxidation with alkaline potassium permanganate producing a green color peaking at 610 nm. Therefore, the reaction mechanism is proposed on the basis of the literature background (39) and our experimental study as shown in Scheme 1.

Scheme 1

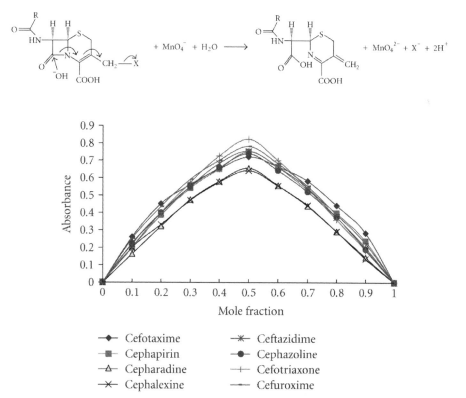

Figure 4. Job's plots of continuous variation between potassium permanganate and the studied drugs.

The opening of the β-lactam ring by the hydroxyl ion proceeds via intermediate and result in the formation of cephalosporoic acid and the intermediate formation is the rate limiting step.

Kinetic of the Reaction

Under the optimum conditions, the absorbance time curves of investigated cephalosporins with potassium permanganate reagent were constructed (Figures 5, 6 for cefotaxime as a representative example). The initial rate of the reaction was determined from the slope of tangents of the absorption time curves. The order of the reaction with respect to permanganate was determined by studying the reaction at different concentrations of permanganate with fixed concentration of investigated cephalosporins. The plot of initial rate ($\Delta A/\Delta t$) against initial absorbance was linear passing through origin indicating that the initial order of the reaction with respect to permanganate was 1. The order with respect to investigated cephalosporins was evaluated by measuring the rate of the reaction at several concentrations of cephalosporins at a fixed concentration of permanganate reagent. This was done by plotting the logarithm of initial rate of the reaction versus logarithm of molar concentration of investigated cephalosporins and was found to be 1. However under the optimized experimental conditions, the concentrations of cephalosporins were determined using relative excess amount of potassium permanganate and sodium hydroxide solutions. Therefore pseudo-zero-order conditions were obtained with respect to their concentrations.

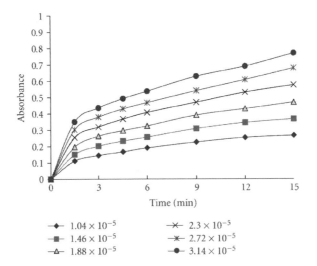

Figure 5. Absorption versus time for the reaction between cefotaxime (different molar concentrations) and $KMnO_4$ (1.8×10^{-3} mol L^{-1}).

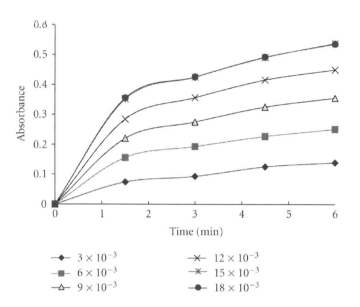

Figure 6. Absorption versus time for the reaction between cefotaxime 3.14×10^{-5} mol L^{-1} and KMnO$_4$ (different molar concentrations).

Quantitation Methods

Initial Rate Method

The initial rate of the reaction would follow pseudo-first-order and were found to obey the following equation:

$$v = \frac{\Delta A}{\Delta t} = K'C'', \qquad (1)$$

where v is the reaction rate, A is the absorbance, t is the measuring time, K′ is the pseudo-first-order rate constant, C is the molar concentration of cephalosporins, and n is the order of the reaction. The logarithmic form of the above equation is written as follows:

$$Logv = \log \frac{\Delta A}{\Delta t} = \log K' + n \log C. \qquad (2)$$

Regression analysis using the method of least square was performed to evaluate the slopes, intercepts, and correlation coefficient.

The analytical parameters and results of regression analysis are given in Table 2. The value of n($\approx$1) in the regression equation confirmed that the reaction of cephalosporins with the potassium permanganate was pseudo-first-order with respect to cephalosporins concentration. The limits of detection (LOD) were calculated and results obtained confirmed good sensitivity of the proposed method and consequently their capabilities to determine low amount of cephalosporins.

Table 2. Analytical parameters for the initial rate method for determination of investigated cephalosporins with alkaline potassium permanganate.

Investigated cephalosporin	Linear range, $M \times 10^{-5}$ ($\mu g\,mL^{-1}$)	Least square equation $\log V = \log K' + n \log C$		Correlation coefficient (r)	LOD $\mu g\,mL^{-1}$
		Intercept ($\log K'$)	Slope (n)		
Cefotaxime	1.04 to 3.14 (5–15)	1.918	1.007	0.9999	0.121
Cephapirin	1.18 to 3.54 (5–15)	1.866	1.010	0.9999	0.117
Cephradine	1.43 to 4.29 (5–15)	1.788	0.9942	0.9997	0.162
Cephalexin	1.43 to 4.31 (5–15)	1.856	1.001	0.9996	0.19
Ceftazidime	0.78 to 2.33 (5–15)	2.044	1.046	0.9995	0.233
Cefazoline	1.1 to 3.3 (5–15)	1.832	0.9918	0.9992	0.28
Ceftriaxone	0.9 to 2.7 (5–15)	1.884	1.009	0.9994	0.241
Cefuroxime	1.17 to 3.53 (5–15)	1.985	1.045	0.9996	0.209

Fixed Time Method

In this method, the absorbance of the reaction solution containing varying amount of cephalosporins was measured at preselected fixed time. Calibration plots of absorbance versus the concentration of cephalosporins at fixed time were established for each investigated cephalosporins. The regression equation, correlation coefficients, and detection limits are given in Table 3. The lowest detection limit was obtained at fixed time of 15 minutes. However the fixed time of 3 minutes showed a wider concentration range for quantification.

According to international conference of harmonization (ICH) guideline for validation of analytical procedures [42], the detection limit is not required to be part of validation procedure for assay. Therefore on the basis of wider concentration range and less time of analysis, the fixed time of 3 minutes was recommended for determination.

Table 3. Analytical parameters for fixed time method of the kinetic spectrophotometric parameter for determination of investigated cephalosporins.

Reaction time	Linear range ($\mu g\,mL^{-1}$)	Intercept (a)	Standard deviation of intercept(S_a)	Slope (b)	Standard deviation of slope (S_b)	Correlation coefficient (r)	LOD ($\mu g\,mL^{-1}$)
			Cefotaxime				
3	5–25	0.00568	0.005827	0.02845	0.0003607	0.9993	0.6144
6	5–19	0.01698	0.005317	0.03472	0.0004140	0.9996	0.4590
9	5–17	0.02618	0.005363	0.04004	0.0004582	0.9997	0.4018
12	5–17	0.02880	0.005112	0.04516	0.0004368	0.9998	0.3395
15	5–17	0.01995	0.005112	0.05102	0.0004368	0.9998	0.3005
			Cephapirin				
3	5–25	0.00105	0.005027	0.02801	0.0003088	0.9995	0.5384
6	5–19	0.03286	0.004621	0.03176	0.0003598	0.9996	0.4364
9	5–17	0.05639	0.003562	0.03525	0.0003043	0.9998	0.30314
12	5–17	0.08327	0.003653	0.03791	0.0003121	0.9998	0.289
15	5–17	0.09054	0.003790	0.04196	0.0003238	0.9999	0.27097
			Cephradine				
3	5–25	0.01464	0.009607	0.03216	0.0005901	0.9985	0.896
6	5–19	0.02004	0.007665	.03856	0.0005967	0.9993	0.596
9	5–17	0.04891	0.005234	0.04184	0.0004471	0.9997	0.375
12	5–17	0.06645	0.004959	0.04566	0.0004237	0.9998	0.3258
15	5–17	0.07721	0.005174	0.04907	0.0004420	0.9998	0.3163
			Cephalexin				
3	5–25	0.01573	0.006484	0.03518	0.0003983	0.9994	0.5529
6	5–19	−0.0067	0.007189	0.04353	0.0005580	0.9995	0.495
9	5–17	−0.0012	0.005694	0.04841	0.0004865	0.9997	0.3528
12	5–17	−0.0040	0.005204	0.05354	0.0004446	0.9998	0.2915
15	5–17	−0.0015	0.005412	0.05821	0.0004624	0.9998	0.2789
			Ceftazidime				
3	5–25	−0.0029	0.003859	0.01905	0.0002371	0.9993	0.6077
6	5–19	0.03039	0.004045	0.02324	0.0003149	0.9994	0.52216
9	5–17	0.04486	0.003966	0.02871	0.0003388	0.9997	0.41442
12	5–17	0.06577	0.003875	0.03270	0.0003311	0.9997	0.3555
15	5–17	0.07993	0.003785	0.03636	0.0003234	0.9998	0.3128
			Cefazoline				
3	5–25	0.00027	0.006894	0.02935	0.0004235	0.9991	0.70466
6	5–19	0.02604	0.006741	0.03439	0.0005246	0.9993	0.588
9	5–17	0.03064	0.006040	0.04064	0.0005161	0.9996	0.4458
12	5–17	0.04073	0.005576	0.04566	0.0004764	0.9997	0.366
15	5–17	0.03973	0.005958	0.05038	0.0005090	0.9997	0.3547
			Ceftriaxone				
3	5–25	−0.0024	0.004203	0.02274	0.0002582	0.9994	0.55448
6	5–19	0.02739	0.004303	0.02770	0.0003329	0.9996	0.466
9	5–17	0.01350	0.004163	0.03307	0.0003557	0.9997	0.377
12	5–17	0.02534	0.004423	0.03741	0.0003779	0.9997	0.3546
15	5–17	0.03982	0.003818	0.03939	0.0003262	0.9998	0.2907
			Cefuroxime				
3	5–25	−0.0065	0.005980	0.02629	0.0003674	0.9991	0.68238
6	5–19	0.02696	0.005472	0.02994	0.00004260	0.9994	0.54829
9	5–17	0.04286	0.005334	0.03471	0.0004557	0.9996	0.4610
12	5–17	0.05480	0.004229	0.03845	0.0003613	0.9998	0.32996
15	5–17	0.06130	0.005614	0.04266	0.0004796	0.9997	0.2747

Validation of the Proposed Method

Concentration range [42] is established by confirming that the analytical procedure provides a suitable degree of precision, accuracy, and linearity when applied to the sample containing amount of analyte within or at the extreme of the specified range of the analytical procedure [43, 44]. In this work, concentrations ranging from 7.8×10^{-6} M to 4.31×10^{-5} M were studied for the investigated drugs in the initial rate method and concentration ranging from 5 to 25 µg mL^{-1} were studied for the investigated drugs in the fixed time method (at preselected fixed time of 3 minutes). The whole set of experiments were carried out through this range to ensure the validation of the proposed procedure. Linear calibration graphs were obtained for all the studied drugs by plotting the logarithm of initial rate of the reaction versus logarithm of molar concentration of analyte in the sample (in initial rate method) within the specified range (Figure 7):

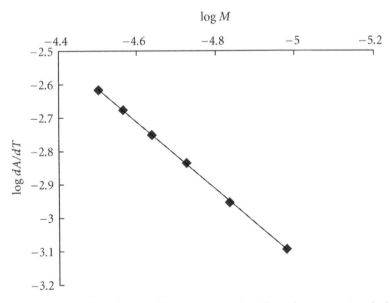

Figure 7. Calibration Plot of logarithm rate of the reaction against logarithm molar concentration of cefotaxime for initial rate method.

$$Logv = \log \frac{\Delta A}{\Delta t} = \log K' + n \log C \qquad (where \; n \approx 1) \qquad (3)$$

or by plotting the absorbance of the studied drugs versus the drug concentration (in fixed time method) within the specified range (Figure 8).

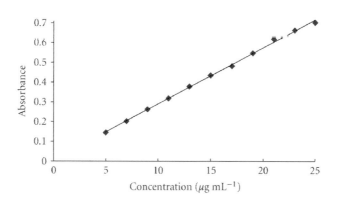

Figure 8. Calibration plot of absorbance versus the concentration of cefotaxime at preselected fixed time of 3 minutes.

Linearity was studied for both initial rate and fixed time method indicated by the values of correlation coefficient (r) and determination coefficient (r2) for both method (Tables 2 and 3).

Accuracy [44] was checked at three concentration levels within the specified range, six replicate measurements were recorded at each concentration levels. The results were recorded as percentage recovery ± standard deviation (Tables 4 and 5).

Table 4. Evaluation of accuracy of the analytical procedure using initial rate method.

| Drug | Recovery %* | | |
	$5.0 \, \mu g \, mL^{-1}$	$9.0 \, \mu g \, mL^{-1}$	$15.0 \, \mu g \, mL^{-1}$
Cefotaxime	99.8 ± 1.061	100.06 ± 0.7136	99.89 ± 0.6435
Cephapirin	100.29 ± 1.071	99.68 ± 0.7181	100.22 + 0.7661
Cephradine	100.74 ± 0.8237	99.98 ± 0.6211	99.87 ± 0.6870
Cephalexin	99.41 ± 0.7567	99.78 ± 0.5612	99.73 ± 0.4031
Ceftazidime	100.14 ± 0.9817	100.16 ± 0.8680	99.18 ± 0.8714
Cefazoline	99.89 ± 1.243	100.87 ± 0.8745	100.4 ± 0.6999
Ceftriaxone	99.31 ± 1.242	99.99 ± 0.9167	100.67 ± 0.7308
Cefuroxime	100.21 ± 1.168	100.18 ± 0.8334	100.69 ± 0.9595

*Mean of 6 replicate ± SD.

Table 5. Evaluation of accuracy of the analytical procedure using fixed time method.

| Drug | Recovery %* | | |
	$5.0 \, \mu g \, mL^{-1}$	$9.0 \, \mu g \, mL^{-1}$	$15.0 \, \mu g \, mL^{-1}$
Cefotaxime	98.75 ± 0.8183	100.98 ± 0.8118	100.75 ± 0.6424
Cephapirin	100.28 ± 1.335	99.35 ± 0.7414	100.36 ± .7051
Cephradine	100.76 ± 1.164	101.87 ± 0.9712	100.58 ± 0.8675
Cephalexin	98.59 ± 0.839	100.09 ± 0.5185	100.77 ± 0.6852
Ceftazidime	99.66 ± 1.485	100.15 ± 0.8789	99.88 ± 0.6548
Cefazoline	99.63 ± 1.262	101.42 ± 0.5617	99.44 ± 0.6155
Ceftriaxone	99.47 ± 0.9185	100.59 ± 0.9242	99.94 ± 0.5474
Cefuroxime	99.57 ± 0.9475	99.05 ± 0.9773	100.85 ± 0.7630

*Mean of 6 replicate ± SD.

Precision [44] was checked at three concentration levels, eight replicate measurements were recorded at each concentration level. The results are summarized in (Table 6). The calculated relative standard deviation were all below 2.2% indicating excellent precision of the proposed procedure at both level of repeatability and intermediate precision.

Table 6. Evaluation of precision of the initial rate and fixed time methods of the proposed kinetic spectrophotometric method for determination of investigated cephalosporins.

Drug	Amount taken ($\mu g\,mL^{-1}$)	Recovery (% ± SD)	
		Initial rate method	Fixed time method
Cefotaxime	5	100.1 ± 1.072	99.82 ± 0.7246
	9	99.88 ± 0.6859	100.9 ± 0.7086
	15	99.95 ± 0.5668	100.6 ± 0.6286
Cephapirin	5	100.16 ± 1.015	100.29 ± 1.144
	9	99.68 ± 0.6840	99.2 ± 0.7160
	15	100.31 ± 0.7624	100.44 ± 0.6225
Cephradine	5	100.61 ± 0.8049	100.53 ± 1.125
	9	99.94 ± 0.5594	101.84 ± 0.9913
	15	99.70 ± 0.6116	100.51 ± 0.9054
Cephalexin	5	99.56 ± 0.7435	98.64 ± 0.7306
	9	99.78 ± 0.5811	100.04 ± 0.4796
	15	99.76 ± 0.4575	100.78 ± 0.6091
Ceftazidime	5	100.2 ± 1.002	99.53 ± 1.309
	9	100.19 ± 0.7958	100.6 ± 0.9820
	15	99.43 ± 0.8901	99.88 ± 0.6205
Cefazoline	5	99.88 ± 1.066	99.64 ± 1.082
	9	101.07 ± 0.8312	101.45 ± 0.4871
	15	100.44 ± 0.7346	99.46 ± 0.5538
Ceftriaxone	5	99.28 ± 1.154	99.34 ± 0.9175
	9	99.99 ± 0.8687	100.54 ± 0.8347
	15	100.76 ± 0.6403	99.98 ± 0.4933
Cefuroxime	5	100.19 ± 1.19	99.52 ± 0.8834
	9	100.4 ± 0.8276	99.23 ± 0.9490
	15	100.39 ± 0.6692	100.94 ± 0.408

Specificity and interference the proposed procedure was performed in visible region away from the UV-absorption region of investigated drugs (228–300 nm), and the interference by reducing sugars in tablets was eliminated by extraction with methanol prior to analysis. However interference was observed from L-arginine formulated with cephradine in velosef vial. Therefore, the proposed procedure could not be used for analysis of cephradine in presence of L-arginine, except after separation by a suitable separation technique.

Limit of detection (LOD) [42] was calculated based on standard deviation of response and the slope of calibration curve [43]. The limit of detection was expressed as [44]

$$LOD = \frac{3\sigma}{S},$$ (4)

where σ is the standard deviation of intercept. S is the slope of calibration curve.

The results were summarized in (Tables 2 and 3) indicating good sensitivity of the proposed method. According to USP XXV validation guidelines [44], the calculated LOD values should be further validated by laboratory experiments. In our work, good results were obtained where the calculated drug concentration by LOD equations were actually detected in these experiments.

Limit of quantitation (LOQ) was calculated based on standard deviation of intercept and slope of calibration curve. In this method, the limit o quantitation is expressed as [44]

$$LOQ = \frac{10\sigma}{S}$$ (5)

The results were summarized in (Tables 2 and 3) indicating good sensitivity of the proposed method. According to USP XXV validation guidelines [45], the calculated LOQ values should be further validated by laboratory experiments. In our work, good results were obtained where the calculated drug concentration by LOQ equations were actually quantitated in these experiments.

Application to Pharmaceutical Dosage Forms

The initial rate and fixed time methods of the proposed kinetic spectrophotometric method for determination of investigated cephalosporins have been tested on commercial pharmaceutical dosage forms. The concentration of investigated cephalosporins was computed from its responding regression equations. The results of proposed method (initial rate and fixed time) were statistically compared with those of reported methods [3–5], in respect to accuracy and precision. The obtained mean recovery values were 99.2–100.67 ± 0.6226–1.69% (Table 7), which ensures that there is no interference of other additives present in the studied formulations.

In the t- and F-tests, no significant differences were found between the calculated and theoretical values of both the proposed and the reported methods at 95% confidence level. This indicates good precision and accuracy in the analysis of investigated cephalosporins in pharmaceutical dosage forms.

Table 7. Determination of studied drugs in their pharmaceutical dosage forms using initial rate and fixed time methods.

Drug	Pharmaceutical dosage form	Proposed methods ± SD ($n = 5$)		Reported methods ± SD ($n = 5$)
		Initial rate	Fixed time	
Cefotaxime	Cefotax vials	100.67 ± 0.9359	99.98 ± 1.146	100.58 ± 0.9753
		$t = 0.1423^*$	$t = 0.8914$	
		$F = 1.086^*$	$F = 1.382$	
Cephapirin	Cefatrexyl vials	100.57 ± 0.9274	99.87 ± 1.112	100.17 ± 1.208
		$t = 0.5901$	$t = 0.4193$	
		$F = 1.698$	$F = 1.180$	
Cephradine	Velosef suspension	99.85 ± 1.274	99.4 ± 1.699	99.89 ± 1.190
		$t = 0.1227$	$t = 0.5303$	
		$F = 1.346$	$F = 2.039$	
Cephradine	Velosef capsuls	100.1 ± 1.071	100.56 ± 0.6226	100.25 ± 0.7994
		$t = 0.2477$	$t = 0.6753$	
		$F = 1.793$	$F = 1.649$	
Cephalexin	Ceporex vials	99.69 ± 1.0	100.43 ± 0.918	100.19 ± 0.9310
		$t = 0.8248$	$t = 0.4069$	
		$F = 1.154$	$F = 1.027$	
Cephalexin	Ceporex tablets	99.93 ± 1.370	100.06 ± 1.647	99.98 ± 1.112
		$t = 0.06844$	$t = 0.08551$	
		$F = 1.518$	$F = 2.194$	
Ceftazidime	Fortum vials	99.2 ± 1.261	99.74 ± 1.212	99.78 ± 0.6750
		$t = 0.9192$	$t = 0.7738$	
		$F = 3.490$	$F = 3.222$	
Cefazoline	Totacef vials	99.69 ± 1.586	99.35 ± 1.059	99.25 ± 1.379
		$t = 0.4724$	$t = 0.1234$	
		$F = 1.321$	$F = 1.695$	
Ceftriaxone	Ceftriaxone vials	99.92 ± 1.292	99.43 ± 1.185	100.04 ± 0.9742
		$t = 0.1658$	$t = 0.8951$	
		$F = 1.760$	$F = 1.479$	
Cefuroxime	Zinnat vials	100.31 ± 1.091	99.82 ± 1.105	100.33 ± 0.7431
		$t = 0.0373$	$t = 0.8533$	
		$F = 2.157$	$F = 2.209$	

*Tabulated value at 95% confidence limit; $t = 2.306$ and $F = 6.388$.

Conclusion

The initial rate and fixed time methods can be easily applied for determination of investigated cephalosporins in pure and dosage forms that do not require elaborate treatment and tedious extraction of chromophore produced. The proposed method (initial rate or fixed time) is sensitive enough to enable determination of lower amounts of drug, these advantages encourage the application of proposed method in routine quality control of investigated cephalosporins in industrial laboratories. Finally our method provides advantages of improving selectivity, avoiding interference of colored and/or turbidity background of samples because it measures the increase in absorbencies with time against blank treated similarly.

References

1. W. A. Remers and J. N. Delgado, Eds., Wilson and Gisvold's Textbook of Organic Medicinal and Pharmaceutical Chemistry, W. A. Remers and J. N. Delgado, Eds., Lippincott-Raven, Philadelphia, Pa, USA, 10th edition, 1998.

2. J. G. Hardman, L. E. Limbird, and A. G. Oilman, Goodman & Gilman's the Pharmacological Basic of Therapeutics, McGraw-Hill, New York, NY, USA, 10th edition, 2001.

3. G. A. Saleh, H. F. Askal, I. A. Darwish, and A.-N. A. El-Shorbagi, "Spectroscopic analytical study for the charge-transfer complexation of certain cephalosporins with chloranilic acid," Analytical Sciences, vol. 19, no. 2, pp. 281–287, 2003.

4. M. M. Ayad, A. A. Shalaby, H. E. Abdellatef, and H. M. Elsaid, "Spectrophotometric and atomic absorption spectrometric determination of certain cephalosporins," Journal of Pharmaceutical and Biomedical Analysis, vol. 18, no. 6, pp. 975–983, 1999.

5. G. A. Saleh, H. F. Askal, M. F. Radwan, and M. A. Omar, "Use of charge-transfer complexation in the spectrophotometric analysis of certain cephalosporins," Talanta, vol. 54, no. 6, pp. 1205–1215, 2001.

6. S. A. Patel, N. M. Patel, and M. M. Patel, "Spectrophotometric methods for the estimation of cephalexin in tablet dosage forms," Indian Journal of Pharmaceutical Sciences, vol. 68, no. 2, pp. 278–280, 2006.

7. H. Salem and H. Askal, "Colourimetric and AAS determination of cephalosporins using Reineck's salt," Journal of Pharmaceutical and Biomedical Analysis, vol. 29, no. 1-2, pp. 347–354, 2002.

8. M. Hefnawy, Y. El-Shabrawy, and F. Belal, "Spectrofluorometric determination of alpha-aminocephalosporins in biological fluids and pharmaceutical preparations," Journal of Pharmaceutical and Biomedical Analysis, vol. 21, no. 4, pp. 703–707, 1999.

9. P. Gutiérez Navarro, A. El Bekkouri, and E. Rodríguez Reinoso, "Spectrofluorimetric study of the degradation of α-amino β-lactam antibiotics catalysed by metal ions in methanol," Analyst, vol. 123, no. 11, pp. 2263–2266, 1998.

10. F. A. Aly, M. M. Hefnawy, and F. Belal, "A selective spectrofluorimetric method for the determination of some α-aminocephalosporins in formulations and biological fluids," Analytical Letters, vol. 29, no. 1, pp. 117–130, 1996.

11. A. F. M. El Walily, A. Abdel-Kader Gazy, F. Belal, and E. F. Khamis, "Selective spectrofluorimetric determination of phenolic β-lactam antibiotics through the formation of their coumarin derivatives," Journal of Pharmaceutical and Biomedical Analysis, vol. 20, no. 4, pp. 643–653, 1999.

12. L. I. Bebawy, K. El Kelani, and L. Abdel Fattah, "Fluorimetric determination of some antibiotics in raw material and dosage forms through ternary complex formation with terbium (Tb3+)," Journal of Pharmaceutical and Biomedical Analysis, vol. 32, no. 6, pp. 1219–1225, 2003.

13. Q. Ma, J. Yang, X. Wu, F. Huang, and L. Sun, "A selective fluorimetric method for the determination of some β-lactamic antibiotics," Analytical Letters, vol. 33, no. 13, pp. 2689–2699, 2000.

14. A. Shalaby, "Simple HPLC method for the analysis of some pharmaceuticals," Journal of Liquid Chromatography and Related Technologies, vol. 21, no. 20, pp. 3161–3171, 1998.

15. V. F. Samanidou, A. S. Ioannou, and I. N. Papadoyannis, "The use of a monolithic column to improve the simultaneous determination of four cephalosporin antibiotics in pharmaceuticals and body fluids by HPLC after solid phase extraction—a comparison with a conventional reversed-phase silica-based column," Journal of Chromatography B, vol. 809, no. 1, pp. 175–182, 2004.

16. N. Ö. Can, G. Altiokka, and H. Y. Aboul-Enein, "Determination of cefuroxime axetil in tablets and biological fluids using liquid chromatography and flow injection analysis," Analytica Chimica Acta, vol. 576, no. 2, pp. 246–252, 2006.

17. M. Becker, E. Zittlau, and M. Petz, "Residue analysis of 15 penicillins and cephalosporins in bovine muscle, kidney and milk by liquid chromatography-tandem mass spectrometry," Analytica Chimica Acta, vol. 520, no. 1-2, pp. 19–32, 2004.

18. M. de Diego Glaría, G. G. Moscciati, and R. G. Ramos, "Determination of ceftriaxone in cerebrospinal fluid by ion-pair liquid chromatography," Journal of AOAC International, vol. 88, no. 2, pp. 436–439, 2005.

19. I. Baranowska, P. Markowski, and J. Baranowski, "Simultaneous determination of 11 drugs belonging to four different groups in human urine samples by reversed-phase high-performance liquid chromatography method," Analytica Chimica Acta, vol. 570, no. 1, pp. 46–58, 2006.

20. S. Al-Rawithi, R. Hussein, D. A. Raines, I. AlShowaier, and W. Kurdi, "Sensitive assay for the determination of cefazolin or ceftriaxone in plasma utilizing LC," Journal of Pharmaceutical and Biomedical Analysis, vol. 22, no. 2, pp. 281–286, 2000.

21. R. V. Oliveira, A. C. De Pietro, and Q. B. Cass, "Quantification of cephalexin as residue levels in bovine milk by high-performance liquid chromatography with on-line sample cleanup," Talanta, vol. 71, no. 3, pp. 1233–1238, 2007.

22. Y.-M. Li, Y. Zhu, D. Vanderghinste, A. Van Schepdael, E. Roets, and J. Hoogmartens, "Micellar electrokinetic capillary chromatography for the separation of cefalexin and its related substances," Electrophoresis, vol. 20, no. 1, pp. 127–131, 1999.

23. H.-H. Yeh, Y.-H. Yang, Y.-W. Chou, J.-Y. Ko, C.-A. Chou, and S.-H. Chen, "Determination of ceftazidime in plasma and cerebrospinal fluid by micellar electrokinetic chromatography with direct sample injection," Electrophoresis, vol. 26, no. 4-5, pp. 927–934, 2005.

24. F. A. Aly, N. A. Alarfaffj, and A. A. Alwarthan, "Permanganate-based chemiluminescence analysis of cefadroxil monohydrate in pharmaceutical samples and biological fluids using flow injection," Talanta, vol. 47, no. 2, pp. 471–478, 1998.

25. Y. Sun, Y. Tang, H. Yao, and X. Zheng, "Potassium permanganate-glyoxal chemiluminescence system for flow injection analysis of cephalosporin antibiotics: cefalexin, cefadroxil, and cefazolin sodium in pharmaceutical preparations," Talanta, vol. 64, no. 1, pp. 156–159, 2004.

26. C. Thongpoon, B. Liawruangrath, S. Liawruangrath, R. A. Wheatley, and A. Townshend, "Flow injection chemiluminescence determination of cephalosporins in pharmaceutical preparations using tris (2,2′-bipyridyl) ruthenium (II)-potassium permanganate system," Analytica Chimica Acta, vol. 553, no. 1-2, pp. 123–133, 2005.

27. Y. Li and J. Lu, "Chemiluminescence flow-injection analysis of β-lactam antibiotics using the luminol-permanganate reaction," Luminescence, vol. 21, no. 4, pp. 251–255, 2006.

28. C. Thongpoon, B. Liawruangrath, S. Liawruangrath, R. A. Wheatley, and A. Townshend, "Flow injection chemiluminescence determination of cefadroxil using potassium permanganate and formaldehyde system," Journal of Pharmaceutical and Biomedical Analysis, vol. 42, no. 2, pp. 277–282, 2006.

29. N. A. Alarfaj and S. A. Abd El-Razeq, "Flow-injection chemiluminescent determination of cefprozil using tris (2,2′-bipyridyl) ruthenium (II)-permanganate system," Journal of Pharmaceutical and Biomedical Analysis, vol. 41, no. 4, pp. 1423–1427, 2006.

30. I. F. Jones, J. E. Page, and C. T. Rhodes, "The polarography of cephalosporin C derivatives," The Journal of Pharmacy and Pharmacology, vol. 20, p. 455, 1968.

31. A. M. M. Ali, "Polarographic determination of cephradine in aqueous and biological media," Bioelectrochemistry and Bioenergetics, vol. 33, no. 2, pp. 201–204, 1994.

32. G. V. S. Reddy and S. J. Reddy, "Estimation of cephalosporin antibiotics by differential pulse polarography," Talanta, vol. 44, no. 4, pp. 627–631, 1997.

33. A. A. Abdel Gaber, M. A. Ghandour, and H. S. El-Said, "Polarographic studies of some metal(II) complexes with cephalosporins selected from the first generation," Analytical Letters, vol. 36, no. 6, pp. 1245–1260, 2003.

34. A. Hilali, J. C. Jiménez, M. Callejón, M. A. Bello, and A. Guiraúm, "Electrochemical reduction of cefminox at the mercury electrode and its voltammetric determination in urine," Talanta, vol. 59, no. 1, pp. 137–146, 2002.

35. S. R. Crouch, T. F. Cullen, A. Scheeline, and E. S. Kirkor, "Kinetic determinations and some kinetic aspects of analytical chemistry," Analytical Chemistry, vol. 70, no. 12, pp. 53–106, 1998.

36. E. M. Hassan and F. Belal, "Kinetic spectrophotometric determination of nizatidine and ranitidine in pharmaceutical preparations," Journal of Pharmaceutical and Biomedical Analysis, vol. 27, no. 1-2, pp. 31–38, 2002.

37. N. Rahman, N. A. Khan, and S. N. H. Azmi, "Kinetic spectrophotometric method for the determination of silymarin in pharmaceutical formulations using potassium permanganate as oxidant," Pharmazie, vol. 59, no. 2, pp. 112–116, 2004.

38. N. Rahman, Y. Ahmad, and S. N. H. Azmi, "Kinetic spectrophotometric method for the determination of norfloxacin in pharmaceutical formulations," European Journal of Pharmaceutics and Biopharmaceutics, vol. 57, no. 2, pp. 359–367, 2004.

39. I. A. Darwish, "Kinetic spectrophotometric methods for determination of trimetazidine dihydrochloride," Analytica Chimica Acta, vol. 551, no. 1-2, pp. 222–231, 2005.

40. P. Job, Advanced Physicochemical Experiments, Oliner and Boyd, Edinburgh, UK, 2nd edition, 1964.

41. P. Job, Annales de Chimie, vol. 6, p. 97, 1936.

42. International Conference on Harmonization, "ICH harmonized tripartite guideline-text on validation of analytical procedures," Federal Register, vol. 60, p. 11260, 1995.

43. "Q2A: text on validation of analytical procedure," in Proceedings of International Conference on Harmonization (ICH '94), Geneva, Switzerland, 1994.

44. "Q2B: validation of analytical procedure, methodology," in Proceedings of International Conference on Harmonization (ICH '96), Geneva, Switzerland, 1996.

45. The United States Pharmacopoeia XXV and NF XX, American Pharmaceutical Association, Washington, DC, USA, 2002.

CITATION

Originally published under the Creative Commons Attribution License or equivalent. Omar MA, Abdelmageed OH, Attia TZ. Kinetic Spectrophotometric Determination of Certain Cephalosporins in Pharmaceutical Formulations. International Journal of Analytical Chemistry, Volume 2009 (2009), Article ID 596379. http://dx.doi. org/10.1155/2009/596379.

Understanding Structural Features of Microbial Lipases — An Overview

John Geraldine Sandana Mala and Satoru Takeuchi

ABSTRACT

The structural elucidations of microbial lipases have been of prime interest since the 1980s. Knowledge of structural features plays an important role in designing and engineering lipases for specifi c purposes. Signifi cant structural data have been presented for few microbial lipases, while, there is still a structure-deficit, that is, most lipase structures are yet to be resolved. A search for 'lipase structure' in the RCSB Protein Data Bank (http://www.rcsb.org/pdb/) returns only 93 hits (as of September 2007) and, the NCBI database (http://www.ncbi.nlm.nih.gov) reports 89 lipase structures as compared to 14719 core nucleotide records. It is therefore worthwhile to consider investigations on the structural analysis of microbial lipases. This review is intended to provide a collection of resources on the instrumental, chemical and bioinformatics approaches for structure analyses. X-ray crystallography is a versatile tool for the structural biochemists and is been exploited till today. The chemical methods of recent interests include molecular modeling and combinatorial designs. Bioinformatics has surged striking interests in protein structural analysis with the advent of innumerable tools. Furthermore, a literature platform of the structural elucidations so far investigated has been presented with detailed descriptions as applicable to microbial lipases. A case study of Candida rugosa lipase (CRL) has also been discussed which highlights important structural features also common to most lipases. A general profile of lipase has been vividly described with an overview of lipase research reviewed in the past.

Keywords: active site, bioinformatics, Candida rugosa lipase, crystallization, lipase structure, structure prediction

Introduction

Lipases (E.C.3.1.1.3) catalyze the hydrolysis of ester linkages in long-chain triacylglycerols with concomitant release of the constituent acid and alcohol moieties. They act at the interface between an insoluble substrate phase and an aqueous phase in which the enzyme is dissolved. Lipases are ubiquitously produced by plants (Bhardwaj et al. 2001), animals (Carriere et al. 1994) and microorganisms (Olempska-Beer et al. 2006). Microbial lipases are the preferred potent sources due to several industrial potentials (Hasan et al. 2006). The world market for lipases has been estimated at approximately US $20 million of the industrial enzyme market (Rahman et al. 2005). Lipases have been intensively investigated for their multiplexity of catalysis with unique specificities (Villeneuve and Foglia, 1997), which have multifold applications in oleochemistry, organic synthesis, detergent formulations and nutrition (Saxena et al. 2003). Also, lipases display useful properties related to their stability as organic solvent-tolerant (Rahman et al. 2005) and thermostable (Li and Zhang, 2005) enzymes. Therefore, microbial lipases have been of recent research interests and a number of lipases have been identified, purified and characterized to date.

In general, microbial lipases are 20–60 kDa proteins, with an active Ser residue of the active site structure Ser-His-Asp. Asp may be replaced by Glu in case of Geotrichum candidum lipases, which have specificity for hydrolysis of fatty acids with cis-unsaturated double bonds. Also, lipases share a consensus sequence of G-X-S-X-G, were X may be any amino acid residue. The lipases belong to the α/β hydrolase family (Ollis et al. 1992) with a central β-sheet, containing the active Ser placed in a loop termed the catalytic elbow. Interfacial activation occurs in presence of a substrate which takes place by the movement of a lid and exposure of the hydrophobic pocket and the active site structure above the critical micellar concentration (CMC) of the substrate (Svendsen, 2000). This interfacial activation is unique to the class of lipases and is also responsible for the versatility of the reactions they catalyse; hydrolysis, esterifi cation, transesterification and interesterification of fats and oils.

The first microbial lipase structure studied was that of the Rhizomucor miehei lipase by Brady et al. (1990) from X-ray crystallographic analysis. It showed that this enzyme had an active site triad as that of the serine proteases. X-ray crystallography was and still continues to be a powerful tool for structure determinations of most biological macromolecules. Recently, especially in this millennium, other approaches also have come into practice for structural analyses, including the use of bioinformatics tools for structure predictions up to the tertiary levels of protein organization. In this review, we provide a basic concept of crystallization and X-crystallographic studies of few microbial lipases. Also, various instrumental and chemical methods of structure analysis have been presented and a description of the structures of microbial lipases and their characteristics studied so far has been discussed.

Table 1. Protein crystallization methods.

Methods of crystallization	Subtypes
Batch	Batch; Microbatch
Seeding	Macroseeding; Microseeding
Nucleation to support growth only	Free interface diffusion; local nucleation
Vapor diffusion	Hanging drop; Sitting drop
Dialysis	-
Lipidic sponge phase crystallization (for membrane proteins)	-

Crystallization and X-ray Crystallographic Analysis

The fundamental approach in X-ray crystallography is crystallization of the molecule under study. This may seem to be a simple task, but the preparation of good quality crystals is a major limiting step in most cases. Several classical methods of crystallization are in practice (Table 1) and a vast literature is available for ready references; however, efficient methods of growing pure crystals suitable for X-ray diffraction analysis are still to be addressed. McPherson (1990) has reviewed different approaches for crystallization of macromolecules and has also emphasized that macromolecular crystallization is still a poorly understood phenomenon. This review presents a wide analysis of crystallization from supersaturated solutions, growth and properties of crystals, various precipitating agents, factors infl uencing protein crystal growth and some useful considerations for an efficient crystallization strategy. A contemporary report by Durbin and Feher (1990) describes the mechanisms of crystal growth of proteins by freeze-etch electron microscopy studies, using lysozyme crystals. The report derives that growth occurs by a lattice defect mechanism at low supersaturation and by two-dimensional nucleation at high supersaturation. Abergel et al. (1991) have analysed the systematic use of an Incomplete Factorial approach for design of protein crystallization experiments. The strategy described by Abergel et al. (1991) can aid other experimentalists to design experiments to crystallize their own proteins. However, this approach hinders the X-ray diffraction analysis of lipases that have significant amounts of carbohydrates. Lipase crystals have also been obtained by nucleation and growth from clarified, concentrated fermented broths by bulk crystallization (Jacobsen et al. 1998). A recent crystallization strategy is reported by Wadsten et al. (2006) for membrane proteins by lipidic sponge phase crystallization. However, classical methods such as hanging drop and sitting drop vapor diffusion methods in presence of saturating amounts of ammonium sulphate and/or polyethylene glycol are in common use in day-to-day laboratories.

Lipase structures have been widely investigated by X-ray crystallography in open or closed conformations. X-ray diffraction analyses of a few microbial lipases are briefl y described.

The crystal structure of Rhizomucor miehei lipase at 1.9 Å resolution using X-ray single crystal diffraction data is reported with refinement of the structure to an R-factor of 0.169 for all available data. Prior to this study, Rhizomucor miehei lipase (RmL) complexed with inhibitors wereanalysed at 3 Å resolution (Brzozowski et al. 1991) and at 2.6 Å resolution (Derewenda et al. 1992a), while, this study presents a detailed analysis of the three-dimensional structure of RmL in its native form (Derewenda et al. 1992b). Lipase I from Rhizopus niveus was crystallized by the hanging drop vapor diffusion with cell dimensions of a = b = 83.7 Å, c = 137.9 Å and the diffraction pattern extended to 2.5 Å resolution (Kohno et al. 1993). Lipase crystals from Staphylococcus hyicus were obtained using dimethyl sulphoxide (DMSO) and isopropanol, with a = 73.31 Å, b = 77.96 Å and c = 169.81 Å and diffracted to 2.8 Å resolution. (Ransac et al. 1995). Lipase crystals from Bacillus stearothermophilus were obtained by hanging drop vapor diffusion method using ammonium sulphate. The unit-cell parameters were a = 118.5 Å, b = 81.23 Å and c = 99.78 Å and diffracted well at 2.2 Å in native form (Sinchaikul et al. 2002). Lipase from Candida rugosa was fi rst determined in an open conformation by X-ray crystallography as reported by Grochulski et al. (1993). In 2003, Mancheno et al. have reported the crystal structure of Lipase 2 isoenzyme of Candida rugosa at 1.97 Å resolution in its closed conformation. Lipase crystals of Penicillium expansum were obtained by the sitting drop vapor diffusion crystallization with unit cell parameters of a = b = 88.09 Å and c = 126.54 Å. Diffraction data were collected to a resolution of 2.08 Å (Bian et al. 2005).

Instrumental Techniques for Structure Analysis

By and large, X-ray crystallography is the powerful tool for most macromolecular structural elucidations. Of recent interests, other instrumentations have emerged with more sophistications and present valuable tools for protein structure analysis. A few important instrumentation methods are reviewed in view of their current applications.

X-ray crystallography is the oldest and most precise method of structure analysis, in which a beam of X-rays is reflected from evenly spaced planes of a single crystal, producing a diffraction pattern of spots called reflections. Each reflection corresponds to one set of evenly spaced planes within the crystal. The density of electrons within the crystal is determined from the position and brightness of the various reflections observed as the crystal is gradually rotated in the X-ray beam; this density, together with supplementary data, allows the atomic positions to be inferred.

Circular dichroism (CD) has become increasingly recognized for examining the structure of proteins in solution. A significant improvement in the provision of CD instrumentation has occurred in recent years. Kelly et al. (2005) have reported a brief summary of the CD technique and its applications with particular reference to the study of proteins. The important practical aspects of performing CD experiments on proteins have been addressed which provide a clear guidance as to how reliable data can be obtained and interpreted. CD instruments, known as spectropolarimeters measure the difference in absorbance between the L (left) and R (right) circularly polarized components in terms of the ellipticity (θ) in degrees. The CD spectrum is obtained when dichroism is a function of wavelength. A CD spectral analysis serves to understand various structural features of proteins. The secondary structure composition such as % helix, sheet, turns from the peptide bond region, tertiary structure fingerprint, integrity of cofactor binding sites, conformational changes in proteins, protein folding and overall structure features of proteins are all attributed to the study of CD spectra of proteins. Interestingly, an integrated software package for CD spectroscopic data processing, analysis and archiving, known as the CD tool, has been developed by Lees et al. (2004). CD tool is a multiplatform graphical user interface (GUI) cross-instrument application package, containing a range of features associated with data handling from initial processing to final storage of data and association with related protein data bank (PDB) crystal structure files. Secondary structures of proteins using vacuum-uv CD spectroscopy has also been studied in the case of lipase from Pseudomonas cepacia and other globular proteins (Matsuo et al. 2005).

Fourier transform infrared (FTIR) spectroscopy is being increasingly used for investigating protein structure and stability (Haris and Severcan, 1999). Different conformational types result in different absorption bands in a FTIR spectrum, which are usually broad and overlapping. To overcome these, Severcan et al. (2004) have successfully reported the use of artificially generated spectral data to improve protein secondary structure prediction from FTIR spectroscopy.

Mass spectroscopy is a versatile tool for protein analysis and has contributed much to the field of proteomics, in conjunction with two-dimensional electrophoresis. Proteomics helps to define the functions and interrelationships of proteins in an organism. As genome sequence information has accumulated, the paradigm has shifted from sequencing to identification of proteins, which has been facilitated by advances in ionization and mass analysis techniques for mass spectrometry. Electrospray ionization (ESI) and Matrix-assisted laser desorption/ionization (MALDI) methods are currently the principal methods for peptide/protein ionization and have been linked to high-throughput sample preparation techniques. Large-scale protein identification has been made possible using mass spectrometry as reviewed by Lin et al. (2003), with emphasis on its methods and applications. Two-dimensional gel electrophoresis (2DGE) is another important technique for proteome analysis by separation in a first dimension using isoelectric focusing (IEF), and then subjected to SDS-PAGE in the second dimension. A 2DGE is very effective for differential analysis of proteins by separation and visualization, but does not explicitly identify proteins, which therefore require a further analytical step for identification or sequencing. Mass spectrometry is now routinely used to identify proteins separated by 2DGE. MALDI-TOF mass spectroscopy is widely used for identification of proteins from 2DGE. Meunier et al. (2005) have investigated data analysis methods for detection of differential protein expression using 2DGE in order to minimize false positives, at the same time, without losing information with false negatives.

Nuclear magnetic resonance (NMR) is another important method in the study of protein structure analysis. Jonas (2002) has reviewed the studies of proteins using high resolution NMR. It is stated that the combination of advanced high resolution NMR technique with high pressure capability is a powerful experimental tool in studies of protein folding. The main advantages of using high resolution and high pressure NMR are the uses of 1-D NMR for determination of structural and dynamic changes in different regions of the protein and the allowance of distance specific information to be obtained between amino acid residues in different regions of the protein.

Electron spin resonance (ESR) spectroscopy in combination with site-directed spin labeling (SDSL) is also an efficient tool for determination of protein structure, dynamics and interactions. Low microwave-amplitude ESR has been used as a novel method especially suitable for studying moderately immobilized spin labels, such as those positioned at exposed sites in a protein (Hedin et al. 2004).

Fluorescence methods are being increasingly used in biochemical characterizations because of its inherent sensitivity. Fluorescence spectroscopy has been employed for monitoring the changes in fl uorescence of Humicola lanuginosa lipase (HLL), by comparison of the conformations of HLL in an aqueous buffer and dissolved in its substrate, triacetin (Jutila et al. 2004). Triacetin is optically transparent and does not impede the use of fluorescence spectroscopy. It has been revealed that Trp89 plays an important role in the structural stability of HLL, and, the carbohydrate moiety attached to Asn33 has only minor effects on the conformational dynamics of the lid. The analysis of differences in frequencies between the two modes of motion augmented in triacetin, indicated that the motion of the Trp89 side chain becomes distinguishable from the motion of the lid. Thus, steady state and time-resolved fluorescence spectroscopy enabled the analysis and identifi cation of specific structural features of the HLL dissolved in its substrate. Spectroscopic methods have also been applied to analyse the thermal stability of proteins as in the case of Chromobacterium viscosum lipase (CVL), whereby, Melo et al. (2000) have studied the CVL thermal stability based on assessment of fluorescence, circular dichroism and static light scattering measurements.

Differential scanning calorimetry revealed unfolding of HLL at 74.4 oC demonstrating significant contribution of Trp residues to the structural stability of the enzyme, when compared with its mutants (Zhu et al. 2001). Small angle X-ray scattering measurements (SAXS) determined the lamellar structure of lipase modified with fatty acids in an aqueous buffer and in n-hexane (Maruyama et al. 2001).

Another major technique is the use of isoelectric focusing electrophoresis to discriminate between closed and open conformations of lipases based on their isoelectric points, as studied by Miled et al. (2005) for HLL and other lipases. They have deduced a significant difference in the isoelectric points between the closed (native) and open (inhibited) conformations, resulting in a distinct electrophoretic pattern, thereby, providing an easy experimental tool for a given lipase.

Chemical Methods of Structure Analysis

As instrumentation is the basic platform for investigations of protein structures, so are chemical modification strategies of proteins due to their primary, secondary and tertiary structural features, hydrophobicity, similarities of motifs and patterns and other structural arrangements.

Molecular modeling plays a key role in structural biology in interpretations of protein structures with experimental observations. Current modeling methods are extremely useful qualitatively and help to predict increased selectivity of biocatalysts by substrate modification or by site-directed mutagenesis. On the contrary, quantitative predictions are still not reliable, while, modeling is also limited by the availability of three-dimensional structures (Kazlauskas, 2000). Molecular modeling has been applied to Rhizopus oryzae lipase (ROL) based on a homology model by Holzwarth et al. (1997) using docking calculations to explain reversals in enantioselectivity. Modeling showed that sn-2 substituent binds in a 'hydrophobic dent'. A flexible β-bond of the substituent avoids clashes with Leu258, but with a rigid β-bond, it can avoid clashes only by turning the substrate so that the sn-3 group is in the hydrolysis site. However, modeling has its limitations in prediction of the degree of enantioselectivity.

Electrostatic potential is one of the critical factors explaining lipase/esterase activity based on its distribution on the molecular surface. Other important factors include the presence and distribution of polar and hydrophobic residues in the active cleft. A negative potential in the active site is correlated with maximum activity towards triglycerides (Peterson et al. 2001).

Modifications of lipase with stearic acid or other fatty acids explained the activation mechanism of modified lipases in relation to structure (Maruyama et al. 2001). Shibamoto et al. (2004) have studied the molecular engineering of ROL using a combinatorial protein library constructed on yeast cell surface, thereby, providing a screening method for novel mutant lipase based on yeast cell-surface displayed mutant library. Modification of lid sequence of lipases show that the lid is a structural and a functional determinant of lipase activity and selectivity. This has been observed by Secundo et al. (2006) for Candida rugosa (CRL), Pseudomonas fragi (PFL) and Bacillus subtilis (BSL) lipases. A CRL chimera enzyme obtained by replacing its lid with that of another CRL isoform was found to be affected in both activity and enantioselectivity in organic solvent. Variants of the PFL protein in which three polar lid residues were replaced with amino acids strictly conserved in homologous lipases displayed altered chain length preference profile and increased thermostability. On the other hand, insertion of lid structures from structurally homologous enzymes into BSL, a lipase that naturally did not possess a lid structure, caused a reduction in the enzyme activity and altered substrate specificity. These results strongly support the concept that the lid plays an important role in modulating not only activity, but also specificity, enantioselectivity and stability of lipase enzymes.

Micellar sodium dodecyl sulfate (SDS) is known to stabilize α-helical conformation in peptides derived from helical regions of proteins, although the precise mechanism remains unclear. This effect has been investigated by Montserret et al. (2000) demonstrating that electrostatic interaction play a significant role in the formation and stabilization of SDS-induced structure.

Use of combinatorial design has earlier been applied to the evolution of increased thermostability, in which a diverse library of proteins is generated and screened for variants with increased stability. Current trends are towards the use of data-driven methods that reduce the library size by using available data to choose areas of the protein to target, without specifying the precise changes. Bommarius et al. (2006) have used high-throughput screening methods for enhancement of protein stability by a combination of these methods which lead to the rapid improvement of protein stability for biotechnological purposes.

Antibodies with enzymatic activities are known as abzymes. Of recent interests, Leong et al. (2007) have studied the selection of lipolytic abzymes from the phage displayed antibody libraries against a transition state analog of lipases. This method is presented as an efficient and convenient means to find new abzymes.

Bioinformatics Approaches for Structure Analysis

Structural knowledge is vital for complete understanding of life at the molecular level. An understanding of structure can lead to derivations of functions and mechanisms of action of proteins. From a practical point of view, the sequence-structure gap is a main factor in motivating the need for predictions of protein structure. A hierarchy of the bioinformatics approaches in protein analysis is represented in Figure 1.

Bioinformatics is a novel approach in recent investigations on sequence analysis and structure prediction of proteins. In general, protein sequence databases may be classified as primary and secondary databases, composite protein pattern databases and structure classifi cation databases. Primary and secondary databases are used to address different aspects of sequence analysis, because they store different levels of protein sequence information. Primary databases are the central repositories of protein sequences, while, secondary databases are based on the analysis of sequences of the primary ones (Table 2). Composite protein pattern databases have been emerged with a view to create a unified database of protein families. ProWeb (Henikoff et al. 1996) is a dedicated protein family website, providing information about individual families through hyperlinks to existing web resources maintained by researchers in their own fields. Protein structure classification databases have been established based on the structural similarities and common evolutionary origins of proteins. SCOP (Structural classification of proteins, Murzin et al. 1995), CATH (Class, Architecture, Topology and Homology, Orengo et al. 1997) and PDBSum (Laskowski et al. 1997) are the major classification schemes. Thus, bioinformatics tools for protein analysis provide a wealth of information related to sequences and structures of proteins. A number of tools are also available for protein structure visualization (Table 3) and protein identification and characterization (Table 4).

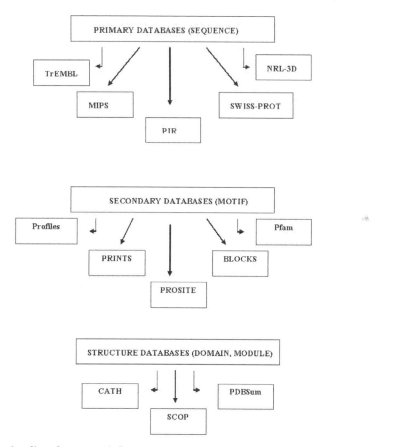

Figure 1. Hierarchy of bioinformatics tools for protein structure analysis.

Earlier, protein secondary structure and active site predictions were obtained by alignment of homologous sequences described by Zvelebil et al. (1987). Another study was the analysis and predictions of different β-turns by Wilmot and

Thornton (1988). More recently, a new database of aligned protein domains known as DOMO has been developed by Gracy and Argos (1998). DOMO can be accessed through the sequence retrieval system (SRS). A form-based query manager allows retrieval of familial domain alignments by identifiers, sequence accession numbers or keywords. The DOMO sequence analysis provides a simple tool for determining domain arrangements, evolutionary relationships and key amino acid residues in a query protein sequence. With the recent revolutions in bioinformatics, new software tools have been designed to meet updated protein information. Bachinsky et al. (2000) have developed Prot_Pat 1.3, an updated database of patterns to detect local similarities, containing patterns of more than 13,000 groups of related proteins in a format similar to that of PROSITE. Simultaneoulsy, Pleiss et al. (2000) have constructed a database exclusively on lipases to understand and exploit sequence-structurefunction relationships. Lipase Engineering Database (LED) serves as a useful tool for protein engineering to help understand the functional role of individual amino acids by reference to annotated aligned sequences and superimposed structures of microbial lipases. The LED is available at http://www.led. uni-stuttgart.de. Of recent interests in bioinformatics for protein structure analysis, Sheehan and Sullivan (2006) have used online resources for homology modeling of milk enzymes. In another report, it has been studied that conformation biases of amino acids play an important role in protein folding, refining domain, structure prediction and structural proteomics, based on the tripeptide microenvironment from PDB (Protein Data Bank) database (Yang et al. 2006). Kartik et al. (2006) have developed a simple web-based computational tool http://www.ccmb.res.in/bioinfo/dsbcp which allows flexible queries to be made on the database in order to retrieve useful information on the disulfide bond containing proteins in the PDB. Thereby, the database may be useful to select suitable protein structure templates in order to model the more distantly related protein homologs/ analogs using the comparative modeling methods. Structural bioinformatics has also been applied to structure prediction of membrane-binding cytosolic peripheral proteins (Bhardwaj et al. 2006).

Table 2. Primary and Secondary databases for protein analysis.

Type of database	Databases	Targets	Web address (URL)
Primary	PIR	Sequence	http://pir.georgetown.edu
	MIPS	Sequence	http://mips.gsf.de
	Swiss-Prot	Sequence	http://www.ebi.ac.uk/swissprot/
Secondary	PROSITE	Patterns	http://www.expasy.ch/prosite/
	PRINTS	Fingerprints	http://www.bioinf.manchester. ac.uk/dbbrowser/PRINTS/
	Pfam	[a]HMMs [b]MSA	http://pfam.sanger.ac.uk/
	BLOCKS	Motifs	http://blocks.fhcrc.org/

a-Hidden Markov Models.
b-Multiple sequence alignments.

Table 3. Protein visualization programs.

Program	Function
RasMol	3-dimensional visualization
Cn3D	3-dimensional visualization, linked to sequence alignments
Chime	3-dimensional visualization
TOPS	Visualization of protein folding topologies
DSSP	Finds secondary structure elements in an input structure
Surfnet	Visualization of protein surface
PROCHECK	Checks stereochemical quality of protein structures
PROMOTIF	Analyses protein structural motifs

Structural Features of Lipase

Lipase structural features are important characteristics for protein engineering to provide efficient biocatalysts and make use of their unique structural specificities for commercial exploitation. In this view, many efforts have been attempted to analyse and characterize significant structural data.

A novel structural approach has been devised to distinguish lipases from esterases (Fojan et al. 2000). Lipases have a lid-like structure which is an important entity for exposing a hydrophobic patch in presence of a substrate. The reaction takes place at the oil-water interface by movement of the lid to allow access of the substrate to the catalytic site, while esterases do not display a lid structure. Akoh et al. (2004) have described the unique family of GDSL hydrolases with multifunctional properties. This new subclass of lipolytic enzymes possesses a distinct GDSL sequence motif different from the G-X-S-X-G motif in many lipases. GDSL motif is a consensus amino acid sequence of Gly, Asp, Ser and Leu around the active site Ser.

Various microbial lipases of bacterial and fungal origins have been investigated for analysis of their structures. We summarize the structural properties of different lipases studied in the recent past. The structures of Pseudomonas lipases have been well conceived since the last decade. Schrag et al. (1997) have studied the open conformations of Pseudomonas cepacia and Pseudomonas glumae lipases suggesting that the conformational changes are important for interfacial activation of these bacterial lipases and that the protein conformation depended strongly on the solution conditions, perhaps by the dielectric constant. Pseudomonas aeruginosa lipase has a single functional disulfide bond, shown by a shift in electrophoretic mobility after treatment with dithiothreitol (DTT) and iodoacetamide. The structural model predicts a catalytic triad consisting of Ser82, Asp229 and His251, with a disulfide bond between Cys813 and Cys235. Residues Asp38 and Glu46 are located on the surface of the enzyme. A striking prediction was the lack of a lid-like α-helical loop structure covering the active site when the substrate existed either as monomeric solutions or aggregates, confirming the absence of interfacial activation (Jaeger et al. 1993). In other studies of Pseudomonas lipases, the role of calcium on the structure and function of a calcium-dependent family 1.3 lipase was characterized and was observed that the C-terminal domain folding was induced by calcium binding (Amada, 2001). In another report, the role of a nine-residue sequence motif in secretion, enzyme activity and protein conformation of a family 1.3 lipase has been described (Kwon et al. 2002). A similar study reported that the repetitive nine-residue sequence motif contributed to the intracellular stability and secretion efficiency of Pseudomonas lipases (Angkawidjaja et al. 2005). Prior to these studies, it has been determined by Liebeton et al. (2001) that the disulfide bond in Pseudomonas aeruginosa lipase stabilized the lipase structure, although it was not required for interaction with its foldase. The complete amino acid sequence of mono- and diacylglycerol lipase from Penicillium camembertii has been determined to consist of 276 amino acid residues with two disulfide linkages and one potential N-glycosylation site (Isobe and Nokihara, 1993). The 3-D structural model of Bacillus stearothermophilus P1 revealed a topological organization of the α/β hydrolase fold. The model structure included both α-helix and extended β-sheet secondary structures in the folded protein, and the β-sheet was in the core region surrounding with α-helix. The helix span between Phe180 to Val197 formed the lid of the model lipase. Ser113, Asp317 and His358 formed the catalytic triad (Sinchaikul et al. 2001). Tyndall et al. (2002) have investigated for the first time the 3-D structure of B. stearothermophilus P1, as a model for thermostable enzymes, with a unique zinc-binding site which may play a role in enhancing thermal stability (Fig. 2). The lipase from Bacillus subtilis showed a single, globular compact domain with dimensions of 35 × 36 × 42 Å. Its fold conformed to the α/β hydrolase fold, although it lacked the β1, β2 strands of the canonical fold. The active site triad consists of Ser78, Asp134 and His157 (Eggert et al. 2002). The primary structure of a novel lipase from Streptococcus sp. N1, showed a consensus sequence containing the active serine [VAGHSIGG], a conserved H-G dipeptide in the N-terminus and a potential site for N-linked glycosylation at amino acid residues 129–131 (Tripathi et al. 2004).

Table 4. Protein identification and characterization programs.

Program	Function
AACompIdent	Identification of amino acid composition
TagIdent	Identification of proteins using mass spectrometric data
PeptIdent	Identification of proteins using peptide mass fingerprinting data
MultiIdent	Identification of proteins using pI, MW, amino acid composition
Propsearch	Find putative protein family
PepSea	Identification of protein by peptide mapping or peptide sequencing
FindPept	Identification of peptides resulting from unspecific cleavage of proteins
TMAP; TMHMM	Prediction of transmembrane helices
ProtParam	Computation of physical and chemical parameters of a protein

Glycosylation is an important feature of eukaryotic lipases, a distinct characteristic of the higher order. Glycosylation is known to contribute to the stability of lipase, but does not affect the enzyme activity (Isobe and Nokihara, 1993). Glycosylation of only a few lipases are reported to due the complexity of its elucidations. However, recently, a number of techniques have evolved to analyse the structural aspects of glycosylation. A strategy for the identification of site-specific glycosylation in glycoproteins using MALDI-TOF mass spectrometry has been described (Mills, 2000). Also, Dell and Morris (2001) have reviewed the various mass spectroscopic techniques as applied for the determination of glycoproteins. More recently, infrared spectroscopy has been used to evaluate the glycosylations of proteins, showing distinct absorption bands for the sugar moiety, the protein amide group and water (Khajehpour et al. 2006). Tang et al. (2001) have reported that glycosylation conferred thermostability to the lipase, while, it did not have any catalytic effect, concluding that glycosylation may effect the structure, stability and movement through the secretory pathways of the lipase.

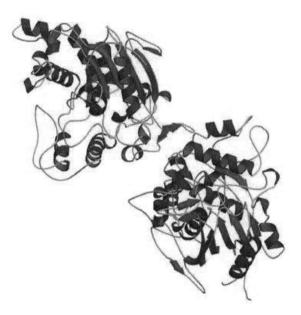

Figure 2. 3-D structure of Bacillus stearothermophilus (Tyndall et al. 2002).

A Case-Study: Candida rugosa Lipase

Candida rugosa lipases (CRL) [exist as isoforms] have been the widely studied lipases in open as well as closed conformations. CRL are of immense significance due to potential applications of commercial interests and have been well-documented as efficient biocatalysts for biotransformations (de Maria et al. 2006). CRL consists of 534 amino acid residue polypeptide chain, with a predicted molecular mass of 60 kDa (Mancheno et al. 2003), showing an α/β hydrolase structure, with a catalytic triad (Ser209-Glu341-His449) and a lid that covers the active site (Akoh et al. 2004). The active site in CRL is covered by an α-helical structure (residues 65–94), composed of variable amino acidic composition of the lid (de Maria et al. 2006). The lid structure is fixed by a disulphide bond (Cys60Cys97) and an ionic interaction between Glu96 and Arg37 (Cygler and Schrag, 1999). Lipases also contain the consensus sequence G-X-S-X-G, where, X = any amino acid residue (Svendsen, 2000). Distinct phenylalanyl-rich region and an aliphatic-rich region have been revealed by structural comparisons of lipase 2 at 1.97 Å resolution in its closed conformation. The aliphatic-rich region is identical to other isoforms, while, the phenylalanyl content is specific for each lipase isoforms, responsible for their varied lipase/esterase characteristics (Mancheno et al. 2003).

Conclusions

In the light of discussion on important instrumental, chemical and bioinformatics approaches, we highlight the basic strategies for structural elucidations of proteins, as in the case of microbial lipases. This review also encompasses the recent advancements in protein science and research for structure analysis. In toto, it is attempted to describe a better understanding of the structural characteristics of proteins with evidences of lipase structural features.

Acknowledgements

The sponsorship of J.Geraldine Sandana Mala by Kikuji Takeuchi and Naomi Takeuchi of TAKENEN, Japan, is gratefully acknowledged.

References

1. Abergel, C., Moulard, M., Moreau, H. et al. 1991. Systematic use of the incomplete factorial approach in the design of protein crystallization experiments. J. Biol. Chem., 266:20131–8.

2. Akoh, C.C., Lee, G.C., Liaw, Y.C. et al. 2004. GSDL family of serine esterases/lipases. Prog. Lipid Res., 43:534–52.

3. Amada, K., Kwon, H.J., Haruki, M. et al. 2001. Ca2+—induced folding of a family 1.3 lipase with repetitive Ca2+ binding motifs at the C-terminus. FEBS Lett., 509:17–21.

4. Angkawidjaja, C., Paul, A., Koga, Y. et al. 2005. Importance of a repetitive nine residue sequence motif for intracellular stability and functional structure of a family 1.3 lipase. FEBS Lett., 579:4707–12.

5. Bachinsky, A.G., Frolov, A.S., Naumochkin, A.N. et al. 2000. PROF_PAT 1.3: Updated database of patterns used to detect local similarities. Bioinf., 16:358–66.

6. Bhardwaj, K., Raju, A. and Rajasekharan, R. 2001. Identification, purification and characterization of a thermally stable lipase from rice bran. A new member of the (phospho) lipase family. Plant Physiol., 127:1728–38.

7. Bhardwaj, N., Stahelin, R.V., Langlois, R.E. et al. 2006. Structural bioinformatics prediction of membrane-binding proteins. J. Mol. Biol., 359:486–95.

8. Bian, C., Yuan, C., Lin, L. et al. 2005. Purification and preliminary crystallographic analysis of a Pencillium expansum lipase. Biochim. Biophys. Acta., 1752:99–102.

9. Bommarius, A.S., Broering, J.M., Chaparro—Riggers, J.F. et al. 2006. High—thoroughput screening for enhanced protein stability. Curr. Opin. Biotechnol., 17:606–10.

10. Brady, L., Brzozowski, A.M., Derewenda, Z.S. et al. 1990. A serine protease triad forms the catalytic centre of a triglyceride lipase. Nature, 343:767–70.

11. Brzozowski, A.M., Derewenda, U., Derewenda, Z.S. et al. 1991. A model for interfacial activation in lipases from the structure of a fungal lipase-inhibitor complex. Nature., 351:491–4.

12. Carriere, F., Thirstrup, K., Hjorth, S. et al. 1994. Cloning of the classical guinea pig pancreatic lipase and comparison with the lipase related protein 2. FEBS Lett., 388:63–8.

13. Cygler, M. and Schrag, J.D. 1999. Structure and conformational flexibility of Candida rugosa lipase. Biochim. Biophys. Acta., 1441:205–14.

14. de Maria, P.D., Sanchez-Montero, J.M., Sinisterra, J.V. et al. 2006. Understanding Candida rugosa lipases : An overview. Biotechnol. Adv., 24:180–96.

15. Dell, A. and Morris, H.R. 2001. Glycoprotein structure determination by Mass spectrometry. Science., 291:2351–6.

16. Derewenda, U., Brzozowski, A.M., Lawson, D.M. et al. 1992. Catalysis at the interface: the anatomy of a conformational change in a triglyceride lipase. Biochem., 31:1532–41.

17. Derewenda, Z.S., Derewenda, U. and Dodson, G.G. 1992. The crystal and molecular structure of the Rhizomucor miehei triacylglyceride lipase at 1.9 Å resolution. J. Mol. Biol., 227:818–39.

18. Durbin, S.D. and Feher, G. 1990. Studies of crystal growth mechanisms of proteins by electron microscopy. J. Mol. Biol., 212:763–74.

19. Eggert, T., van Pouderoyen, G., Pencreac'h, G. et al. 2002. Biochemical properties and three-dimensional structures of two extracellular lipolytic enzymes from Bacillus subtilis. Coll. Surf. B: Biointerf., 26:37–46.

20. Fojan, P., Jonson, P.H., Peterson, M.T.N. et al. 2000. What distinguishes an esterase from a lipase: A novel structural approach. Biochimie., 82:1033–41.

21. Gracy, J. and Argos, P. 1998. DOMO: a new database of aligned protein domains. TIBS, 23:495–7.

22. Grochulski, P., Li, Y., Schrag, J.D. et al. 1993. Insights into interfacial activation from an open structure of Candida rugosa lipase. J. Biol. Chem., 268:12843–7.

23. Haris, P.I. and Severcan, F. 1999. FTIR. spectroscopic characterization of protein structure in aqueous and non-aqueous media. J. Mol. Catal. B: Enz.,

7:207–21. Hasan, F., Shah, A.A. and Hameed, A. 2006. Industrial applications of microbial lipases. Enz. Microb.Technol., 39:235–51.

24. Hedin, E.M.K., Hult, K., Mouritsen, O.G. et al. 2004. Low microwave-amplitude ESR. spectroscopy : Measuring spin—relaxation interactions of moderately immobilized spin labels in proteins. J. Biochem. Biophys. Methods, 60:117–38.

25. Henikoff, S., Endow, S.A. and Greene, E.A. 1996. Connecting protein family resources using the proweb network. TIBS, 21:444–5.

26. Holzwarth, H.C., Pleiss, J. and Schmid, R.D. 1997. Computer-aided modeling of stereoselective triglyceride hydrolysis catalysed by Rhizopus oryzae lipase. J. Mol. Catal. B., 3:73–82.

27. Isobe, K. and Nokihara, K. 1993. Primary structure determination of mono-and diacylglycerol lipase from Pencillium camembertii. FEBS Lett., 320:101–6.

28. Jacobsen, C., Garside, J. and Hoare, M. 1998. Nucleation and growth of microbial lipase crystals from clarified concentrated fermentation broths. Biotechnol. Bioeng., 57:666–75.

29. Jaeger, K.E., Ransac, S., Koch, H.B. et al. 1993. Topological characterization and modeling of the 3D structure of lipase from Pseudomonas aeruginosa. FEBS Lett., 232:143–9.

30. Jonas, J. 2002. High-resolution nuclear magnetic resonance studies of proteins. Biochim. Biophys. Acta., 1595:145–59.

31. Jutila, A., Zhu, K., Tuominen, E.K.J. et al. 2004. Fluorescence spectroscopic characterization of Humicola lanuginose lipase dissolved in its substrate. Biochim. Biophys. Acta., 1702:181–9.

32. Kartik, V.J., Lavanya, T. and Guruprasad, K. 2006. Analysis of disulfide bond connectivity patterns in protein tertiary structure. Int.J. Biol. Macromol., 38:174–9.

33. Kazlauskas, R.J. 2000. Molecular modeling and biocatalysis: explanations, predictions, limitations, and opportunities. Curr. Opin. Chem. Biol., 4:81–8.

34. Kelly, S.M., Jess, T.J. and Price, N.C. 2005. How to study proteins by circular dichroism. Biochim. Biophys. Acta., 1751:119–39.

35. Khajehpour, M., Dashnau, J.L. and Vanderkooi, J.M. 2006. Infrared spectroscopy used to evaluate glycosylation of proteins. Anal. Biochem., 348:40–8.

36. Kohno, M., Kugimiya, W., Hashimoto, Y. et al. 1993. Preliminary investigation of crystals of lipase I from Rhizopus niveus. J. Mol. Biol., 229:785–6.

37. Kwon, H.J., Haruki, M., Morikawa, M. et al. 2002. Role of repetitive nine—residue sequence motifs in secretion, enzymatic activity and protein conformation of a family 1.3 lipase. J. Biosc. Bioeng., 93:157–64.

38. Laskowski, R.A., Hutchinson, E.G., Michie, A.D. et al. 1997. PDBSum: a web based database of summaries and analysis of all PDB. structures. TIBS, 22:488–90.

39. Lees, J.G., Smith, B.R., Wein, F. et al. 2004. CD tool—an integrated software package for circular dichroism spectroscopic data processing, analysis, and archiving. Anal. Biochem., 332:285–9.

40. Leong, M.K., Chen, C., Shar, K.C. et al. 2007. Selection and characterization of lipase abzyme from phage displayed antibody libraries. Biochem. Biophys. Res. Commun., 361:567–73.

41. Li, H. and Zhang, X. 2005. Characterization of thermostable lipase from thermophilic Geobacillus sp. TW1. Protein Expr. Purif., 42:153–9.

42. Liebeton, K., Zacharias, A. and Jaeger, K.E. 2001. Disulphide bond in Pseudomonas aeruginosa lipase stabilizes the structure but is not required for interaction with its foldase. J. Bacteriol., 183:597–603.

43. Lin, D., Tabb, D.L. and Yates III, J.R. 2003. Large-scale protein identification using mass spectrometry. Biochim. Biophys. Acta., 1646:1–10.

44. Mancheno, J.M., Pernas, M.A., Martinez, M.J. et al. 2003. Structural insights into the lipase/esterase behavior in the Candida rugosa lipases family: Crystal structure of the lipase2 isoenzyme at 1.97 Å resolution. J. Mol. Biol., 332:1059–69.

45. Maruyama, T., Nakajima, M., Ichikawa, S. et al. 2001. Structural study of lipase modified with fatty acids. Biochem. Eng. J., 9:185–91.

46. Matsuo, K., Yonehara, R. and Gekko, K. 2005. Improved estimation of the secondary structures of proteins by vacuum-ultraviolet circular dichroism spectroscopy. J.Biochem., 138:79–88.

47. McPherson, A. 1990. Current approaches to macromolecular crystallization. Eur. J. Biochem., 189:1–23.

48. Melo, E.P., Taipa, M.A., Castellar, M.R. et al. 2000. A spectroscopic analysis of thermal stability of the Chromobacterium viscosum lipase. Biophys. Chem., 87:111–20.

49. Meunier, B., Bouley, J., Piec, I. et al. 2005. Data analysis methods for detection of differential protein expression in two-dimensional gel electrophoresis. Anal. Biochem., 340:226–30.

50. Miled, N., Riviere, M., Cavalier, J.F. et al. 2005. Discrimination between closed and open forms of lipases using electrophoretic techniques. Anal. Biochem., 338:171–8.

51. Mills, K., Johnson, A.W., Diettrich, O. et al. 2000. A strategy for the identification of site-specific glycosylation in glycoproteins using MALDI TOF MS. Tet. Asymm., 11:75–93.

52. Montserret, R., McLeish, M.J., Bockmann, A. et al. 2000. Involvement of electrostatic interactions in the mechanism of peptide folding induced by sodium dodecyl sulfate binding. Biochem., 39:8362–73.

53. Murzin, A.G., Brenner, S.E., Hubbard, T. et al. 1995. Scop: a structural classification of proteins database for the investigation of sequences and structures. J. Mol. Biol., 247:536–40.

54. Olempska-Beer, Z.S., Merker, R.I., Ditto, M.D. et al. 2006. Food processing enzymes from recombinant microorganisms-a review. Reg. Toxicol. Pharmacol.., 45:144–58.

55. Ollis, D.L., Cheah, E., Cygler, M. et al. 1992. The alpha/beta hydrolase fold. Protein Eng., 5:197–221.

56. Orengo, C.A., Michie, A.D., Jones, D.T. et al. 1997. CATH—a hierarchic classification of protein domain structures. Structure, 5:1093–108.

57. Peterson, M.T.N., Fojan, P. and Peterson, S.B. 2001. How do lipases and esterases work: the electrostatic contribution. J. Biotechnol., 85:115–47.

58. Pleiss, J., Fischer, M., Peiker, M. et al. 2000. Lipase engineering database: Understanding and exploiting sequence—structure—function relationships. J. Mol. Cat. B: Enz., 10:491–508.

59. Rahman, R.N.Z.R.A., Baharum, S.N., Basri, M. et al. 2005. High-yield purification of an organic solvent-tolerant lipase from Pseudomonas sp. strain S5. Anal. Biochem., 341:267–74.

60. Ransac, S., Blaauw, M., Dijkstra, B.W. et al. 1995. Crystallization and preliminary X-ray analysis of a lipase from Staphylococcus hyicus. J. Struct. Biol., 114:153–5.

61. Saxena, R.K., Sheoran, A., Giri, B. et al. 2003. Purification strategies for microbial lipases. J. Microbiol. Methods, 52:1–18. Schrag, J.D., Li, Y., Cygler, M. et al. 1997. The open conformation of a Pseudomonas lipase. Structure, 5:187–202.

62. Secundo, F., Carrea, G., Tarabiono, C. et al. 2006. The lid is a structural and functional determinant of lipase activity and selectivity. J. Mol. Cat. B: Enz., 39:166–70.

63. Severcan, M., Haris, P.I. and Severcan, F. 2004. Using artifi cially generated spectral data to improve protein secondary structure prediction from Fourier transform infrared spectra of proteins. Anal. Biochem., 332:238–44.

64. Sheehan, D. and Sullivan, S.O. 2006. Homology modeling of milk enzymes using on-line resources : Insights to structure—function and evolutionary relationships. Int. Diary J., 16:701–6.

65. Shibamoto, H., Matsumoto, T., Fukuda, H. et al. 2004. Molecular engineering of Rhizopus oryzae lipase using a combinatorial protein library constructed on the yeast cell surface. J. Mol. Catal. B: Enz., 28:235–9.

66. Sinchaikul, S., Sookkheo, B., Phutrakul, S. et al. 2001. Structural modeling and characterization of a thermostable lipase from Bacillus stearothermophilus P1. Biochem. Biophys. Res. Commun., 283:868–75.

67. Sinchaikul, S., Tyndall, J.D.A., Fothergill-Gilmore, L.A. et al. 2002. Expression, purification, crystallization and preliminary crystallographic analysis of a thermostable lipase from Bacillus stearothermophilus P1. Acta. Cryst., D58:182–5.

68. Svendsen, A. 2000. Lipase protein engineering. Biochim. Biophys. Acta., 1543:233–8.

69. Tang, S.J., Shaw, J.F., Sun, K.H. et al. 2001. Recombinant expression and characterization of the Candida rugosa lip4 lipase in Pichia pastoris: Comparison of glycosylation, activity, and stability. Arch. Biochem. Biophys., 387:93–8.

70. Tripathi, M.K., Roy, U. and Jinwal, U.K. 2004. Cloning, sequencing and structural features of a novel Streptococcus lipase. Enz. Microb. Technol., 34:437–45.

71. Tyndall, J., Sinchaikul, S., Gilmore, L., Taylor, P. and Walkinshaw, M. 2002. Crystal structure of a thermostable lipase from Bacillus stearothermophilus P1. J. Mol. Biol., 323:859–69.

72. Villeneuve, P. and Foglia, T.A. 1997. Lipase specificities: Potential application in lipid bioconversions. Inform., 8:640–50.

73. Wadsten, P., Wohri, A.B., Snijder, A. et al. 2006. Lipidic sponge phase crystallization of membrane proteins. J. Mol. Biol., 364:44–53.

74. Wilmot, C.M. and Thornton, J.M. 1988. Analysis and prediction of the different types of β-turn in proteins. J. Mol. Biol., 203:221–32.

75. Yang, J., Dong, X.C. and Leng, Y. 2006. Conformation biases of amino acids based on tripeptide microenvironment from PDB. database. J. Theoret. Biol., 240:374–84.

76. Zhu, K., Jutila, A., Tuominen, E.K.J. et al. 2001. Impact of the tryptophan residues of Humicola lanuginosa lipase on its thermal stability. Biochim. Biophys. Acta., 1547:329–38.

77. Zvelebil, M.J., Barton, G.J., Taylor, W.R. et al. 1987. Prediction of protein secondary structure and active sites using the alignment of homologous sequences. J. Mol. Biol., 195:957–61.

CITATION

Significance Analysis of Microarray for Relative Quantitation of LC/MS Data in Proteomics

Bryan A. P. Roxas and Qingbo Li

ABSTRACT

Background

Although fold change is a commonly used criterion in quantitative proteomics for differentiating regulated proteins, it does not provide an estimation of false positive and false negative rates that is often desirable in a large-scale quantitative proteomic analysis. We explore the possibility of applying the Significance Analysis of Microarray (SAM) method (PNAS 98:5116-5121) to a differential proteomics problem of two samples with replicates. The quantitative proteomic analysis was carried out with nanoliquid chromatography/ linear iron trap-Fourier transform mass spectrometry. The biological sample model included two Mycobacterium smegmatis unlabeled cell cultures grown at pH 5 and pH 7. The objective was to compare the protein relative abundance between the two unlabeled cell cultures, with an emphasis on significance analysis of protein differential expression using the SAM method. Results using the SAM method are compared with those obtained by fold change and the conventional t-test.

Results

We have applied the SAM method to solve the two-sample significance analysis problem in liquid chromatography/mass spectrometry (LC/MS) based quantitative proteomics. We grew the pH5 and pH7 unlabelled cell cultures in triplicate resulting in 6 biological replicates. Each biological replicate was mixed with a common ^{15}N-labeled reference culture cells for normalization prior to SDS/PAGE fractionation and LC/MS analysis. For each biological replicate, one center SDS/PAGE gel fraction was selected for triplicate LC/MS analysis. There were 121 proteins quantified in at least 5 of the 6 biological replicates. Of these 121 proteins, 106 were significant in differential expression by the t-test (p < 0.05) based on peptide-level replicates, 54 were significant in differential expression by SAM with Δ = 0.68 cutoff and false positive rate at 5%, and 29 were significant in differential expression by the t-test (p < 0.05) based on protein-level replicates. The results indicate that SAM appears to overcome the false positives one encounters using the peptide-based t-test while allowing for identification of a greater number of differentially expressed proteins than the protein-based t-test.

Conclusion

We demonstrate that the SAM method can be adapted for effective significance analysis of proteomic data. It provides much richer information about the protein differential expression profiles and is particularly useful in the estimation of false discovery rates and miss rates.

Background

Fold change is commonly used in quantitative proteomic analysis where proteins differing by more than an arbitrary cut-off value in abundance are considered to be differentially expressed [1-5]. A fold change test is equivalent to a global t-test assuming homogenous variance between different proteins. Although it is a convenient and cost effective way to evaluate protein expression level differences between two conditions, fold change alone is not a statistical test that can indicate the level of confidence in differential expression of proteins.

Rapid development of liquid chromatography-mass spectrometry (LC/MS) based proteomics has led to gradual replacement of the traditional 2D gel approach by the LC/MS approach. Accordingly, variation and quality control of quantitation by LC/MS has been actively explored [6-12]. The importance of significance analysis for biomarker discovery has also been stressed [13]. Molina et al. simultaneously measured three states of Hela cells in response to stimuli using SILAC labeling for quantitation [6]. Fold changes were evaluated at protein and peptide level by analysis of variance performed in the statistical program R. The authors demonstrated the capability of detecting 1.8-fold change at a significance level of 95%. No significance score was assigned to individual proteins, however. Piening et al. proposed Mass Deviance, a quality control metric, for assessing the accuracy of peptide detection in Saccharomyces cerevisiae [7]. This approach was rigorous at validating peptide identification in LC/MS but not yet directly applicable for quantifying relative abundance. Meng et al. used the differential mass spectrometry (dMS) method for label-free LC/MS profiling, demonstrating detection of peptides with a change as small as 1.5-fold with ~20% relative errors in peptide relative abundance in a processed plasma background [8]. Andreev et al. developed Q-MEND algorithm for label-free quantitation of relative protein abundances across multiple complex E. coli proteome samples, achieving 7% quantitation accuracy and mean precision of 15% [9]. Wang et al. reported the algorithm Quoil for label-free quantitation measurements across repeated LC/MS runs with Student's t-test after applying the step-down adjustment of probability threshold [14]. Most recently, reproducibility assessment of differential quantitation by SILAC, ICAT, and label-free methods was reported [11]. In this study, a ratio distribution analysis was applied to common peptides between samples to remove outliers until a normal distribution was obtained. Using the filtered common peptides, it was assessed that 95% of the total common peptides have intensities within a ~2-fold change for a pair of cultures of T47D human breast cancer cells, with SILAC analysis having the best summary statistics. These conclusions were drawn from peptide-level quantitation in combination with a ratio distribution analysis. Earlier efforts in statistical and computational methods for quantitative proteomics by LC/MS was reviewed by Listgarten and Emili [12]. These studies showed a great deal of efforts and progress in statistical analysis of LC/MS data in proteomics. Currently, few have systematically assessed significance of analysis at a systems level along with estimation of false positive and negative rates.

In this work, we explore the use of the Significance Analysis of Microarray (SAM) method [15] for analysis of a two-sample significance problem in LC/MS quantitative proteomics. We used Mycobacterium smegmatis cells grown at pH 5 and pH 7 in unlabeled media as the two-sample model. We also grew one 15N-labeled M. smegmatis cell culture and used it as the internal standard to normalize the protein abundance in the pH5 and pH7 unlabeled cells by the popular 14N/15N quantitation method [16]. Cell protein extracts were first fractionated by SDS/PAGE. Then a high resolution nanoliquid chromatography/linear ion trap-Fourier transform mass spectrometry (nanoLC/LTQ-FTMS) system was used for peptide separation and identification. The LC/MS data was further quantified by the previously described algorithm [10]. We report the results of quantifying the protein relative abundance between the pH5 and pH7 unlabeled cells, with an emphasis on significance analysis of protein differential expression using the SAM method in comparison with fold change and conventional t-test methods.

Results and Discussion

In this study, we report using the SAM method to solve the two-sample significance analysis problem in LC/MS based quantitative proteomics. We compare the SAM method with the conventional fold change test and t-tests.

SAM was originally developed for microarray analysis by Tusher et al. [15]. Development of this method was initially propelled by the need to resolve the issue of multiplicity of testing in conventional t-tests when a large number of genes were studied simultaneously. As Tusher et al. stated, "SAM identifies genes with statistically significant changes in expression by assimilating a set of gene specific t-tests. Each gene is assigned a score on the basis of its change in gene expression relative to the standard deviation of repeated measurements for that gene. Genes with scores greater than a threshold are deemed potentially significant with an assigned q-value." SAM incorporated q-value as a measurement of the significance of a gene based on the work of Storey [17]. Each time when the threshold is adjusted, a false discovery rate is estimated for the resulting set of genes with significant differential expression. Low-level data processing in the LC/MS measurements is typically very different from that in DNA microarray experiments. However, at the higher level of protein differential expression determination, we treated the protein abundance data the same as the gene abundance data and used the SAM method without modification.

For statistical analysis in this work, the pH5 and pH7 M. smegmatis cell cultures were grown in triplicate resulting in total 6 biological replicates. We also grew a 15N-labeled M. smegmatis culture and used it as the reference for normalizing the 6 unlabeled biological replicates. Each of the 6 unlabeled biological replicates was first mixed with the 15N-labeled reference. It was then processed for protein quantitation by the widely used 14N/15N relative abundance measurement method [10,16]. After all 6 unlabeled biological replicates were normalized to the 15N-labeled reference, they were analyzed either by fold change test, t-test or the SAM method.

In the following sections, we discuss the experimental layout of sample replicates, fold change analysis, random fluctuation of measurements, conventional t-tests, SAM analysis, and differentially expressed proteins.

Sample Replicates

In DNA microarray experiments, arrays are often spotted with gene probes in replicates. The typical practice is to average the replicates for each probe before assessing the differential expression of the gene. Since the geometrical arrangement of gene probes on an array is known before an experiment, the replicates for a gene can be known a priori within an array and across multiple arrays. In proteomics, assignment and cross-reference of peptides and proteins across multiple LC/MS analysis is not as straightforward. In a typical LC/MS based proteomics experiment, a protein is digested with an enzyme into multiple peptides. The mixture of peptides from multiple proteins is injected into a LC/MS instrument for separation and peptide identification by MS/MS scan. Protein relative abundance is assessed from the quantitation of one or more peptides originating from the protein. This process is analogous to the quantitation of genes based on multiple gene probes.

Contrary to DNA microarray experiments, one distinct characteristic of LC/MS based quantitative proteomics is that the number of peptides being quantified is usually not known a priori. There are different reasons for this. For example, if a peptide is highly hydrophobic or highly negatively charged, the chance of identifying this peptide by LC/MS is significantly reduced. There are also LC/MS instrument related considerations. A data-dependent acquisition algorithm is employed in most LC/MS instrument methods for peptide identification [18]. Due to the limited speed with which a mass spec can acquire MS/MS spectra for peptide identification, only a limited number of precursor ions with the highest intensities in one MS scan will be selected for MS/MS identification. Currently the sampling rate of a typical LC/MS instrument can easily be overwhelmed by the complexity of a protein sample. Saturation of sampling a complex protein mixture requires more than just a few replicate runs. For this reason, a common way to increase the number of identified and quantified peptides for a sample is to pool the peptides identified by MS/MS from replicate runs of the same sample [5].

Callister et al. [19] demonstrated an accurate mass and time tag (AMT) approach to overcome the above mentioned limitation. In this approach, the LC/MS scanning process is decoupled from the MS/MS peptide sequencing process.

This is done by first accumulating enough MS/MS identification of peptides followed by high-throughput LC/MS analysis. This powerful AMT approach skips the rate limiting MS/MS step. It thus avoids the random sampling effect of the MS/MS peptide identification process. However, successful application of this approach relies upon a database containing the AMTs accumulated from multiple LC/MS/MS runs. This requires precise control of the LC/MS/MS and LC/MS operation parameters to ensure the reliability of AMTs.

Because the primary focus of this work was to compare several statistical significance analysis methods, we opted to take a straight-forward approach by only quantifying those peptides with confident MS/MS identification. These were the peptides assigned a probability of misidentification smaller than 0.01 by the Bio-Works software based on a MS/MS spectrum database search. A probability of 0.01 implies one misidentification out of 100 by chance. A peptide may be identified at different charge states typically ranging from +1 to +4. The most often observed charge state is +2 or +3 in the nanoLC/LTQ-FTMS system. BioWorks assigns a probability for a peptide detected at each charge state. Accordingly, a peptide detected at a particular charge state is called a peptide charge state (PCS) [10].

In this study, we grew the pH5 and pH7 unlabelled cell cultures in triplicate resulting in a total of 6 biological replicates. Each biological replicate was mixed with the 15N-labeled reference sample prior to SDS/PAGE fractionation and LC/MS analysis. Each biological replicate was analyzed by nanoLC/LTQ-FTMS with triplicate runs. The PCS's with $p < 0.01$ from the triplicate runs of a biological replicate were combined for calculating the protein and peptide relative abundance [10]. Statistical analysis of relative abundance between the pH5 and pH7 unlabelled cells was performed for the proteins quantified in at least 5 of the 6 replicates. The 6 biological replicates were designated as pH5A, pH5B, pH5C, pH7A, pH7B, and pH7C (Table 1). The average of the pH5 biological triplicates was named pH5av. Similarly, the average of the pH7 biological triplicates was named pH7av.

Table 1. Sample names. The first number in parenthesis is the number of quantified proteins for the sample replicate name preceding the two numbers in the parenthesis. The second number in the same parenthesis is the average number of PCS's quantified for a protein in the same sample replicate. See main text for more details.

Cell sample	Culture triplicates	*in silico* pooled replicate	Average of culture triplicates
pH5	pH5A (119, 15) pH5B (120, 14) pH5C (121, 15)	pH5p (174, 41)	pH5av
pH7	pH7A (112, 14) pH7B (110, 14) pH7C (113, 13)	pH7p (174, 31)	pH7av

In addition, we pooled the PCS's from the pH5 biological triplicates to calculate the protein relative abundance for the in silico pooled replicate for the pH5 cells, which we called pH5p. Similarly, we also pooled the PCS's from the pH7 biological triplicates to calculate the protein relative abundance for pH7p, the in silico pooled replicate for the pH7 cells (Table 1).

The protein mixture of each biological replicate was fractionated into 5 fractions by SDS/PAGE. Only the center fraction was further processed for nanoLC/LTQ-FTMS analysis. Although it was desirable to analyze all the fractions in triplicate LC/MS analysis, we chose to only analyze the center fraction for each biological replicate for two reasons. First, focusing on one common fraction for all 6 biological replicates is sufficient for demonstrating the principle of statistical analysis we investigated in this work. Second, we were conservative about the cost of analyzing all 5 fractions for all 6 replicates because it would have required 90 LC/MS runs lasting for more than 135 hrs. This estimation was based on 5 SDS/PAGE fractions per biological replicate, triplicate runs per SDS/PAGE fraction, and 90 min per run (see Methods).

With only the center fraction analyzed, there were 121 proteins quantified in at least 5 of the 6 biological replicates of the pH5 and pH7 samples (Table 1). Ninety were quantified in all 6 replicates, and 31 in 5 replicates. Figure 1a shows the CV boxplots for these proteins in the 6 biological replicates and the 2 in silico pooled replicates. Meanwhile, there were 174 proteins found in common between pH5p and pH7p. The CV boxplots for these 174 proteins are shown in Figure 1. The complete set of protein and peptide data for statistical significance analysis is summarized in Table 2. Table 2 shows the protein relative abundance, standard deviation, number of unique peptides and number of PCS's for each protein in the sample replicates pH5A, pH5B, pH5C, pH7A, pH7B, pH7C, pH5p, and pH7p. Results of the fold change test, the 2 t-tests, and the SAM analysis to be described in a later section are shown to the right of Table 2.

Figure 1 indicates that more than 75% of proteins in every sample replicate have CV less than 30%. The mean CV and median CV for all sample replicates was less than 21% and 15%, respectively. It was noticed that the CV summary statistics were improved only slightly for the 121 proteins in Figure 1a compared to the 174 proteins in Figure 1b. The average of the average CV's for the biological triplicates was $18 \pm 2\%$ for the pH5 sample, and $18 \pm 3\%$ for the pH7 sample. For the 174 proteins common between pH5p and pH7p, the average CV was 19% and 21% for pH5p and pH7p, respectively. These results indicate that the sample replicates have consistent CV summary statistics. They are suitable for use in subsequent analysis to compare several significance analysis methods.

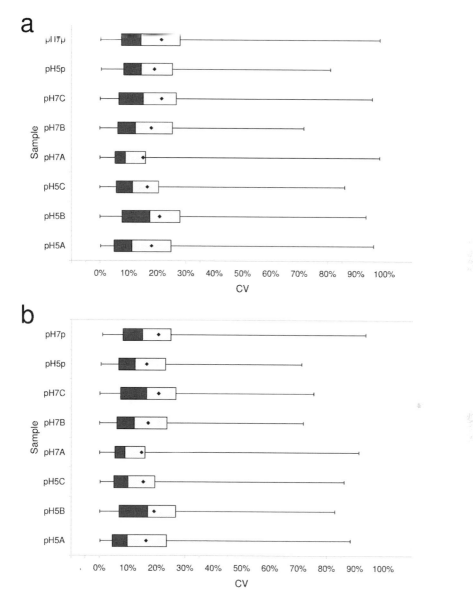

Figure 1. CV summary statistics. Boxplots displaying the summary statistics of the coefficient of variance (CV) of protein relative abundances for the pH5 culture triplicates (pH5A, pH5B, and pH5C), the pH7 culture triplicates (pH7A, pH7B, and pH7C), and the in silico pooled replicates (pH5p and pH7p). A boxplot summarizes the minimum, 25 percentile, 50 percentile, 75 percentile, and maximum CV's of a sample. a) Boxplots are plotted for the 174 proteins quantified between pH5p and pH7p. These 174 proteins include all those proteins quantified in pH5A(159), pH5B(159), pH5C(161), pH7A(131), pH7B(124), and pH7C(134). The numbers in parenthesis indicate the number of protein in a sample. b) Boxplots are plotted for the 121 proteins quantified in at least 5 of the 6 replicates of the pH5 and pH7 culture samples. The diamond dots indicate the mean CV's.

Fold Change Analysis

Since pH5p and pH7p represent the average of both analytical and biological replicates for the pH5 and pH7 unlabeled culture samples respectively, we examine the number of differentially expressed proteins between the two samples by the fold change test. Within this context, fold change refers to the ratio of relative abundance of a protein between the pH5 and pH7 unlabeled samples. It has a value greater than or equal to 1. This definition is consistent with that of the SAM.

Based on the simple 2- and 3-fold change tests, 55 and 29 proteins were respectively found to be differentially expressed between pH5p and pH7p (Table 2). As discussed earlier, the fold change threshold alone is not a statistical test that can indicate the level of confidence about differentially expressed proteins. It does not reveal the random fluctuation inherent in protein differential expression levels. It would be of interest to test the level of such random fluctuation. As described below, we took a simple approach to test if random fluctuation was confined within a 2- or even 3-fold change boundary.

Random Fluctuation

To test random fluctuation, the number of quantified PCS's of a protein was plotted against the log ratio between the average of relative abundance of its biological triplicates (A_{av}, representing either A_{pH5av} or A_{pH7av}) and the relative abundance of its in silico pooled replicate (A_p, representing either A_{pH5p} or A_{pH7p}), as shown in Figure 2. In addition, the histogram for each sample was also plotted based on protein number and $\log_2(A_{av}/A_p)$. We reasoned that pH5av versus pH5p or pH7av versus pH7p represents a form of permutation for the biological triplicates of the pH5 or pH7 sample. The distribution of $\log_2(A_{av}/A_p)$, as summarized by the histograms, should therefore reveal some random errors in protein relative abundance quantitation. We chose the number of PCS's for plotting against $\log_2(A_{av}/A_p)$ because it was interesting to examine the effect on random errors.

From Figure 2, it was noted that most of the proteins clustered within 1.5 fold change, or ± 0.585 on the log base 2 scale. The 95% interval was -0.52 to 0.48 for the pH5 sample (light gray trace) and -0.53 to 0.80 for the pH7 sample (dark gray trace). There were 3 proteins in the pH5 sample and 4 in the pH7 sample falling outside the 2-fold boundary. There was 1 protein in the pH5 sample and 1 in the pH7 sample falling outside the 3-fold boundary. There were 3 proteins in the pH5 sample and 2 in the pH7 sample falling within the range of 2- to 3-fold change. There were 2 proteins in the pH5 sample and 6 in the pH7 sample falling within 1.5- to 2-fold change range.

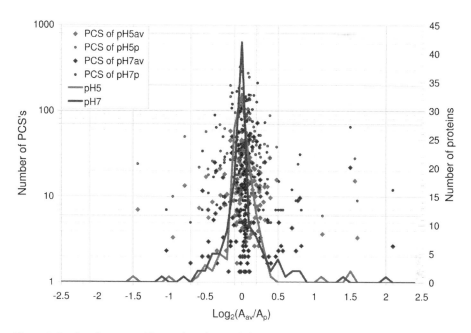

Figure 2. Random fluctuation. The number of quantified PCS's of a protein is plotted against the log ratio of the average relative abundance of its biological triplicates (A_{av}, representing either A_{pH5av} or A_{pH7av}) over the relative abundance of its in silico pooled replicate (A_p, representing either A_{pH5p} or A_{pH7p}) for the pH5 and pH7 samples respectively. The light gray diamonds and the small light gray dots represent the pH5 sample. The dark gray diamonds and the small dark gray dots represent the pH7 sample. The diamond symbols represent the average number of PCS's of a protein of the biological triplicates. The small dots represent the number of total PCS's of a protein in the in silico pooled replicates. The histograms based on the number of quantified proteins are also plotted, with the light gray trace representing the pH5 sample and the dark gray trace representing the pH7 sample.

These results suggested that the random errors could occur outside a 2- or even 3-fold change boundary. In addition, the random errors shown in Figure 2 were not limited to those proteins that had a very low number (< 5) of PCS's, even though the trend was that random errors mostly occurred below 25 PCS's for pH5av and pH7av or below 70 for pH5p and pH7p.

To evaluate the influence of random fluctuation on the confidence of measured protein differential expression, we performed 2 t-tests as described in the following. One t-test was based on peptide-level replicates. The other was based on protein-level replicates.

T-Tests

In general, a t-test is used to evaluate whether the means of control and experiment groups are statistically different. The t-value is the ratio between the difference of group means and the variability of groups. The standard deviation of the t distribution is determined by the number of degrees of freedom derived from the sample sizes. The number of degrees of freedom need not be the same for the control and the experiment groups. For the same z score, a falling sample size will make the t distribution take on an increasingly larger standard deviation. Increased standard deviation of the t distribution has the tendency to incur a higher false negative rate. On the other hand, a very large number of degrees of freedom may allow a higher false positive rate. In this analysis, the number of degrees of freedom may be very high for pH5p and pH7p for some proteins when it is based on the number of PCS's detected for each protein, i.e., peptide-level replicates. However, the number of degrees of freedom is no more than 3 when protein-level replicates are used. In either case, the t-tests do not require equal number of degrees of freedom between control and experiment.

For simplicity in describing proteins found to have statistically significant differential expression, the term "significant protein" is used hereafter with the meaning of "protein with significant differential expression."

T-Test with Peptide-Level Replicates

To test if the observed differential expression of these proteins was significant, a two-sample t-test assuming equal variances was performed on the 174 proteins using peptide-level relative abundance information. We adopted the t-test which was previously demonstrated by Wu et al. [20] in quantitative proteomic analysis of mammalian organisms. To compute the two-sample t-test, a pooled standard deviation was first calculated from the standard deviations of the protein relative abundance of the 2 samples. The pooled standard deviation was between the 2 standard deviations with greater weight given to the standard deviation of the sample with larger number of PCS's detected. The mathematical formula for the t-test was fully described by Wu et al. [20]. Since all of the PCS's were pooled to calculate the protein relative abundance in pH5p and pH7p, there was only one protein relative abundance value for each protein in pH5p or pH7p. The t-test for comparing pH5p and pH7p was thus performed using peptide-level replicates without protein-level replicates. This means that the number of degrees of freedom for measuring a protein was represented by multiple PCS measurements for that protein. This t-test with peptide-level replicates is different from that described later with protein-level replicates.

We used the volcano plot in Figure 3 to visualize the proteins categorized as up- or down-regulated based on the simple 2- and 3-fold change thresholds, and to display their statistical significance based on the t-test with peptide-level replicates. In the volcano plot, the t-test p value was plotted against the relative abundance ratio between pH5p and pH7p on a logarithmic scale. The t-test rejected one of the proteins found upregulated in pH5p with greater than 3-fold change. This resulted in a total of 53 proteins having greater than 2-fold change with t-test significance (p < 0.05). Of these 53 proteins, 25 had fold change between 2 and 3, and 28 had greater than a 3-fold change. Of the remaining 120 proteins that had less than a 2-fold change, 32 were not significant (p > = 0.05), and 88 were significant (p < 0.05).

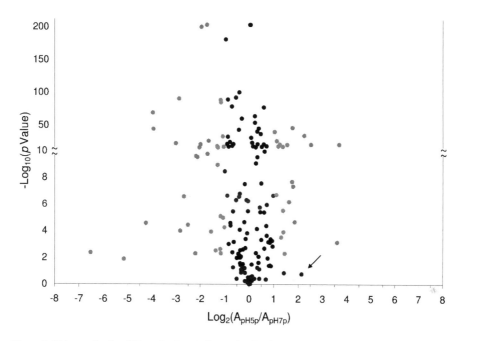

Figure 3. Volcano plot for pH5p and pH7p. Volcano plot for the in silico pooled replicates pH5p and pH7p. The arrow indicates the protein with 3-fold change but found not significant by the t-test. See text for more details.

T-Test with Protein-Level Replicates

The above t-test with peptide-level replicates utilized the PCS's of a protein quantified in pH5p and pH7p. Each PCS should be an independent event for a protein. This assumption is complicated by several factors. In LC/MS based proteomic experiments, detection of a PCS depends not only on its concentration but also on the composition of the peptide mixture. Ion suppression effect in electrospray ionization, space charge effect in FT mass spectrometer, LC column separation efficiency for complex samples, and data-dependant acquisition, etc., can directly or indirectly affect the quantitation of a PCS. Therefore, a conventional t-test performed on such data requires cautious interpretation.

For comparison, we performed the second t-test at a protein level. This means that only protein relative abundance values were used without referring to the PCS information as for the t-test shown in Figure 3. Basically, pH5A, pH5B, and pH5C represented the triplicates for the pH5 sample. pH7A, pH7B, and pH7C represent the triplicates for the pH7 sample. Using the 2 sets of protein-level triplicates, we calculated their respective average pH5av and pH7av. To perform the t-test, the 2 sets of protein-level replicates were input as two arrays in the Microsoft Excel TTEST function with option selection of two-sample equal variance, two-tailed, and type of homoscedastic. Figure 4 shows the volcano plot for the protein relative abundance ratios between the pH5 and pH7 samples based on the triplicate protein relative abundances for each sample. Compared to Figure 3, one apparent difference is that a smaller number of significant proteins (55%) had fold change less than 1.5 (0.585 on log base 2 scale). The percentage of significant proteins was also reduced to 24% in Figure 4 compared to 81% in Figure 3. Of the 14 proteins with greater than 3-fold change, 9 (69%) were tested significant, compared to 28 out of 29 (97%) in Figure 3.

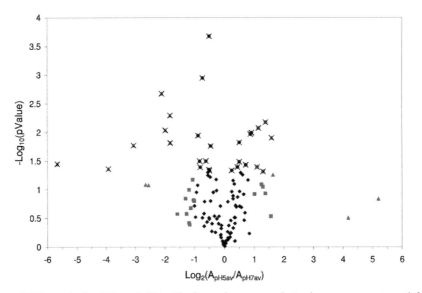

Figure 4. Volcano plot for pH5av and pH7av. The diamonds, squares, and triangles represent proteins with fold change of less or equal to 2, greater than 2 but less or equal to 3, and greater than 3 respectively. The x marks indicate that a protein is found significant ($p < 0.05$) by the t-test based on triplicate protein relative abundances without referring to peptide information. See text for more details.

There is an apparent discrepancy between the t-test results shown in Figure 3 and Figure 4 based on respective peptide- and protein-level information.

This suggests that we need a third method to examine whether the two conventional t-test methods are overly aggressive or conservative. To do so, we need to assess not only the individual protein significance but also the false positive and false negative rates for the group of proteins under significance testing. A similar issue has been extensively investigated in DNA microarray data analysis. SAM is one of the widely accepted methods for such analysis in DNA microarray. In the following, we explore the applicability of the SAM method towards our proteomics problem.

Significance Analysis with SAM

As described earlier, SAM is a statistical technique originally developed for finding genes with significant differential expression in a set of microarray experiments [21]. SAM is capable of taking input from different response variables. For our proteomics problem of two-sample significance analysis between the pH5 and pH7 cell cultures, the response variable is equivalent to a grouping of untreated (pH7) and treated (pH5) samples (unpaired). For each sample, at least two replicates are required by SAM. Using the protein-level replicates from the pH5 and pH7 samples, SAM calculates observed and expected scores for each protein. The observed score represents the relative difference of a protein between the pH5 and pH7 samples. The expected score represents the random fluctuation when there is no difference between the two samples. When the difference between the observed and expected scores is beyond a certain threshold, the protein is called significant in differential expression.

To perform the SAM analysis, the protein relative abundance data from the pH5 and pH7 biological triplicate samples were input as two-class unpaired response type into SAM, with 600 permutations, t-statistic test, 1% fixed percentile for estimation of s0 factor for denominator, and K-nearest neighbors imputer as the imputation engine with 5 neighbors for filling missing values. Due to the space limit here, we will not repeat further detail description of the SAM software and its operation. The users guide and technical documents for SAM are readily available elsewhere by Chu et al. [21]. We will instead focus on the result output and interpretation of the method.

Figure 5 shows the results from the SAM analysis. The SAM plotsheets are presented with slight graphical modification, with the Δ value and fold change inserted into the upper right corner for convenience of comparison. Each SAM plotsheet contains all the proteins plotted by their observed scores and expected scores.

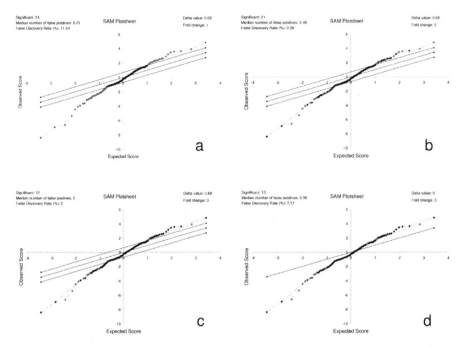

Figure 5. SAM plotsheet outputs. SAM plotsheet outputs under the four sets of criteria: a) Δ = 0.68, fc = 1; b) Δ = 0.68, fc = 2; c) Δ = 0.68, fc = 3; d) Δ = 0, fc = 3, which are indicated at the upper right corner of each plotsheet. The dots represent upregulated, downregulated, and insignificant proteins respectively. The upper and lower 45° degree lines indicate the Δ threshold boundaries. Proteins with Δ = 0 would fall on the 45° line through the origin. The number of significant proteins, median number of false positives, and false discovery rate are indicated at the upper left corner of each plotsheet.

The observed score is the relative difference [15] in protein expression. It is calculated by dividing the difference between protein relative abundances in the pH5 and pH7 samples by the pooled standard error of repeated measurements of that protein in the pH5 and pH7 samples [15]. The expected score is calculated using the large set of permutations of protein relative abundance data of the 6 biological replicates from the pH5 and pH7 samples.

The observed score provides a control over random fluctuation, while the expected score allows assignment of statistical significance. The correlation of these two scores is used for identifying proteins with potentially significant differential expression as shown in Figure 5. If a protein has absolutely no differential expression, the observed relative difference would be the same as the random fluctuation that is represented by the expected score. The data point of such a protein in the SAM plotsheet would fall on the 45° line through the origin. Data points representing differentially expressed proteins will deviate from this 45° line. The point displacement of a protein from the 45° line through the origin is quantitatively measured by a Δ value in SAM. Proteins with Δ values beyond a certain threshold are called significant. The 45° upper and lower Δ lines indicate the boundary defined by a selected Δ value.

SAM provides an estimation of false discovery rate (FDR) for the proteins called significant by each Δ value. A Δ value can be set together with a fold change threshold. FDR is calculated from the average number of falsely significant proteins in all the permutations divided by the number of proteins called significant above that Δ threshold.

Figure5a presents the results with Δ = 0.68. This Δ value results in 6.23 estimated false positives out of the 121 proteins under testing, equivalent to a 5.1% false positive rate. This is the same as the nominal false positive rate defined by a p < 0.05 threshold in a conventional t-test. Fifty-four proteins are called significant with a FDR of 11.54%, with 22 upregulated and 32 downregulated in the pH5 versus the pH7 sample.

Combination of Δ = 0.68 with 2-fold change reduces the number of significant proteins to 21 with a FDR of 2.28% (Figure 5b). When Δ = 0.68 is used together with 3-fold change, the number of significant proteins are further reduced to 12, with a FDR of 0 (Figure 5c) When the 3-fold change criterion is used alone, there are 13 proteins called significant with a FDR of 7.37% (Figure 5d). The 13th protein (MSMEG4520) increases the total significant proteins determined by SAM using the four different criteria to 55, as shown in Figure 5. For comparison, Table 3 lists these 55 proteins with their analysis output by SAM and the 2 conventional t-tests.

Of the 55 proteins, 26 were found significant by the t-test (p < 0.05) performed on the triplicates of the pH5 and pH7 samples (Figure 4), and all were found significant by the t-test (p < 0.05) performed on the in silico pooled replicates of the pH5 and pH7 samples (Table 2). Of the 13 proteins with greater than 3-fold change in Table 3, only 9 were found significant by the t-test shown in Figure 4. It is noted that there are 14 proteins with greater than 3-fold change shown in Figure 4. This extra 14th protein (MSMEG2382) with greater than 3-fold change in Figure 4 has a significant fold change of 2.8 calculated by SAM after imputation. This protein was not found significant under the t-test in Figure 4, even though it showed a fold change of 3.1 in Figure 4. Of the 121 proteins analyzed by SAM, 106 are called significant by the t-test (p < 0.05) shown in Figure 3, whereas only 54 are called significant by SAM with Δ = 0.68 cutoff which controls false positive rate at 5% and FDR at 13.1%. Therefore, the t-test for the in silico pooled replicates pH5p and pH7p shown in Figure 3 is overly aggressive while the t-test in Figure 4 appears to be overly conservative. These results indicate that SAM provides more reasonable results. The resampling approach used by SAM appears to overcome the false positives one encounters using the peptide-based t-test while allowing for identification of a greater number of differentially expressed proteins than the protein-based t-test.

Most importantly, for each significant protein, SAM assigns a q-value that represents the minimum FDR of the list of proteins having Δ values and/or fold changes equal to or greater than that at which the protein is called significant in differential expression. Therefore, q-value quantitatively measures how significantly the protein is differentially expressed. This is the lowest FDR at which the protein is called significant [17,21]. As further explained by Chu et al. [21], it is like the familiar 'p-value' but adapted to the analysis of a large number of genes. In other words, it is the p-value at which proteins with Δ values and/or fold changes smaller than the significant threshold are actually differentially expressed. The q-values for proteins called significant under different Δ and/or fold change criteria are presented in Table 3 in comparison with conventional t-tests.

Thirty-four (63%) of the 54 proteins called significant with the Δ = 0.68 threshold have fold change between 1.2 and 2.0. Of these 34 proteins, 22 (65%) have a q-value greater than 5%. For the 20 (37%) proteins with greater than 2-fold change, 3 (15%) have a q-value greater than 5% (Table 3). This illustrates that q-value properly predicts the significance of protein differential expression. While conventional t-tests provide an estimation of probability for individual proteins, the distribution of errors is not known.

Combination of Δ = 0.68 and 2-fold change results in 21 significant proteins, of which 15 have a q-value of 0 and 6 have a q-value between 2.3 and 3.0. Combination of Δ = 0.68 and 3-fold change results in 12 significant proteins all of which have a q-value of 0. Using the 3-fold change criterion alone generates 13 significant proteins that include the 12 proteins called significant by Δ = 0.68 and 3-fold change. The 13th additional protein (MSMEG4520) has a q-value of 7.4%. The other 12 proteins all have a q-value of 0. SAM predicts that 1 out of the 13 proteins (13 × 7.4% [congruent with] 1) would be a false positive. Since MSMEG4520 has the lowest observed scored = 1.2 and a q-value of 7.4%, by definition, MSMEG4520 is the one most likely to be falsely called significant by the 3-fold change criterion. The t-test performed in Figure 3 identifies this protein as significant with a p value of 7.2 × e-4 which is not the lowest among the proteins with greater than 3-fold change (Table 3). MSMEG4520 was originally annotated as nitrite reductase (NirA), but is recently re-annotated as sulfite reductase (SirA) [22]. SirA is essential for growth of mycobacteria on sulfite or sulfate as the sole sulfur source. It does not appear to have an apparent role in acid stress response.

SAM also generates a miss rate table for each Δ and/or fold change threshold. The miss rate is equivalent to a false negative rate for the proteins that are between specified score cut points and do not make the list of significant proteins. The contents of the miss rate tables for the four conditions shown in Figure 5 are presented graphically in Figure 6. The general feature is that the proteins in the

0.25–0.75 quantile range tend to have the lower miss rate, and the proteins at the two tails tend to have a higher miss rate. This is totally as expected. Comparison of panels a and b does not reveal apparent difference in the overall miss rates, suggesting that a combination of Δ = 0.68 and 2-fold change can reduce FDR without increasing miss rates compared to either Δ = 0.68 or 2-fold change alone. Thus, this combination is a more optimum criterion. Panel c shows increase in miss rates. This is expected when the 3-fold change threshold is applied in combination with Δ = 0.68. When only the 3-fold change threshold is used, the overall miss rate decreases (Panel d). The miss rate for the upregulated proteins decreases more than those for the downregulated ones. This suggests the 3-fold change threshold does not work equally for the up- and down-regulated proteins. This may be because fold change cutoff alone assumes a normal distribution, while SAM does not impose this restriction. Asymmetrical cutoff is preferred because the observed scores for up- and down-regulated proteins may behave differently in some biological experiments [15]. The samples analyzed in this study appear to be such a case.

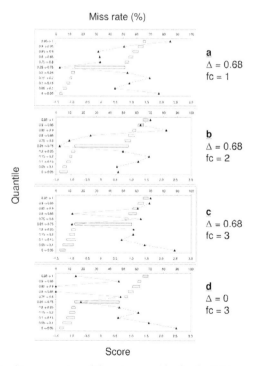

Figure 6. Miss rates. Graphic presentation of the miss rate tables for the SAM outputs shown in Figure 5. Each horizontal open bar represents the observed score cut points of a quantile of the proteins not making to the significant list. The triangle symbols represent the miss rates for the quantiles. In each panel, the left vertical category axis represents the quantiles. The top value axis represents the miss rates. The bottom value axis represents the observed scores. See text for more details.

Differentially Expressed Proteins

By the 2-fold cutoff and FDR of 2.28% with $\Delta = 0.70$ (Figure5b), SAM found 9 induced and 12 repressed proteins in the pH5 versus the pH7 samples (Table 3). There were more repressed than induced proteins. This trend was similar to that observed in a microarray study of 15 min acid shocked M. tuberculosis by Fisher et al. [23], in which 20 genes were found induced while 58 were found repressed by SAM with a 1.5 fold cutoff and 2.86% FDR. Similarly, more genes were also repressed than induced when Shewanella oneidensis was exposed to acidic pH [24].

Of the 9 induced proteins, 2 (MSMEG1600 and MSMEG5766) are involved in purine ribonucleotide biosynthesis, 3 (MSMEG0772, MSMEG1024, and MSMEG5516) in energy metabolism, 2 (MSMEG0366 and MSMEG5709) in fatty acid and phospholipid degradation, 1 (MSMEG2382) in glutamyl-tRNA aminoacylation, and 1 (MSMEG4283) in protein degradation. Inosine-5-monophosphate (IMP) dehydrogenase (MSMEG1600;GuaB) is an important enzyme involved in guanine nucleotide synthesis, catalyzing the oxidation of IMP to xanthosine 5'-monophosphate with the concomitant reduction of NAD to NADH. The enzyme was identified as a DNA binding protein [25]. It has been reported that protein GuaB was induced by acid in E. coli K-12 [26], consistent with our result here that GuaB was induced in M. smegmatis grown at pH 5.

Genome analysis of mycobacteria has revealed an array of genes involved in lipid metabolism [27]. It has been suggested that mycobacteria grown in vivo are largely lipolytic rather than lipogenic due to the variety and quantity of lipids available within mammalian cells and the tubercle [28]. The acidic growth condition probably triggers induction of fatty acid degradation related proteins such as MSMEG0366 and MSMEG5709, even though there was no fatty acid supplied in the growth media for M. smegmatis in this study.

Cytosolic protein degradation is central to regulating various aspects of cell biology, including stress response [29]. Proteins targeted for degradation are unfolded and cleaved to release large peptides in an ATP-dependant manner. These peptides are further cleaved or degraded by endopeptidases such as aminopeptidases in an ATP-independent manner. This general scheme of cytosolic protein degradation is conserved in all organisms. While most of the enzymes involved in the upstream ATP-dependant proteolysis are more organism-specific, the enzymes involved in the downstream ATP-independent proteolysis, including leucine aminopeptidase (PepA), are present in most organisms. Induction of cytosol aminopeptidase (MSMEG4283;PepA) in pH5 grown M. smegmatis is consistent with the putative function of PepA in stress response.

Of the 12 repressed proteins, 2 (MSMEG3082 and MSMEG3837) have roles in biosynthesis of cofactors, prosthetic groups, and carriers. One (MSMEG3166) is an enzyme involved in central intermediary metabolism. The remaining 9 proteins are involved in energy metabolism including the ATP synthase F1 beta subunit (MSMEG4921;AtpD). Decrease in ATP synthesis and downshift of metabolism is commonly observed in cells under stressful conditions.

Schnappinger et al. used SAM as the significance analysis program for transcriptional analysis of adaptation by M. tuberculosis in phagosomal environment [30]. The results indicated that all the seven ATP synthase subunit genes (atpBEF-HAGD) were repressed for intraphagosomal M. tuberculosis, consistent with the stressful condition within phagosomes. Similarly, in a gene expression analysis of Corynebacterium glutamicum in response to acid adaptation at pH 5.7, the seven F0F1-type ATP synthase subunits (NCgl1159-1165) were repressed [31]. In our recent study of protein turnover in M. smegmatis [32], AtpD was found to have lower protein turnover when logarithmically growing cells were shifted to acidic (pH5) or low iron medium, suggesting down regulation of AtpD under both stress conditions. This result further supports our finding here that AtpD was repressed in pH5 grown M. smegmatis cells. Since only 1 of the 5 SDS/PAGE fractions was analyzed in this study, it is reasonable to expect that other ATP synthase subunits could be found repressed as well if all the SDS/PAGE fractions were analyzed [32]. This expectation is based on the transcriptional analysis of M. tuberculosis and C. glutamicum under stress [30,31], as well as our work on protein turnover analysis of M. smegmatis in which three detected ATP synthase subunits (MSMEG4920, MSMEG4921, and MSMEG4926) had lower protein turnover when the M. smegmatis cells encountered an acidic or low iron condition [32].

Conclusion

We have shown that the SAM method for DNA microarray data analysis can be adapted for significance analysis in LC/MS based quantitative proteomics. SAM assigns a significance value, a false discovery rate, and a miss rate for differential expression of individual proteins and groups called significant or insignificant. Such information is not readily available by conventional t-test or fold change test alone. The SAM method provides richer information and is more adaptive to different biological experiments that may have asymmetrical distribution of differential protein expression profiles.

One limitation of applying the SAM method for quantitative proteomics is that it requires sample replicates. Such data sets require more effort to obtain them in proteomics than in microarray analysis due to the limited MS/MS sampling speed in LC/MS analysis. In this work, we performed multiple runs for each biological replicate to cover as many proteins as possible so that enough proteins were commonly quantified between replicates. In on-going work, we will incorporate the cross-reference method that has already been developed by other research groups to align peptides between runs based on accurate mass and elution time information [6,8,9,19]. This will allow a peptide identified by MS/MS scan in one run to also be quantified in another run, even if the peptide is missed by MS/MS scan in the other run. Implementation of this cross-reference method will also make it possible to perform time course study using SAM [21]. Storey et al. showed that "an actual time course analysis offers a sizable increase in statistical power over a static design analysis" [33]. Measuring differential expression over time with single sampling at each time point will likely be a more sensitive study design than a typical static design even if replicates are sampled at the single time point. Once the issue of protein cross-reference between samples is addressed for quantitation of LC/MS data, it is more desirable to perform a time course study for quantitative proteomics than a single time point design with replicates. SAM is a suitable statistical analysis software for such a time course study [33].

Methods

Chemicals and Bacterial Strain

Dextrose, Tween 80, citric acid, biotin, pyridoxine, NaCl, Na_2HPO_4, KH_2PO_4, $MgSO_4 \cdot 6H_2O$, $CuSO_4 \cdot 5H_2O$, $ZnSO_4 \cdot 6H_2O$, $CaCl_2 \cdot 2H_2O$, ferric ammonium citrate, ammonium bicarbonate, and acetonitrile were purchased at certified ACS or reagent grade from Fisher Scientific (Pittsburgh, PA). 7H9 broth base and 99At% $(^{15}NH_4)_2SO_4$ were purchased from Sigma (St. Louis, MO). At% denotes atomic percent. Sequencing grade trypsin was obtained from Promega (Madison, WI). M. smegmatis strain mc2 155 was obtained from the American Type Culture Collection (ATCC; Rockville, Md). BCA Protein Assay kit was obtained from Pierce (Rockford, IL).

Cell Culturing

Two unlabeled (i.e. 14N labeled) M. smegmatis culture samples were grown for study, one at pH 5 and the other at pH 7. Each culture sample was grown in triplicate and harvested at approximately the same OD during the exponential growth phase. During exponential growth, the cells are at the same physiological state so that the only difference is the pH value of the cultures. It is more important to ensure that cells are collected in the exponential phase rather than at the same OD [34] because cell cultures under different stresses may grow to different maximum OD. For quantitative proteomic analysis by isotope ratios, one single 15N labeled culture was grown as the common reference for all the replicates of the pH5 and pH7 culture samples. Since this 15N labeled culture was used as the reference for comparing the 14N labeled pH5 and pH7 cultures, we chose to collect this culture at OD 1.1 in the late exponential phase for a high cell yield.

The medium for growing the unlabeled cells was prepared with Sigma 7H9 base plus 0.05% Tween80 and 0.2% glucose. The medium pH was adjusted to 7.0 or 5.0 by titrating with 1 M sodium hydroxide or 2 M hydrochloric acid. The six unlabeled culture replicates were grown at 100 ml in loosely capped 250-ml nephelo culture flasks under shaking at 37°C. Growth was monitored by measuring turbidity in a Spec20 spectrometer (Thermo Fisher Scientific, Waltham, MA) at 600 nm. The triplicates of the pH5 culture were collected at OD 0.71, 0.69, and 0.67 and named pH5A, pH5B, and pH5C respectively. Similarly, the triplicates of the pH7 culture were sampled at OD 0.77, 0.74, and 0.76 and named pH7A, pH7B, and pH7C respectively. Only 30 ml from each culture was collected, allowing the rest of the culture to continue to grow until stationary phase for recording the complete growth curves.

The medium for growing 15N labeled cells consisted of (g/L) 99At% $(15NH4)2SO4$: 0.5; glucose: 2; Tween 80: 0.5; citric acid: 0.094; biotin: 0.0005; pyridoxine: 0.001; NaCl: 0.1; $Na2HPO4$: 2.5; $KH2PO4$: 1; $MgSO4{\cdot}6H2O$: 0.1; $CuSO4{\cdot}5H2O$: 0.001; $ZnSO4{\cdot}6H2O$: 0.002; $CaCl2{\cdot}2H2O$: 0.0007; ferric ammonium citrate: 0.04; pH5.0. The single 15N labeled cell culture was grown at 50 ml in a loosely capped 250-ml nephelo culture flask under shaking at 37°C. Thirty milliliter of the 15N labeled reference culture was collected at OD 1.1 in the late-log phase. All the collected 30 ml cultures were centrifuged at 4000 rpm in a 5810R refrigerated Eppendorf centrifuge (Fisher Scientific, Pittsburgh, PA) for 10 min at 4°C to collect the cell pellets.

Sample Preparation

Proteins were extracted from each cell pellet by bead beating using a protein extraction buffer that consisted of 100 mM ammonium bicarbonate. A protease inhibitor cocktail (Pierce) was added at 1× as recommended by manufacturer into the mixtures of cell pellet and extraction buffer during protein extraction. The mixtures were vigorously agitated for total 2 min at maximum speed in a Mini-BeadBeater™ (BioSpec, Bartlesville, OK) with 30 sec of ice cooling at the 1 min intermittent. The resulted mixtures were cleared by centrifugation at 13,000 g at 4°C for 30 min. The protein concentrations were determined with the BCA

Protein Assay kit according to the standard protocol. The protein extract concentrations were 3.2, 3.2, 3.3, 3.9, 3.7, and 3.4 mg/ml for pH5A, pH5B, pH5C, pH7A, pH7B, and pH7C respectively. The concentration of the protein extract of the ^{15}N labeled reference was 6.3 mg/ml.

The quantified six unlabeled protein extracts were respectively spiked with an equal amount of the 15N labeled reference protein extract. The six spiked protein extracts were separated by 1D-SDS/PAGE. One hundred micrograms of total proteins of a spiked protein extract was loaded for separation in each lane on a 10% Tris-HCl SDS-PAGE gel (Pierce) of 5-cm length. Gel bands were revealed by Imperial Protein Stain (Pierce) and destained overnight in water. Each lane of the gel was divided into 5 fractions. The band cutting pattern was maintained the same across all the lanes.

Only the 3rd fraction from each of the lanes was processed for mass spectrometry analysis. Gel pieces were minced to 1-mm3 cubes, washed, and processed for in-gel digestion and peptide extraction as previously described [35]. The final peptide extract for each spiked protein extract was concentrated to near dryness in an Eppendorf Vacufuge concentrator (Fisher Scientific) and reconstituted to 25 µl with 5% formic acid for mass spectrometry analysis as below.

Mass Spectrometry Analysis and Data Processing

Samples were submitted for analysis at the Mass Spectrometry Laboratory of Research Resource Center at University of Illinois at Chicago. The resulted raw data files were processed with the BioWorks software (Finnigan, San Jose, CA) licensed to the facility.

The peptide extracts of all the six spiked protein extracts were analyzed by the nanoLC/LTQ-FTMS system. The LTQ-FTMS is the Finnigan hybrid mass spectrometer consisting of a linear ion trap and a Fourier transform ion cyclotron resonance instrument as a second mass analyzer manufactured by Thermo Finnigan (San Jose, CA). Each peptide extract was analyzed in triplicate runs. The instrument was operated in 24-hr unattended service mode with samples injected from an auto-sampler.

For each run, about 5 µl of peptide extract solution was loaded for separation on a 150 mm × 75 µm ZORBAX C18 reverse phase column (Agilent, Germany) with a 5–35% acetonitrile (v/v) gradient in 0.1% TFA over 60 min and detected by the LTQ-FTMS. The 60-min gradient was followed by a step-gradient elution program with 80% acetonitrile in 0.1% TFA, resulting in 90 min per run. The LTQ-FTMS was operated in a data-dependant acquisition mode with up to 10 MSMS spectra acquired after each FTMS scan. The acquired RAW data files were searched against the M. smegmatis strain mc2 155 NCBI database in two separate searches by BioWorks, one corresponding to 14N labeling and the other 15N labeling. The precursor ion tolerance was set to ± 1.5Da and digestion enzyme was designated as trypsin with 2 missed cleavages allowed. Peptide and protein probabilities were calculated in BioWorks.

Protein quantitation procedure was based upon the previously described QN algorithm [10]. The program was kindly provided by Prof. Barry L. Karger's laboratory at Northeastern University and was modified in Matlab v7.2 environment to accommodate using peptide probabilities calculated by BioWorks. Relative abundance was calculated for every identified PCS [10] with p < 0.01. The relative abundance of a peptide is expressed as the ratio of the unlabeled sample isotopologue intensity and the 15N labeled reference isotopologue intensity for this peptide.

To compute the protein relative abundances, the peptide lists of replicate runs for each spiked protein extract were combined. Outliers were filtered by Dixon's Q-test (95% confidence level) before being used to calculate the protein relative abundance. Protein relative abundance refers to the ratio of the abundance of an unlabeled protein relative to that of the 15N labeled reference. It was calculated by averaging the qualified peptide relative abundances of a protein. Relative abundance was calculated only for proteins with at least two qualified PCS identifications [10,34]. The resulted protein relative abundances were then normalized by median.

We also generated an in silico pooled replicate for the pH5 and pH7 culture samples respectively. To do so, the combined peptide relative abundances for each biological replicate were first normalized by the median of these peptide relative abundances. For each sample, i.e. pH5 or pH7, the normalized peptide relative abundances from the biological triplicates were combined for computing the protein relative abundances. The protein relative abundances were finally normalized by the median of the protein relative abundances. The in silico pooled replicates for the pH5 and pH7 samples were named pH5p and pH7p respectively.

Significance Analysis

The significance analysis was carried out with the software Significance Analysis of Microarray (academic version 3.0 for Windows XP) obtained from Stanford University [21]. The software functions as an add-in in Microsoft Excel.

Authors' Contributions

BAPR performed cell culturing, sample processing, database search, and some data interpretation. QL performed data interpretation. Both authors read and approved the manuscript.

Acknowledgements

This work was supported by the start-up fund from University of Illinois at Chicago. The authors appreciate the constructive comments and suggestions from the reviewers for this manuscript.

References

1. Hendrickson EL., Xia Q., Wang T., Leigh JA., Hackett M. Comparison of spectral counting and metabolic stable isotope labeling for use with quantitative microbial proteomics. Analyst. 2006;131:1335–1341.

2. Xia Q., Hendrickson EL., Zhang Y., Wang T., Taub F., Moore BC., Porat I., Whitman WB., Hackett M., Leigh JA. Quantitative proteomics of the archaeon Methanococcus maripaludis validated by microarray analysis and real time PCR. Mol Cell Proteomics. 2006;5:868–881.

3. Park Y., Downing SR., Kim D., Hahn WC., Li C., Kantoff PW., Wei LJ. Simultaneous and exact interval estimates for the contrast of two groups based on an extremely high dimensional variable: application to mass spec data. Bioinformatics. 2007;23:1451–1458.

4. Cho SH., Goodlett D., Franzblau S. ICAT-based comparative proteomic analysis of non-replicating persistent Mycobacterium tuberculosis. Tuberculosis (Edinb). 2006;86:445–460.

5. Li L., Li Q., Rohlin L., Kim U., Salmon K., Rejtar T., Gunsalus RP., Karger BL., Ferry JG. Quantitative proteomic and microarray analysis of the archaeon Methanosarcina acetivorans grown with acetate versus methanol. J Proteome Res. 2007;6:759–771.

6. Molina H., Parmigiani G., Pandey A. Assessing reproducibility of a protein dynamics study using in vivo labeling and liquid chromatography tandem mass spectrometry. Anal Chem. 2005;77:2739–2744.

7. Piening BD., Wang P., Bangur CS., Whiteaker J., Zhang H., Feng LC., Keane JF., Eng JK., Tang H., Prakash A., McIntosh MW., Paulovich A. Quality control metrics for LC-MS feature detection tools demonstrated on Saccharomyces cerevisiae proteomic profiles. J Proteome Res. 2006;5:1527–1534.

8. Meng F., Wiener MC., Sachs JR., Burns C., Verma P., Paweletz CP., Mazur MT., Deyanova EG., Yates NA., Hendrickson RC. Quantitative analysis of complex peptide mixtures using FTMS and differential mass spectrometry. J Am Soc Mass Spectrom. 2007;18:226–233.

9. Andreev VP., Li L., Cao L., Gu Y., Rejtar T., Wu SL., Karger BL. A new algorithm using cross-assignment for label-free quantitation with LC-LTQ-FT MS. J Proteome Res. 2007;6:2186–2194.

10. Andreev VP., Li L., Rejtar T., Li Q., Ferry JG., Karger BL. New algorithm for 15N/14N quantitation with LC-ESI-MS using an LTQ-FT mass spectrometer. J Proteome Res. 2006;5:2039–2045.

11. Kim YJ., Zhan P., Feild B., Ruben SM., He T. Reproducibility assessment of relative quantitation strategies for LC-MS based proteomics. Anal Chem. 2007;79:5651–5658.

12. Listgarten J., Emili A. Statistical and computational methods for comparative proteomic profiling using liquid chromatography-tandem mass spectrometry. Mol Cell Proteomics. 2005;4:419–434.

13. Veenstra TD. Global and targeted quantitative proteomics for biomarker discovery. J Chromatogr B Analyt Technol Biomed Life Sci. 2007;847:3–11.

14. Wang G., Wu WW., Zeng W., Chou CL., Shen RF. Label-free protein quantification using LC-coupled ion trap or FT mass spectrometry: Reproducibility, linearity, and application with complex proteomes. J Proteome Res. 2006;5:1214–1223.

15. Tusher VG., Tibshirani R., Chu G. Significance analysis of microarrays applied to the ionizing radiation response. Proc Natl Acad Sci U S A. 2001;98:5116–5121.

16. de Godoy LM., Olsen JV., de Souza GA., Li G., Mortensen P., Mann M. Status of complete proteome analysis by mass spectrometry: SILAC labeled yeast as a model system. Genome Biol. 2006;7:R50.

17. Storey JD. The positive discovery rate: a Bayesian interpretation and the q-value. The Annuals of statistics. 2003;31:2013–2035.

18. Liu H., Sadygov RG., Yates JR. 3rd A model for random sampling and estimation of relative protein abundance in shotgun proteomics. Anal Chem. 2004;76:4193–4201.

19. Callister SJ., Dominguez MA., Nicora CD., Zeng X., Tavano CL., Kaplan S., Donohue TJ., Smith RD., Lipton MS. Application of the accurate mass and time tag approach to the proteome analysis of sub-cellular fractions obtained from Rhodobacter sphaeroides 2.4.1. Aerobic and photosynthetic cell cultures. J Proteome Res. 2006;5:1940–1947.

20. Wu CC., MacCoss MJ., Howell KE., Matthews DE., Yates JR. 3rd Metabolic labeling of mammalian organisms with stable isotopes for quantitative proteomic analysis. Anal Chem. 2004;76:4951–4959.

21. http://www-stat.stanford.edu/~tibs/SAM/. http://www-stat.stanford.edu/~tibs/SAM/. http://www-stat.stanford.edu/~tibs/SAM/

22. Pinto R., Harrison JS., Hsu T., Jacobs WR., Jr., Leyh TS. Sulfite reduction in mycobacteria. J Bacteriol. 2007;189:6714–6722.

23. Fisher MA., Plikaytis BB., Shinnick TM. Microarray analysis of the Mycobacterium tuberculosis transcriptional response to the acidic conditions found in phagosomes. J Bacteriol. 2002;184:4025–4032.

24. Leaphart AB., Thompson DK., Huang K., Alm E., Wan XF., Arkin A., Brown SD., Wu L., Yan T., Liu X., Wickham GS., Zhou J. Transcriptome profiling of Shewanella oneidensis gene expression following exposure to acidic and alkaline pH. J Bacteriol. 2006;188:1633–1642.

25. Matsuno K., Miyamoto T., Yamaguchi K., Abu Sayed M., Kajiwara T., Hatano S. Identification of DNA-binding proteins changed after induction of sporulation in Bacillus cereus. Biosci Biotechnol Biochem. 1995;59:231–235.

26. Yohannes E., Barnhart DM., Slonczewski JL. pH-dependent catabolic protein expression during anaerobic growth of Escherichia coli K-12. J Bacteriol. 2004;186:192–199.

27. Trivedi OA., Arora P., Sridharan V., Tickoo R., Mohanty D., Gokhale RS. Enzymic activation and transfer of fatty acids as acyl-adenylates in mycobacteria. Nature. 2004;428:441–445.

28. Wheeler PR., Ratledge C. In: Tuberculosis: Pathogenesis, Protection, and Control. Bloom BR, editor. Washington DC , Am. Soc. Microbiol.; 1994. pp. 353–385.

29. Chandu D., Nandi D. PepN is the major aminopeptidase in Escherichia coli: insights on substrate specificity and role during sodium-salicylate-induced stress. Microbiology. 2003;149:3437–3447.

30. Schnappinger D., Ehrt S., Voskuil MI., Liu Y., Mangan JA., Monahan IM., Dolganov G., Efron B., Butcher PD., Nathan C., Schoolnik GK. Transcriptional Adaptation of Mycobacterium tuberculosis within Macrophages: Insights into the Phagosomal Environment. J Exp Med. 2003;198:693–704.

31. Jakob K., Satorhelyi P., Lange C., Wendisch VF., Silakowski B., Scherer S., Neuhaus K. Gene expression analysis of Corynebacterium glutamicum subjected to long-term lactic acid adaptation. J Bacteriol. 2007;189:5582–5590.

32. Rao PK., Roxas BA., Li Q. Determination of global protein turnover in stressed mycobacterium cells using hybrid-linear ion trap-fourier transform mass spectrometry. Anal Chem. 2008;80:396–406.

33. Storey JD., Xiao W., Leek JT., Tompkins RG., Davis RW. Significance analysis of time course microarray experiments. Proc Natl Acad Sci U S A. 2005;102:12837–12842.

34. Li Q., Li L., Rejtar T., Lessner DJ., Karger BL., Ferry JG. Electron transport in the pathway of acetate conversion to methane in the marine archaeon Methanosarcina acetivorans. J Bacteriol. 2006;188:702–710.

35. Zhang X., Guo Y., Song Y., Sun W., Yu C., Zhao X., Wang H., Jiang H., Li Y., Qian X., Jiang Y., He F. Proteomic analysis of individual variation in normal livers of human beings using difference gel electrophoresis. Proteomics. 2006;6:5260–5268.

CITATION

Originally published under the Creative Commons Attribution License or equivalent. Roxas BA, Li Q. Significance analysis of microarray for relative quantitation of LC/MS data in proteomics. BMC Bioinformatics. 2008 Apr 10;9:187. doi:10.1186/1471-2105-9-187.

Determination of Key Intermediates in Cholesterol and Bile Acid Biosynthesis by Stable Isotope Dilution Mass Spectrometry

**Tadashi Yoshida, Akira Honda, Hiroshi Miyazaki
and Yasushi Matsuzaki**

ABSTRACT

For more than a decade, we have developed stable isotope dilution mass spectrometry methods to quantify key intermediates in cholesterol and bile acid biosynthesis, mevalonate and oxysterols, respectively. The methods are more sensitive and reproducible than conventional radioisotope (RI), gas-chromatography (GC) or high-performance liquid chromatography (HPLC) methods, so that they are applicable not only to samples from experimental animals but also to small amounts of human specimens. In this paper, we review the development of stable isotope dilution mass spectrometry for quantifying mevalonate and oxysterols in biological materials, and demonstrate the usefulness of this technique.

Keywords: isotope dilution mass spectrometry, biomarker, cholesterol synthesis, bile acid synthesis, mevalonate, oxysterol

Pathways for Cholesterol and Bile Acid Biosynthesis

Cholesterol homeostasis in human is maintained by two input pathways, comprised of dietary absorption and de novo synthesis, and two output pathways, comprised of direct secretion from liver to bile and conversion into bile acids (Everson, 1992). The rate-limiting step in the de novo cholesterol synthesis is the conversion of 3-hydroxy-3-methylglutaryl-CoA (HMG-CoA) into mevalonic acid (MVA) by HMG-CoA reductase (HMGCR) (Dietschy and Brown, 1974). In contrast, the bile acid biosynthetic pathway is initiated by either hepatic 7α-hydroxylation or hepatic and extrahepatic 27-hydroxylation of cholesterol. The former is catalyzed by microsomal cholesterol 7α-hydroxylase (CYP7A1), the first and rate-limiting enzyme in the classic pathway, while the latter is catalyzed by mitochondrial sterol 27-hydroxylase (CYP27A1), a key enzyme in the alternative pathway (Vlahcevic et al. 1992). Bile acid synthesis by the classic pathway accounts for more than 90% of total bile acids in humans (Duane and Javitt, 1999) while less than 50% of total bile acids is produced by this pathway in rats (Vlahcevic et al. 1997) and mice (Schwarz et al. 1996). Therefore, the measurement of CYP7A1 activity is more important than that of CYP27A1 activity for the evaluation of bile acid biosynthesis in humans.

Direct and Indirect Assays of HMGCR and CYP7A1 Activities

Since HMGCR and CYP7A1 are crucial enzymes in understanding whole body cholesterol metabolism, a great deal of effort has been made to develop suitable assay methods for these enzyme activities. The primary methods have the great disadvantage that an invasive tissue biopsy is necessary for direct determination of these enzyme activities in humans. To overcome this problem, plasma biomarkers for evaluation of these enzyme activities has been explored.

Plasma levels of MVA, the immediate product of HMGCR, were positively correlated with HMGCR activities in rat liver (Popjak et al. 1979). In humans, the plasma MVA concentrations reflected (i) increased rates of whole-body cholesterol synthesis by treatment with cholestyramine resin, and (ii) decreased rates of whole-body sterol synthesis after consumption of a cholesterol-rich diet (Parker et al. 1982 and 1984). In addition, plasma concentration of lathosterol, an intermediate in the late cholesterol biosynthetic pathway, was reported to reflect hepatic HMGCR activity (Björkhem et al. 1987a) as well as whole body cholesterol synthesis (Kempen et al. 1988) in humans.

As for bile acid biosynthesis, Björkhem et al. (Björkhem et al. 1987b) demonstrated that serum levels of 7α-hydroxycholesterol (7A) correlated well with the activities of CYP7A1 in patients with gallstones treated with cholestyramine. In addition, Axelson et al. (Axelson et al. 1988) measured serum concentrations of 7α-hydroxy-4-cholesten-3-one (C4), the product of the next reaction following 7α-hydroxylation of cholesterol, and showed that it was a good marker for CYP7A1 activity in humans (Axelson et al. 1991). It was subsequently reported that serum concentrations of 7A (Hahn et al. 1995) and C4 (Sauter et al. 1996) reflected not only CYP7A1 activities but also bile acid synthesis in humans.

The Methods for the Quantification of MVA

Table 1 summarizes the previously described methods for the quantification of MVA in the liver (enzyme assay), plasma or urine. The primary methods for assaying HMGCR activity have utilized a RI technique that measures the radioactivity in [^{14}C]MVA produced from [^{14}C]HMG-CoA (Shapiro et al. 1969; Goldfarb and Pitot, 1971; Shefer et al. 1972). The methods have been used for the direct determination of enzyme activity but they are not applicable to the quantification of plasma or urinary MVA. In contrast, the following methods, i.e. radioenzymatic assay, enzyme immunoassay, gas chromatography-mass spectrometry (GC-MS), high-performance liquid chromatography (HPLC), liquid chromatography-mass spectrometry (LC-MS) and liquid chromatography-tandem mass spectrometry (LC-MS/MS), can measure not only enzyme activity but also MVA concentrations in plasma and urine.

Radioenzymatic Assay

Radioenzymatic assay of the plasma MVA concentration was reported by Popjak et al. (Popjak et al. 1979). The method depends on the phosphorylation of MVA with [γ-^{32}P]ATP and MVA kinase to 5-[^{32}P]phospho-MVA, and the subsequent isolation of the 5-[^{32}P]phospho-MVA together with known amounts of added 5-phospho[^{14}C]MVA by ion-exchange chromatography. The detection limit of their radioenzymatic assay was 1–2 pmol (148–296 pg) indicating that it was not adequate for determining MVA in small amounts of plasma. In addition, there was a safety concern due to the handling of radioactive materials.

Enzyme Immunoassay

In 1998, Hiramatsu et al. developed an enzyme immunoassay for urinary MVA using a specific monoclonal antibody against MVA (Hiramatsu et al. 1998). This method is not only simpler than the previously described radioenzymatic assay but also completely avoids the risk of radiation hazards. However, the limit of detection was not better than that of the radioenzymatic assay.

GC-MS

In 1972, Hagenfeldt and Hellström attempted to determine MVA concentration in rat blood by using GC-MS (Hagenfeldt and Hellström, 1972). In this procedure, the MVA was extracted from the acid aqueous phase as the lactone. The lactonization increased the hydrophobicity of MVA, so that they could extract it into organic phase. The resulting extract was treated with diazomethane to convert the coexisting fatty acids into their methyl esters. The unchanged mevalonolactone (MVL) with diazomethane in the extract was quantified by GC-MS in electron ionization mode (GC-EI-MS). The peak corresponding to the retention time of MVL appeared large due to interfering materials, such as fatty acids. However, the MVL could be quantified selectively because MVL exhibited an intensive peak at m/z 71 in the spectrum, whereas all fatty acid methyl esters gave rise to the inherent peak at m/z 74 produced by the McLafferty rearrangement ion of the methyl ester. Since then, urinary MVA has been successfully quantified by a similar GC-EI-MS method described above (Woollen et al. 2001).

In the 1970s, the RI technique was the standard method for assaying HMGCR activity, but the handling of radiolabeled materials was a great disadvantage of this method. In 1978, Miyazaki et al. developed a new non-RI method for assaying HMGCR activity in rat liver microsomes or liver slices using [2H3]HMG-CoA as a substrate and GC-MS in chemical ionization mode (GC-CI-MS) (Miyazaki et al. 1978). In this method, the resulting [2H3]MVL was derivatized to the corresponding n-propylamide-n-butylboronate, and deuterium labeled [2H7]MVL was first used as an internal standard.

Table 1. Methods for quantification of MVA in biological samples.

Author	Year	Method (ionization mode)	Derivatization	Lower limit of detection	Intra-assay variation	Inter-assay variation	Recovery	Application
Hagenfeldt et al.	1972	GC-MS (P-EI)	MVL	NA	NA	6.2%	87% ± 4%	blood
Miyazaki et al.	1978	GC-MS (P-CI)	MVL- PABB	NA	NA	NA	NA	liver
Popjak et al.	1979	radioenzymatic assay	5-[^{32}P]phospho-MVA	150–300 pg	NA	NA	100%	plasma
Cighetti et al.	1981	GC-MS (P-EI)	MVL-TMS	NA	1.5%	6.1%	NA	liver
Del Puppo et al.	1989	GC-HR-MS (P-EI)	MVL-TMS	NA	6.5%	NA	101% ± 4%	plasma urine
Honda et al.	1991	GC-HR-MS (P-EI)	MVL-B-DMES	800 fg	4.9%	7.8%	96%–100%	liver
Scoppola et al.	1991	GC-MS (N-CI)	MVA-TFB-TMS	10 pg	5.1%	7.7%	NA	plasma
Yoshida et al.	1993	GC-HR-MS (P-EI)	MVL-B-DMES	NA	2.8%	5.6%	91%–96%	plasma
Ishihama et al.	1994	GC-MS (P-CI)	MVL	NA	2.2%	4.5%	101%–103%	plasma
Siavoshian et al.	1995	GC-MS (P-CI)	MVL-TMS	NA	4.0%	8.0%	70% ± 2%	urine
Saisho et al.	1997	GC-MS (N-CI)	MVA-PFB-CB	NA	2.0%	7.5%	100%–107%	plasma
Hiramatsu et al.	1998	enzyme immunoassay	MVA	195 pg	3.4%	5.2%	102% ± 7%	urine
Woollen et al.	2001	GC-MS (P-EI)	MVL	NA	<13.7%	<9.8%	82%–110%	urine
Park et al.	2001	LC-MS (P-ESI)	MVL	6.5 pg	4.1%	9.4%	95% ± 4%	liver
Ndong-Akoume et al.	2002	LC-MS/MS (P-ESI)	MVL	NA	<1.0%	NA	98%–99%	liver
Abrar et al.	2002	LC-MS/MS (P-ESI)	MVL	NA	4.1%–15%	13%–16%	89%–114%	plasma
Jemal et al.	2003	LC-MS/MS (N-ESI)	MVA	NA	<4.5%	<3.3%	98%–103%	plasma urine
Buffalini et al.	2005	HPLC-UV	MVL	741 ng	NA	<3.0%	97%–103%	liver
Saini et al.	2006	LC-MS/MS (N-ESI)	MVA	NA	1%–17%	3%–12%	99%–108%	plasma
Honda et al.	2007	LC-MS/MS (P-ESI)	MV-PLEA	31 fg	1.8%	3.2%	93%–96%	liver

Abbreviations: P-EI: positive electron ionization; NA: not available; P-CI: positive chemical ionization; N-CI: negative chemical ionization; MVL-PABB: MVL n-propylamide-n-butylboronate; MVL-TMS: trimethylsilyl ether of MVL; GC-HR-MS: high-resolution GC-MS; MVL-B-DMES: dimethylethylsilyl ether of mevalonylbenzylamide; MVA-TFB-TMS: trimethylsilyl ether of bis(trifluoromethyl)benzyl ester of MVA; MVA-PFB-CB: cyclic boronate-pentafluorobenzyl ester of MVA; P-ESI: positive electrospray ionization: N-ESI: negative electrospray ionization; HPLC-UV: high-performance liquid chromatography equipped with an ultraviolet detector; MV-PLEA: MV-(2-pyrrolidin-1-yl-ethyl)-amide.

In 1991, Scoppola et al. (Scoppola et al. 1991) extended this approach, and quantifi ed plasma MVA concentrations. The MVA was lactonized, extracted with [2H3]MVL and reconverted to the free acid. The resulting MVA was then converted to 3,5-bis(trifluoromethyl)benzyl ester followed by its trimethylsilyl (TMS) ether derivative. The quantification method was based on GC-CI-MS using ammonia as a reagent gas and the detection limit of MVA in plasma was 100 pg/mL. The GC-CI-MS method for the quantification of plasma MVA was subsequently improved by Ishihama et al. (Ishihama et al. 1994) and Saisho et al. (Saisho et al. 1997), and the method for the determination of urinary MVA was developed by Siavoshian et al. (Siavoshian et al. 1995). However, the GC-CI-MS methods have one disadvantage in that they required frequent cleaning of the CI ion source to maintain the high sensitivity.

To eliminate the aforementioned tedious operations in GC-CI-MS, another approach by gas chromatography-electron ionization-mass spectrometry (GC-EI-MS) was also developed. Cighetti et al. (Cighetti et al. 1981; Galli Kienle, 1984) assayed HMGCR activity by GC-EI-MS after conversion of MVL into the corresponding trimethylsilyl (TMS) ether. They used the ions at m/z 187 (M–15) for MVL-TMS and m/z 150 (M–15– CH_2CO) for [2H5]MVL-TMS because these ions were not influenced by interfering peaks in extracts from liver microsomes. In 1989, the same group improved their original method by using GC-high-resolution (HR)-EI-MS (Del Puppo et al. 1989). This group lactonized plasma and urinary MVA into MVL using a cation exchange resin, and extracted with organic solvent after the addition of [2H5]MVL as an internal standard. The extracted MVL was then converted into the TMS ether derivative, and quantified by GC-HR-EI-MS with a mass spectral resolution of 5,000. The ions at m/z 145.0685 for MVL-TMS and m/z 150.0965 for [2H5]MVL-TMS were used for selected ion monitoring (SIM).

We also developed new assay methods to measure hepatic HMGCR activity (Honda et al. 1991) and plasma MVA concentration (Yoshida et al. 1993) by GC-HR-EI-MS. These methods made it possible to simultaneously quantify not only MVA but also 7A. Other features of these methods are described below.

(i) A purification procedure was developed by the serial use of commercially available solid-phase extraction cartridges, which provided high recovery and reproducibility. In brief, plasma MVA was extracted by an anion exchange Bond Elut SAX cartridge, and then eluted as MVL with 0.6 M HCl. The MVL was further purified by a reversed phase Bond Elut C18 and a normal phase Bond Elut CN cartridges. In addition, an excess benzylamine was removed by another Bond Elut CN cartridge after derivatization into mevalonylbenzylamide. The recovery of spiked MVA through the purification procedures using these cartridges was 94.1%, and the relative standard deviations between sample preparations and between measurements by this method were 5.6% and 2.8%, respectively (Yoshida et al. 1993).

(ii) [2H7]MVL was used as an internal standard. This hepta-deuterated variant of MVL provided both good linearity of the calibration curve and easiness to distinguish between MVL peak and interfering peaks even if the MVL peak was small.

(iii) MVL was easily converted into mevalonylbenzylamide without any catalyst under mild conditions followed by its dimethylethylsilyl (DMES) ether derivative. This amidation via MVL from MVA is a characteristic reaction for γ-hydroxyfatty acids, such as MVA, however, the free fatty acids also present in the extract did not react without catalysts. The resulting derivative gave a [M-C2H5]+ ion at m/z 380.2077 with a prominent intensity in the high mass region, which was a great advantage in the elimination of interfering peaks originating from endogenous substances in the extract by GC-EI-MS.

(iv) The DMES ether derivative was much more stable than the TMS ether derivative.

(v) The MVL derivative was quantified by GC-HR-EI-MS with a mass spectral resolution of 10,000, which was also useful to eliminate peaks of unknown substances that could interfere with the monitoring.

(vi) Trace amounts, less than 1 pg, of MVA could be detected by this method, and the lower limit of quantification in plasma sample was 180 pg/mL.

(vii) Using these methods, it was validated that there was a highly significant correlation between the hepatic HMGCR activities and plasma concentrations of MVA in ten patients (r = 0.83, P < 0.01) (Yoshida et al. 1993).

(viii) The GC-EI-MS method did not require frequent cleaning. This indicated that the GC-EI-MS method was suitable for clinical applications, in which it is necessary to assay a large number of samples at once.

LC-MS and LC-MS/MS

Since the early 2000s, LC-MS or LC-MS/MS have been used more extensively than GC-MS to analyze relatively polar compounds, such as MVA or MVL, because LC-MS and LC-MS/MS do not generally require a derivatization step.

Park et al. (Park et al. 2001) and Ndong-Akoume et al. (Ndong-Akoume et al. 2002) assessed HMGCR activity by measuring MVL with LC-MS and LC-MS/MS using the positive electrospray ionization (P-ESI) mode. Plasma and urinary MVA concentrations were quantified by LC-P-ESI-MS/MS after conversion into MVL (Abrar and Martin, 2002), as well as directly by LC-negative (N)-ESI-MS/MS without lactonization (Jemal et al. 2003; Saini et al. 2006). The detection limit of MVL by the LC-MS method was 6.5 pg (Park et al. 2001), and the lower limit of quantification of plasma MVA by the LC-MS/MS methods were 200–500 pg/mL, which were similar to those obtained using GC-MS methods.

Recently, we developed a highly-sensitive method to assess HMGCR activity by LC-MS/MS (Honda et al. 2007a). In this method, MVA was extracted as MVL and its detection sensitivity was enhanced through derivatization (Fig. 1). The features of this method are described below.

(i) The P-ESI mode was selected to quantify MVA because the positive mode provides more abundant ions than the negative mode (Hiraoka and Kudaka, 1992).

(ii) To select the most suitable derivative of MVA for P-ESI mode, the amidation reaction from MVA via MVL, a characteristic reaction for γ-hydroxy fatty acids such as MVA, was conducted using seven types of primary alkylamines with a tertiary amine moiety to promote protonation. Of these amide derivatives, mevaonyl-2-pyrrolidin-1-yl-ethyl)-amide (MV-PLEA) was the best derivative for the LC-P-ESI-MS/MS method.

(iii) The detection limit of this MV-PLEA was about 30 fg (signal-to-noise ratio (S/N) = 3), indicating that this is the most sensitive method at present for the detection of MVL.

(iv) [2H7]MVL was used as an internal standard. The recovery of spiked MVA was 94.6%, and the relative standard deviations between sample preparations and between measurements by this method were 3.2% and 1.8%, respectively.

(v) MV-PLEA was determined by selected reaction monitoring (SRM) using m/z 245 (M+H) as a precursor ion and m/z 227 (M+H–H2O) as a production, which almost completely eliminated the interfering peaks on the SRM chromatogram.

(vi) Hepatic HMGCR activities in 11 normal rats were measured by both the RI and LC-P-ESIMS/MS methods. The HMGCR activities obtained by the present method correlated well with those obtained by the conventional RI method (r = 0.93, P < 0.0001). In the RI method, [14C]HMG-CoA is usually used as 30 dpm/pmol = 33.3 fmol/dpm. When the standard deviation of background noise is 2 dpm, the signal would be 6 dpm when the S/N = 3. Therefore, the detection limit of the conventional RI method is calculated to be 200 fmol (S/N = 3). In comparison, the detection limit of the LC-P-ESI-MS/MS method is 240 amol (S/N = 3), ~800 times more sensitive than that of the conventional RI method.

HPLC

In 2005, Buffalini et al. reported a new method for the determination of HMGCR activity by HPLC (Buffalini et al. 2005). In this method, MVL produced from unlabeled HMG-CoA was extracted and quantified by HPLC with a fixed ultraviolet (UV) detector (200 nm). This method does not require very expensive equipment, such as a mass spectrometer, but the detection limit of MVL is at least 100,000 times less than that by mass spectrometry.

The Methods for the Quantification of 7A

CYP7A1 activity has previously been assayed by measuring the radioactivity of 7A produced from exogenously added [^{14}C]cholesterol by incubation with liver microsomes (Shefer et al. 1968). However, the extent of equilibration of exogenous labeled cholesterol with the endogenous cholesterol pool under different conditions still remains to be elucidated. To overcome this problem, several methods, i.e. a radioisotope derivative method, and GC-MS and HPLC methods, have been developed. These methods are able to measure the net amount of 7A produced from endogenous and exogenous cholesterol. Table 2 summarizes the previously reported methods for the diresct determination of the mass 7A in the liver (enzyme assay) or plasma.

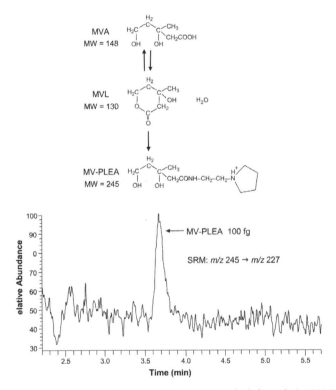

Figure 1. Representative chromatogram of mevalonyl-(2-pyrrolidin-1-yl-ethyl)-amide (MV-PLEA) by positive ESI-SRM at m/z 245 → m/z 227. Authentic standard of MV-PLEA (100 fg) was injected into the HPLC. LC-MS/MS conditions have been described previously (Honda et al. 2007a).

Radioisotope Derivative Method

This technique can measure the net amount of 7A produced from exogenous [^{14}C]cholesterol and endogenous unlabeled cholesterol (Mitropoulos et al. 1972; Shefer et al. 1975). The resultant 7A was extracted, acetylated with [^{3}H]acetic anhydride and purifi ed by thin layer chromatography (TLC). The mass of 7A was calculated from the amount of radioactivity in the acetylated product based upon the specifi c radioactivity of the reagent.

Table 2. Methods for quantification of 7A in biological samples.

Author	Year	Method (ionization mode)	Derivatization	Lower limit of detection	Intra-assay variation	Inter-assay variation	Recovery	Application
Mitropoulos et al.	1972	radioisotope derivative method	acetylated 7A	NA	NA	NA	NA	liver
Björkhem et al.	1974	GC-MS (P-EI)	7A-TMS	NA	NA	2.2%	95%	liver
Sanghvi et al.	1981	GC-MS (P-EI)	7A-TMS	50 pg	3.5%	2.3%–4.7%	NA	liver
Noshiro et al.	1985	NP-HPLC-UV	7A	NA	NA	NA	>85%	liver
Ogishima et al.	1986	NP-HPLC-UV	C4	NA	NA	NA	NA	liver
Björkhem et al.	1987	GC-MS (P-EI)	7A-TMS	1–2 ng/mL	NA	4%–8%	105%	serum
Hylemon et al.	1989	RP-HPLC-UV	C4	8 ng	NA	NA	NA	liver
Yamashita et al.	1989	GC-MS (P-EI)	7A-TMS	NA	3.8%	4.6%	92%–99%	liver
Oda et al.	1990	GC-MS (P-EI)	7A-TMS	NA	NA	3%	97%–109%	serum
Honda et al.	1991	GC-HR-MS (P-EI)	7A-DMES	1.6 pg	7.9%	7.0%	94%–102%	liver
Yoshida et al.	1993	GC-HR-MS (P-EI)	7A-DMES	NA	4.2%	2.6%	93%–95%	serum

Abbreviations: NA: not available; P-EI: positive electron ionization; 7A-TMS: trimethylsilyl ether of 7α-hydroxycholesterol; HPLC-UV: high-performance liquid chromatography equipped with an ultraviolet detector; NP: normal-phase; C4: 7α-hydroxy-4-cholesten-3-one; RP: reversed-phase; GC-HR-MS: high-resolution GC-MS; 7A-DMES: dimethylethylsilyl ether of 7α-hydroxycholesterol.

GC-MS

In 1974, Björkhem and Danielsson developed a method for the assay of hepatic CYP7A1 activity by GC-MS (Björkhem and Danielsson, 1974). Their method was based on stable isotope dilution-mass spectrometry using [^{2}H$_3$]7A as an internal standard. In this method, 7A produced from endogenous microsomal cholesterol was extracted in organic solvent, purified by TLC, converted to the TMS ether derivative, and analyzed by GC-MS. In 1981, Sanghvi et al. reported an alternative method by GC-MS in which 7A produced from microsomal cholesterol was extracted with 5α-cholestane as an internal standard by organic solvent, converted to the TMS ether derivative, and quantified by SIM (Sanghvi et al. 1981). Meanwhile, Yamashita et al. measured hepatic CYP7A1 activity by GC-SIM using 5α-cholestane-3β,7β-diol as an internal standard (Yamashita et al. 1989).

We also developed a new assay method for hepatic CYP7A1 activity by GC-HR-SIM (Honda et al. 1991). As mentioned in the previous MVA section, this method made it possible to quantify simultaneously not only 7A but also MVA. [2H7]7A was used as an internal standard and 7A was converted into its DMES ether derivative before analysis by GC-HR-MS. This DMES ether derivative was not only more stable but also much advantageous compared with the TMS ether derivative for the separation of 7A from contaminated cholesterol on GC chromatograms.

The concentration of 7A in human serum was first quantified by Björkhem et al. using GC-SIM (Björkhem et al. 1987b). They also showed that serum free (unesterified) 7A reflected hepatic CYP7A1 activities in humans. In contrast, Oda et al. quantified human serum free and esterified 7A concentrations by GC-SIM and reported that the hepatic CYP7A1 activities correlated better with the serum esterified 7A than with the free 7A (Oda et al. 1990).

In 1993, we applied our GC-HR-SIM method to the determination of human serum 7A concentrations and confirmed that there was a significant correlation product of the next oxidative enzymatic reaction ($r = 0.76$, $p < 0.05$) between serum free 7A after CYP7A1. In fact, CYP7A1 activities correlated concentrations and hepatic CYP7A1 activities in humans (Yoshida et al. 1993). However, neither the esterified 7A ($r = 0.45$, $p > 0.05$) nor the total (free + esterified) 7A concentrations ($r = 0.51$, $p > 0.05$) correlated significantly with CYP7A activities.

HPLC

The assay method for hepatic CYP7A1 activity by HPLC wasfirst reported by Noshiro et al. (Noshiro et al. 1985). The 7A produced from microsomal cholesterol was extracted and separated by normal phase HPLC. Although the absorption maximum of 7A was lower than 200 nm, they monitored 7A at 214 nm because there was an interference due to absorption of oxygen and/or solvent impurities at lower wavelengths.

In 1986, the same group improved the assay method by converting the produced 7A into C4 by incubating with cholesterol oxidase (Ogishima et al. 1986). Because C4 exhibits an intense absorption at 240 nm and there are fewer interfering peaks at this wavelength than at 214 nm, this improved method exhibited a more than 10-fold increase in the sensitivity compared with the previous one

(Noshiro et al. 1985). In 1989, Hylemon et al. modified Ogishima's method by using reverse-phase HPLC and adding 7β-hydroxycholesterol as an internal standard (Hylemon et al. 1989).

The Methods for the Quantification of C4

Another plasma or serum marker for the evaluation of hepatic CYP7A1 activities is C4, which is a product of the next oxidative enzymatic reaction after CYP7A1. In fact, CYP7A1 activities correlated better with serum C4 levels compared with those of 7A irrespective of the esterification (Yoshida et al. 1994). Table 3 shows the previously described methods for the quantification of serum C4 concentrations by HPLC, GC-MS, and LC-MS/MS.

HPLC

In 1988, Axelson et al. (Axelson et al. 1988) reported a method for the quantification of plasma C4 using normal-phase HPLC with UV detection, and demonstrated that plasma C4 concentration reflected bile acid biosynthesis in humans. In addition, they reported that there was a strong positive correlation between the plasma levels of C4 and the activities of CYP7A1 in patients treated with cholestyramine, chenodeoxycholic acid, or ursodeoxycholic acid (Axelson et al. 1991). However, their method required the addition of 3H-labeled 25-hydroxyvitamin D_3 as an internal standard. On the other hand, Pettersson et al. (Pettersson and Eriksson, 1994) and Gälman et al. (Gälman et al. 2003) used unlabeled 7β-hydorxy-4-cholesten-3-one as an internal standard and analyzed C4 levels using HPLC with a reversed-phase column. The detection limits of C4 by these HPLC-UV methods were nearly 1 ng, so that at least 1 mL of plasma was required for each assay.

Table 3. Methods for quantification of serum C4 concentration.

Author	Year	Method (ionization mode)	Derivatization	Lower limit of detection	Intra-assay variation	Inter-assay variation	Recovery	Application
Axelson et al.	1988	NP-HPLC-UV	C4	0.5–1.5 ng/mL	NA	5%	82%–106%	plasma
Pettersson et al.	1994	RP-HPLC-UV	C4	3 ng	3.2%	3.8%	96%–105%	serum
Yoshida et al.	1994	GC-HR-MS (P-EI)	C4-MO-DMES	1 pg	2.54%	5.16%	94%–100%	plasma
Gälman et al.	2003	RP-HPLC-UV	C4	500 pg	4.4%	5.6%	NA	blood
Honda et al.	2007	LC-MS/MS (P-ESI)	C4-picolinate	30 fg	3.9%	5.7%	92%–94%	serum

Abbreviations: HPLC-UV: high-performance liquid chromatography equipped with an ultraviolet detector; NA: not available; NP: normal-phase; RP: reversed-phase; GC-HR-MS: high-resolution GC-MS; P-EI: positive electron ionization; C4-MO-DMES: methyloxime dimethyl-ethylsilyl ether of C4; P-ESI: positive electrospray ionization.

GC-MS

In 1994, we developed a more sensitive method for the quantification of plasma C4 by GC-HR-MS using [^{2}H$_7$]C4 as an internal standard (Yoshida et al. 1994). C4 was extracted from 200 μL of plasma by a salting-out extraction, and then purified by serial solid-phase extractions. The extract was treated with O-methylhydroxylamine hydrochloride and then dimethylethylsilylated. The resulting methyloxime-DMES ether derivative was quantified by GC-HR-SIM. This method was very sensitive as well as specific, and a lower limit of detection of 1 pg was achieved.

We compared the relationships between hepatic CYP7A1 activity and plasma concentrations of C4 and free 7A in humans using our GC-HR-SIM methods (Yoshida et al. 1994). Both biomarkers correlated significantly with hepatic CYP7A1 activity (C4: r = 0.84, p < 0.001; free 7A: r = 0.73, p < 0.01), and C4 correlated better with CYP7A1 activity compared with free 7A. However, perhaps these plasma markers do not precisely reflect hepatic CYP7A1 activities in some patients with markedly changed concentrations of plasma lipoproteins. Because plasma oxysterols including C4 and 7A are transported in lipoproteins, the concentrations of oxysterols can be affected by the half-life of the lipoproteins. This hypothesis was supported by another study by ourselves (Honda et al. 2004), in which plasma C4 concentrations and hepatic CYP7A1 activities were compared in New Zealand white rabbits that were fed a high cholesterol diet and/or a bile fistula was constructed. Feeding cholesterol markedly increased and bile drainage reduced plasma cholesterol concentrations. Initially, in these models there was no correlation between plasma C4 concentrations and hepatic CYP7A1 activities (r = -0.24, p > 0.05). Cholesterol feeding was associated with downregulated CYP7A1 activities, while plasma C4 concentrations were elevated in the presence of increased plasma cholesterol levels. However, this discrepancy was overcome and a significant correlation was observed (r = 0.73, p < 0.05) by expressing C4 levels relative to cholesterol. These results suggested that plasma C4 relative to cholesterol was a better marker for hepatic CYP7A1 activity than the absolute concentration when plasma cholesterol concentrations were changed markedly.

LC-MS/MS

HPLC with UV detection is a more convenient method than GC-MS for the measurement of plasma C4 concentrations. However, the sensitivity is not sufficient to quantify C4 in limited amounts of human serum. Therefore, we recently developed a highly-sensitive new method by LC-MS/MS (Honda et al. 2007b). After the addition of [^{2}H$_7$]C4 as an internal standard, C4 was extracted from human serum (2–50 μL) by a salting-out procedure, derivatized into the picolinoyl ester (C4-7α-picolinate), and then purified using a disposable C_{18} cartridge. The resulting picolinoyl ester derivative of C4 was quantified by LC-P-ESI-MS/MS (Fig. 2). LC-MS/MS method do not always require a derivatization step. However, it is also true that the introduction of charged moieties enhances the ionization efficiency of neutral steroids in ESI and atmospheric pressure chemical ionization processes. In P-ESI mode, the picolinoyl ester of C4 exhibited an $[M+H]^+$ ion at m/z 506 as the base peak. In the MS/MS spectrum, the $[M-C_6H_5O_2N]^+$ ion was observed at m/z 383 as the most prominent peak. The SRM was conducted using m/z 506 →m/z 383 for the C4-7α-picolinate and m/z 513 → m/z 390 for the $[^2H_7]$ variant. The detection limit of the C4-7α-picolinate was 30 fg (S/N = 3), which was more than 1,000 times more sensitive than that of C4 with a conventional HPLC-UV method. The recovery of spiked C4 was 93.4%, and the relative standard deviations between sample preparations and between measurements by this method were 5.7% and 3.9%, respectively. Thus, this LC-MS/MS method is not only the most sensitive method at present for the detection of C4 but it is also highly reliable and reproducible.

Applications to Clinical Studies

The quantification of MVA, 7A or C4 in human blood has made it possible to monitor in vivo cholesterol and bile acid synthesis without the need for an invasive liver biopsy. Therefore, these methods are very useful for basic or clinical time-course studies of cholesterol metabolism (Table 4).

Diurnal Cycle

In 1982, Parker et al. observed the diurnal cycle of the MVA concentrations in human plasma (Parker et al. 1982). At the peak of the cycle (between midnight and 3 a.m.), the MVA concentrations were 3–5 times greater than those at the nadir (between 9 a.m. and noon). Pappu et al. also reported that the plasma concentrations of MVA exhibited a diurnal cycle in normal subjects and patients with abetalipoproteinemia, and the highest levels were observed between midnight and 4 a.m. (Pappu and Illingworth, 1994).

On the other hand, the diurnal cycle of bile acid biosynthesis in the human liver was reported (deletion) by Duane et al. (Duane et al. 1983). They used a radioisotope technique and demonstrated for the first time that humans with an intact enterohepatic circulation exhibited a diurnal cycle of bile acid synthesis with an amplitude of ± 35%–55% around mean synthesis, and an acrophase at about 9 a.m. The same group also reported in 1988 that neither chenodeoxycholic acid nor ursodeoxycholic acid administration significantly altered the circadian rhythm of bile acid synthesis in humans (Pooler and Duane, 1988).

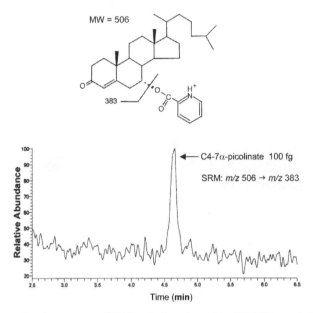

Figure 2. Representative chromatogram of C4-7α-picolinate by positive ESI-SRM at m/z 506 → m/z 383. Authentic standard of C4-7αpicolinate (100 fg) was injected into HPLC. LC-MS/MS conditions have been described previously (Honda et al. 2007b).

We investigated the diurnal cycle of bile acid biosynthesis by using plasma C4 and 7A concentrations (Yoshida et al. 1994). Plasma was obtained every 2 hours from three normal volunteers and the C4 and 7A concentrations were determined using our GC-HR-MS method (Fig. 3). These levels were fitted to a cosine curve as reported in the previous studies using the isotope kinetic method (Duane et al. 1983; Pooler and Duane, 1988). The amplitudes of C4 and free 7A averaged 45% and 32%, respectively, and the acrophases of C4 and free 7A averaged 5:35 a.m. and 5:39 a.m., respectively, which was compatible with the previous results obtained using the radioisotope technique (Fig. 4). In contrast, total 7A and esterified 7A did not exhibit any significant diurnal cycle.

In 2005, Gälman et al. also reported the diurnal cycle of C4 and lathosterol, another biomarker of cholesterol biosynthesis (Gälman et al. 2005). They concluded that bile acid synthesis in humans exhibits a diurnal cycle with 2 peaks during the daytime, which is opposite from the circadian rhythm of cholesterol biosynthesis. These results were different from previous studies. Further investigations will be required to elucidate the reason for the discrepancy.

Table 4. Applications of MVA and C4 as biomarkers for cholesterol and bile acid biosynthesis.

	MVA	C4	References
HMGCR inhibitors	decrease	–	(Nozaki, 1996; Naoumova, 1996; Yoshida, 1997; Naoumova, 1997; Pfohl, 1998; O'Neill, 2001; Pappu, 2002)
	–	no effect	(Yoshida, 1997; Naoumova, 1999; O'Neill, 2001)
HMGCR inhibitors with partial ileal resection	decrease	decrease	(Naoumova, 1999)
Insulin	decrease	–	(Lala, 1994; Scoppola, 1995; Naoumova, 1996)
Growth hormone	no effect	–	(Boyle, 1992)
	–	no effect	(Leonsson, 1999; Lind, 2004)
Thyroid hormone	–	no effect	(Sauter, 1997)
CDCA	–	decrease	(Einarsson, 2001)
DCA	–	decrease	(Einarsson, 2001)
UDCA	–	increase	(Sauter, 2004)
Rifampin	–	increase	(Lutjohann, 2004)
Gallstone	–	increase	(Muhrbeck, 1997; Gälman, 2004)
Liver cirrhosis	no change	decrease	(Yoshida, 1999)
Diarrhea	–	increase	(Eusufzai, 1993; Sauter, 1999)

HMGCR Inhibitors

The measurement of plasma MVA concentration is very useful to evaluate the in vivo effects of HMGCR inhibitors. Pfohl et al. reported that the HMGCR inhibitor, simvastatin, rapidly down-regulated cholesterol biosynthesis, which was then up-regulated when the drug was withdrawn (Pfohl et al. 1998). Nozaki et al. investigated the difference in the effect of another HMGCR inhibitor, pravastatin, on cholesterol biosynthesis between the morning and the evening. They administered pravastatin to the same patients in the morning or evening, and found that morning and evening administrations of pravastatin elicited equivalent reductions in the plasma and urinary MVA concentrations (Nozaki et al. 1996). Pappu and Illingworth demonstrated that patients with familial hypercholesterolemia exhibited a diurnal pattern in plasma MVA levels similar to that reported previously in controls (Pappu and Illingworth, 2002). In addition, they reported that the administration of lovastatin in the evening reduced the nocturnal increases in MVA levels, and the administration of simvastatin completely abolished the nighttime rise. Naoumova et al. treated familial hypercholesterolemia patients with 3 different HMGCR inhibitors, pravastatin, simvastatin, and atorvastatin, and showed that the patients who responded well to statins exhibited higher basal plasma levels of MVA (Naoumova et al. 1996).

We investigated the short-term effects of pravastatin on cholesterol and bile acid biosynthesis by measuring MVA and C4 as biomarkers (Yoshida et al. 1997). Six male volunteers were administered 40 mg of pravastatin, and the plasma MVA and C4 levels were measured every 2 hours. The plasma MVA levels 2 hours after the administration of pravastatin were decreased compared with those in controls. While the decrease in MVA concentrations continued for 8 hours, the plasma C4 concentrations did not change during the initial 6 hours and then decreased 8 hours after the administration. Three-way analysis of this study indicated that the MVA level was influenced significantly by both pravastatin treatment and the time-course. In contrast, C4 level was affected significantly by both inter-individual differences and time-course, but not by pravastatin treatment. These results indicated that cholesterol biosynthesis was inhibited by pravastatin treatment, but bile acid biosynthesis was not influenced in normal subjects (Yoshida et al. 1997). Naoumova et al. treated familial hypercholesterolemia patients with atorvastatin and partial ileal bypass (Naoumova et al. 1999). Atorvastatin decreased the rate of bile acid synthesis only when bile acid synthesis was up-regulated by partial ileal bypass or bile acid sequestrants, presumably by limiting the supply of newly synthesized free cholesterol.

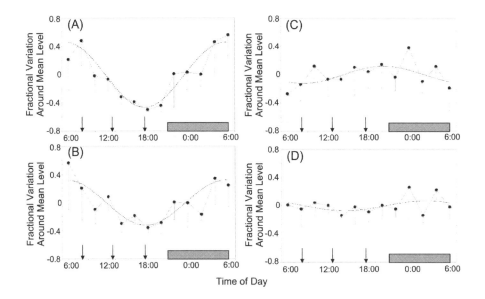

Figure 3. The circadian rhythm of the plasma levels of C4 and 7A in three normal volunteers. The volunteers consumed a normal hospital diet three times a day (shown by arrows), and slept from 21:00 on the fi rst day to 6:00 on the second day (shown by the shaded box). The values are expressed as fractional variations around the mean levels (mean ± SD). Dashed lines represent the curves of best fi t. (A) C4; (B) free 7A; (C) esterifi ed 7A; (D) total 7A. Reprinted with minor modifi cation from our previous paper (Yoshida et al. 1994), Copyright (1994), with permission from Elsevier.

Hormones

There are several reports that show the effects of hormones e.g. insulin, growth hormone and thyroid hormone, on in vivo cholesterol metabolism. Euglycemic hyperinsulinemia acutely decreased the circulating levels of MVA (Lala et al. 1994), which indicated that insulin could decrease cholesterol biosynthesis. Naoumova et al. also investigated the effects of hyperinsulinemia on the plasma MVA concentrations and reported that acute hyperinsulinemia decreased cholesterol biosynthesis less in the subjects with non-insulin-dependent diabetes mellitus compared with non-diabetic subjects, which suggests that the patients with non-insulin-dependent diabetes mellitus exhibit insulin resistance (Naoumova et al. 1996).

Because plasma growth hormone levels and cholesterol biosynthesis are both increased during sleep, Boyle et al. speculated that growth hormone might stimulate de novo cholesterol biosynthesis (Boyle et al. 1992). However, the peak nocturnal and fasting MVA concentrations did not correlate with the growth hormone levels, and they concluded that nocturnal growth hormone secretion was not related to the stimulation of cholesterol production during sleep.

Patients with hypothyroidism exhibit hypercholesterolemia, while those with hyperthyroidism exhibit hypocholesterolemia. Sauter et al. measured serum C4 concentrations before and after treatment for hypo- and hyperthyroidism and showed that in humans, thyroid hormones influenced the serum cholesterol concentrations by mechanisms other than through modification of the CYP7A1 activity (Sauter et al. 1997).

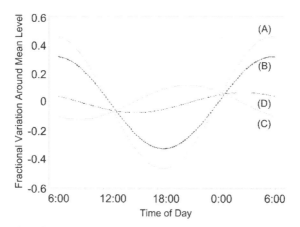

Figure 4. The curves of best fit for C4 and 7A. (A) C4, y = 0.46 cos ((2π/24)t -1.46) (p < 0.005); (B) Free 7A, y = 0.32 cos ((2π/24)t -1.48) (p < 0.005); (C) Esterified 7A, 0.12 cos ((2π/24)t + 0.97) (NS); (D) Total 7A, 0.07 cos ((2π/24)t -0.65) (NS). Statistical significance was evaluated by a Zero-amplitude test by Nelson et al. (Nelson et al. 1979). Reprinted with minor modification from our previous paper (Yoshida et al. 1994), Copyright (1994), with permission from Elsevier.

Bile Acids

Bile acids, particularly chenodeoxycholic acid (CDCA) and deoxycholic acid (DCA), are physiological ligands for the farnesoid X receptor (FXR, NR1H4). They inhibit bile acid biosynthesis through activation of this nuclear receptor. In fact, Einarsson et al. reported that the treatment of healthy subjects with CDCA or DCA reduced the serum concentrations of C4 (Einarsson et al. 2001). They also found that CDCA reduced cholesterol biosynthesis while DCA did not when they evaluated in vivo cholesterol biosynthesis by measuring the serum 7-dehydrocholesterol concentrations. In contrast, UDCA treatment for 40 days did not affect cholesterol synthesis, as evaluated by urinary excretion of MVA, but the same treatment significantly increased bile acid biosynthesis determined by serum C4 concentrations (Sauter et al. 2004).

Hepatobiliary Diseases

Cholesterol gallstone disease is caused by abnormal cholesterol and bile acid metabolism. The formation of cholesterol supersaturated bile is one of the important factors in the pathogenesis of this disease. Shoda et al. proposed an estimated biliary cholesterol saturation index $(CSI)_E = 1[MVL] + 0.7[C4]$ that was calculated by multivariate linear regression analysis using the plasma MVA and C4 concentrations of patients with hyperlipoproteinemia and demonstrated that this convenient calculation of $(CSI)_E$ corresponded well to actual biliary CSI (Shoda et al. 1997). However, the hypersecretion of biliary cholesterol in patients with gallstones does not seem to be due to increased hepatic synthesis of cholesterol or decreased catabolism of cholesterol to bile acids. This could be because the plasma levels of lathosterol were not significantly different between gallstone subjects and controls and the C4 levels were about 40% higher in the gallstone subjects compared with the controls (Muhrbeck et al. 1997). The increased bile acid biosynthesis determined by the plasma C4 levels, corrected for plasma cholesterol, was also reported in gallstone subjects and gallstone high-risk Mapuche Indians (Gälman et al. 2004).

Conversely, some hepatobiliary diseases affect cholesterol and bile acid metabolism. In patients with liver cirrhosis (LC), the blood cholesterol levels are relatively preserved, despite other markers, including the serum albumin levels, show liver dysfunction. We studied the association between hepatic cholesterogenesis and bile acid synthesis in hepatocellular impairment using the plasma levels of MVA and C4 (Yoshida et al. 1999). There were no significant differences in the plasma MVA levels between chronic hepatitis (CH), LC and control groups. In contrast, plasma C4 levels were significantly lower in LC compared with the CH and control groups. Although the MVA levels did not correlate with the Child-Pugh's score, which reflects the severity of liver damage (Albers et al. 1989), there was a significant correlation between the C4 level and Child-Pugh's score. In addition, plasma C4 levels in the control subjects correlated positively with the MVA levels, but there was no significant correlation between these biomarkers in CH and LC patients. Therefore, it was concluded that in the patients with chronic liver disease, there was a tendency for hepatic cholesterogenesis to be sustained in the face of hepatocellular dysfunction, while bile acid synthesis declined in parallel with the severity of impairment.

Perspectives

Biological specimens contain many types of organic acids and sterols. While fatty acids and cholesterol are relatively abundant compounds, MVA and oxysterols (7A and C4) are minor components. To quantify the concentrations of such minor components, stable isotope dilution mass spectrometry (GC-MS or LC-MS/MS) is an ideal method because of its high sensitivity, specificity and accuracy.

Recently, LC-MS/MS has come to be used more readily than GC-MS. Because MS/MS is more specific than MS, sample preparation process for the elimination of interfering materials can be simplified. In addition, LC-MS/MS does not require a derivatization step, which is also advantageous for high-throughput analyses. However, simple and rapid procedures do not always produce good results for the microanalysis of biological samples. A careful sample purification can increase the sensitivity of an analyte by reducing matrix effect (Jemal et al. 2003). Derivatization is useful not only to increase the sensitivity by enhancing the ionization efficiency but also to give a prominent ion in the high mass region, which makes it possible to avoid interfering peaks and to increase the specificity. A thorough chromatographic separation is also important to distinguish between similar biological compounds, e.g. hydroxycholesterols that have the same molecular weight and a virtually identical MS/MS spectrum. Thus, the importance of basic analytical techniques, i.e. sample purification, derivatization and chromatographic separation will not be denied even if the performance of mass spectrometer is improved further.

In conclusion, the methods for the quantification of key intermediates in cholesterol and bile acid biosynthetic pathways using stable isotope dilution mass spectrometry exhibit superior accuracy and sensitivity. By using this technique, the MVA and oxysterols in blood were established as biomarkers for cholesterol and bile acid biosynthesis. The use of these biomarkers has made it possible to monitor in vivo cholesterol and bile acid synthesis without the need for invasive liver biopsy, which is very useful for basic or clinical studies of cholesterol metabolism in humans.

References

1. Abrar, M. and Martin, P.D. 2002. Validation and application of an assay for the determination of mevalonic acid in human plasma by liquid chromatography tandem mass spectrometry. J. Chromatogr. B., 773:103–11.

2. Albers, I., Hartmann, H., Bircher, J. et al. 1989. Superiority of the Child-Pugh classification to quantitative liver function tests for assessing prognosis of liver cirrhosis. Scand. J. Gastroenterol., 24:269–76.

3. Axelson, M., Aly, A. and Sjövall, J. 1988. Levels of 7α-hydroxy-4-cholesten3-one in plasma reflect rates of bile acid synthesis in man. FEBS Lett, 239:324–8.

4. Axelson, M., Björkhem, I., Reihnér, E. et al. 1991. The plasma level of 7α-hydroxy-4-cholesten-3-one reflects the activity of hepatic cholesterol 7α-hydroxylase in man. FEBS Lett., 284:216–8.

5. Björkhem, I. and Danielsson, H. 1974. Assay of liver microsomal cholesterol 7α-hydroxylase using deuterated carrier and gas chromatography-mass spectrometry. Anal. Biochem., 59:508–16.

6. Björkhem, I., Miettinen, T., Reihnér, E. et al. 1987a. Correlation between serum levels of some cholesterol precursors and activity of HMG-CoA reductase in human liver. J. Lipid Res., 28:1137–43.

7. Björkhem, I., Reihnér, E., Angelin, B. et al. 1987b. On the possible use of the serum level of 7α-hydroxycholesterol as a marker for increased activity of the cholesterol 7α-hydroxylase in humans. J. Lipid Res., 28:889–94.

8. Boyle, P.J., Avogaro, A., Smith, L. et al. 1992. Role of GH in regulating nocturnal rates of lipolysis and plasma mevalonate levels in normal and diabetic humans. Am. J. Physiol., 263:E168–72.

9. Buffalini, M., Pierleoni, R., Guidi, C. et al. 2005. Novel and simple high-performance liquid chromatographic method for determination of 3-hydroxy-3-methylglutaryl-coenzyme A reductase activity. J. Chromatogr. B., 819:307–13.

10. Cighetti, G., Santaniello, E. and Galli, G. 1981. Evaluation of 3-hydroxy-3-methylglutaryl-CoA reductase activity by multiple-selected ion monitoring. Anal. Biochem., 110:153–8.

11. Del Puppo, M., Cighetti, G., Galli Kienle, M. et al. 1989. Measurement of mevalonate in human plasma and urine by multiple selected ion monitoring. Biomed. Environ. Mass Spectrom., 18:174–6.

12. Dietschy, J.M. and Brown, M.S. 1974. Effect of alterations of the specific activity of the intracellular acetyl CoA pool on apparent rates of hepatic cholesterogenesis. J. Lipid Res., 15:508–16.

13. Duane, W.C., Levitt, D.G., Mueller, S.M. et al. 1983. Regulation of bile acid synthesis in man. Presence of a diurnal rhythm. J. Clin. Invest., 72:1930–6.

14. Duane, W.C. and Javitt, N.B. 1999. 27-hydroxycholesterol: production rates in normal human subjects. J. Lipid Res., 40:1194–9.

15. Einarsson, C., Hillebrant, C.G. and Axelson, M. 2001. Effects of treatment with deoxycholic acid and chenodeoxycholic acid on the hepatic synthesis of cholesterol and bile acids in healthy subjects. Hepatology., 33:1189–93.

16. Eusufzai, S., Axelson, M., Angelin, B. et al. 1993. Serum 7α-hydroxy-4-cholesten3-one concentrations in the evaluation of bile acid malabsorption in patients with diarrhoea: correlation to SeHCAT test. Gut., 34:698–701.

17. Everson, G.T. 1992. Bile acid metabolism and its role in human cholesterol balance. Semin. Liver Dis., 12:420–8.

18. Galli Kienle, M., Galli, G., Bosisio, E. et al. 1984. Evaluation of enzyme activities by gas chromatography-mass spectrometry: HMGCoA reductase and cholesterol 7α-hydroxylase. J. Chromatogr., 289:267–76.

19. Gälman, C., Arvidsson, I., Angelin, B. et al. 2003. Monitoring hepatic cholesterol 7α-hydroxylase activity by assay of the stable bile acid intermediate 7α-hydroxy-4-cholesten-3-one in peripheral blood. J. Lipid Res., 44:859–66.

20. Gälman, C., Miquel, J.F., Perez, R.M. et al. 2004. Bile acid synthesis is increased in Chilean Hispanics with gallstones and in gallstone high-risk Mapuche Indians. Gastroenterology, 126:741–8.

21. Gälman, C., Angelin, B. and Rudling, M. 2005. Bile acid synthesis in humans has a rapid diurnal variation that is asynchronous with cholesterol synthesis. Gastroenterology, 129:1445–53.

22. Goldfarb, S. and Pitot, H.C. 1971. Improved assay of 3-hydroxy-3methylglutaryl coenzyme A reductase. J. Lipid Res., 12:512–5.

23. Hagenfeldt, L. and Hellström, K. 1972. Blood concentration and turnover of circulating mevalonate in the rat. Life Sci., 11:669–76.

24. Hahn, C., Reichel, C. and von Bergmann, K. 1995. Serum concentration of 7α-hydroxycholesterol as an indicator of bile acid synthesis in humans. J. Lipid Res., 36:2059–66.

25. Hiramatsu, M., Hayashi, A., Hidaka, H. et al. 1998. Enzyme immunoassay of urinary mevalonic acid and its clinical application. Clin. Chem., 44:2152–7.

26. Hiraoka, K. and Kudaka, I. 1992. Nagative-mode electrospray-mass spectrometry using nonaqueous solvents. Rapid Commun. Mass Spectrom., 6:265–8.

27. Honda, A., Shoda, J., Tanaka, N. et al. 1991. Simultaneous assay of the activities of two key enzymes in cholesterol metabolism by gas chromatography-mass spectrometry. J. Chromatogr., 565:53–66.

28. Honda, A., Yoshida, T., Xu, G. et al. 2004. Significance of plasma 7αhydroxy-4-cholesten-3-one and 27-hydroxycholesterol concentrations as markers for hepatic bile acid synthesis in cholesterol-fed rabbits. Metabolism, 53:42–8.

29. Honda, A., Mizokami, Y., Matsuzaki, Y. et al. 2007a. Highly-sensitive assay of HMG-CoA reductase activity by LC-ESI-MS/MS. J. Lipid Res., 48:1212–20.

30. Honda, A., Yamashita, K., Numazawa, M. et al. 2007b. Highly-sensitive quantifi cation of 7α-hydroxy-4-cholesten-3-one in human serum by LC-ESI-MS/MS. J. Lipid Res., 48:458–64.

31. Hylemon, P.B., Studer, E.J., Pandak, W.M. et al. 1989. Simultaneous measurement of cholesterol 7α-hydroxylase activity by reverse-phase high-performance liquid chromatography using both endogenous and exogenous [4–14C]cholesterol as substrate. Anal. Biochem., 182:212–6.

32. Ishihama, Y., Mano, N., Oda, Y. et al. 1994. Simple and sensitive quantitation method for mevalonic acid in plasma using gas chromatography/mass spectrometry. Rapid Commun. Mass Spectrom., 8:377–80.

33. Jemal, M., Schuster, A. and Whigan, D.B. 2003. Liquid chromatography/ tandem mass spectrometry methods for quantitation of mevalonic acid in human plasma and urine: method validation, demonstration of using a surrogate analyte, and demonstration of unacceptable matrix effect in spite of use of a stable isotope analog internal standard. Rapid Commun. Mass Spectrom., 17:1723–34.

34. Kempen, H.J.M., Glatz, J.F.C., Leuven, J.A.G. et al. 1988. Serum lathosterol concentration is an indicator of whole-body cholesterol synthesis in humans. J. Lipid Res., 29:1149–55.

35. Lala, A., Scoppola, A., Ricci, A. et al. 1994. The effects of insulin on plasma mevalonate concentrations in man. Ann. Nutr. Metab., 38:257–62.

36. Leonsson, M., Oscarsson, J., Bosaeus, I. et al. 1999. Growth hormone (GH) therapy in GH-deficient adults influences the response to a dietary load of cholesterol and saturated fat in terms of cholesterol synthesis, but not serum low density lipoprotein cholesterol levels. J. Clin. Endocrinol. Metab., 84:1296–303.

37. Lind, S., Rudling, M., Ericsson, S. et al. 2004. Growth hormone induces low-density lipoprotein clearance but not bile acid synthesis in humans. Arterioscler. Thromb. Vasc. Biol., 24:349–56.

38. Lutjohann, D., Hahn, C., Prange, W. et al. 2004. Influence of rifampin on serum markers of cholesterol and bile acid synthesis in men. Int. J. Clin. Pharmacol. Ther., 42:307–13.

39. Mitropoulos, K.A. and Balasubramaniam, S. 1972. Cholesterol 7α-hydroxylase in rat liver microsomal preparations. Biochem. J., 128:1–9.

40. Miyazaki, H., Koyama, M., Hashimoto, M. et al. 1978. Assay for 3-hydroxy3-methylglutaryl coenzyme A reductase by selected ion monitoring. In Stable Isotopes: Proceedings of the Third International Conference, Oak Brook, IL, May 23–26, 1978. Klein, ER. and Klein PD, editors. 267–73.

41. Muhrbeck, O., Wang, F.H., Björkhem, I. et al. 1997. Circulating markers for biosynthesis of cholesterol and bile acids are not depressed in asymptomatic gallstone subjects. J. Hepatol., 27:150–5.

42. Naoumova, R.P., Cummings, M.H., Watts, G.F. et al. 1996. Acute hyperinsulinaemia decreases cholesterol synthesis less in subjects with non-insulin-dependent diabetes mellitus than in non-diabetic subjects. Eur. J. Clin. Invest., 26:332–40.

43. Naoumova, R.P., Marais, A.D., Mountney, J. et al. 1996. Plasma mevalonic acid, an index of cholesterol synthesis in vivo, and responsiveness to HMG-CoA reductase inhibitors in familial hypercholesterolaemia. Atherosclerosis, 119:203–13.

44. Naoumova, R.P., Dunn, S., Rallidis, L. et al. 1997. Prolonged inhibition of cholesterol synthesis explains the efficacy of atorvastatin. J. Lipid Res., 38:1496–500.

45. Naoumova, R.P., O'Neill, F.H., Dunn, S. et al. 1999. Effect of inhibiting HMG-CoA reductase on 7α-hydroxy-4-cholesten-3-one, a marker of bile acid synthesis: contrasting findings in patients with and without prior up-regulation of the latter pathway. Eur. J. Clin. Invest., 29:404–12.

46. Ndong-Akoume, M.Y., Mignault, D., Perwaiz, S. et al. 2002. Simultaneous evaluation of HMG-CoA reductase and cholesterol 7α-hydroxylase activities by electrospray tandem MS. Lipids, 37:1101–7.

47. Nelson, W., Tong, Y.L., Lee, J.K. et al. 1979. Methods for cosinorrhythmometry. Chronobiologia, 6:305–23.

48. Noshiro, M., Ishida, H., Hayashi, S. et al. 1985. Assays for cholesterol 7α-hydroxylase and 12α-hydroxylase using high performance liquid chromatography. Steroids, 45:539–49.

49. Nozaki, S., Nakagawa, T., Nakata, A. et al. 1996. Effects of pravastatin on plasma and urinary mevalonate concentrations in subjects with familial hypercholesterolaemia: a comparison of morning and evening administration. Eur. J. Clin. Pharmacol., 49:361–4.

50. Oda, H., Yamashita, H., Kosahara, K. et al. 1990. Esterified and total 7αhydroxycholesterol in human serum as an indicator for hepatic bile acid synthesis. J. Lipid Res., 31:2209–18.

51. Ogishima, T. and Okuda, K. 1986. An improved method for assay of cholesterol 7α-hydroxylase activity. Anal. Biochem., 158:228–32.

52. O'Neill, F.H., Patel, D.D., Knight, B.L. et al. 2001. Determinants of variable response to statin treatment in patients with refractory familial hypercholesterolemia. Arterioscler. Thromb. Vasc. Biol., 21:832–7.

53. Pappu, A.S. and Illingworth, D.R. 1994. Diurnal variations in the plasma concentrations of mevalonic acid in patients with abetalipoproteinaemia. Eur. J. Clin. Invest., 24:698–702.

54. Pappu, A.S. and Illingworth, D.R. 2002. The effects of lovastatin and simvastatin on the diurnal periodicity of plasma mevalonate concentrations in patients with heterozygous familial hypercholesterolemia. Atherosclerosis, 165:137–44.

55. Park, E.J., Lee, D., Shin, Y.G. et al. 2001. Analysis of 3-hydroxy-3-methylglutaryl-coenzyme A reductase inhibitors using liquid chromatography-electrospray mass spectrometry. J. Chromatogr. B. Biomed. Sci. Appl., 754:327–32.

56. Parker, T.S., McNamara, D.J., Brown, C. et al. 1982. Mevalonic acid in human plasma: relationship of concentration and circadian rhythm to cholesterol synthesis rates in man. Proc. Natl. Acad. Sci. U.S.A., 79:3037–41.

57. Parker, T.S., McNamara, D.J., Brown, C.D. et al. 1984. Plasma mevalonate as a measure of cholesterol synthesis in man. J. Clin. Invest., 74:795–804.

58. Pettersson, L. and Eriksson, C.G. 1994. Reversed-phase high-performance liquid chromatographic determination of 7α-hydroxy-4-cholesten-3one in human serum. J. Chromatogr. B. Biomed. Appl., 657:31–6.

59. Pfohl, M., Naoumova, R.P., Kim, K.D. et al. 1998. Use of cholesterol precursors to assess changes in cholesterol synthesis under nonsteady-state conditions. Eur. J. Clin. Invest., 28:491–6.

60. Pooler, P.A. and Duane, W.C. 1988. Effects of bile acid administration on bile acid synthesis and its circadian rhythm in man. Hepatology, 8:1140–6.

61. Popjak, G., Boehm, G., Parker, T.S. et al. 1979. Determination of mevalonate in blood plasma in man and rat. Mevalonate "tolerance" tests in man. J. Lipid Res., 20:716–28.

62. Saini, G.S., Wani, T.A., Gautam, A. et al. 2006. Validation of LC-MS/MS method for quantification of mevalonic acid in human plasma and an approach to differentiate recovery and matrix effect. J. Lipid Res., 47:2340–5.

63. Saisho, Y., Kuroda, T. and Umeda, T. 1997. A sensitive and selective method for the determination of mevalonic acid in dog plasma by gas chromatography/

negative ion chemical ionization-mass spectrometry. J. Pharm. Biomed. Anal., 15:1223 30.

64. Sanghvi, A., Grassi, E., Bartman, C. et al. 1981. Measurement of cholesterol 7α-hydroxylase activity with selected ion monitoring. J. Lipid Res., 22:720–4.

65. Sauter, G., Berr, F., Beuers, U. et al. 1996. Serum concentrations of 7αhydroxy-4-cholesten-3-one reflect bile acid synthesis in humans. Hepatology, 24:123–6.

66. Sauter, G., Weiss, M. and Hoermann, R. 1997. Cholesterol 7α-hydroxylase activity in hypothyroidism and hyperthyroidism in humans. Horm. Metab. Res., 29:176–9.

67. Sauter, G.H., Munzing, W., von Ritter, C. et al. 1999. Bile acid malabsorption as a cause of chronic diarrhea: diagnostic value of 7α-hydroxy4-cholesten-3-one in serum. Dig. Dis. Sci., 44:14–9.

68. Sauter, G.H., Thiessen, K., Parhofer, K.G. et al. 2004. Effects of ursodeoxycholic acid on synthesis of cholesterol and bile acids in healthy subjects. Digestion, 70:79–83.

69. Schwarz, M., Lund, E.G., Setchell, K.D. et al. 1996. Disruption of cholesterol 7α-hydroxylase gene in mice. II. Bile acid defi ciency is overcome by induction of oxysterol 7α-hydroxylase. J. Biol. Chem., 271:18024–31.

70. Scoppola, A., Maher, V.M., Thompson, G.R. et al. 1991. Quantitation of plasma mevalonic acid using gas chromatography-electron capture mass spectrometry. J. Lipid Res., 32:1057–60.

71. Scoppola, A., Testa, G., Frontoni, S. et al. 1995. Effects of insulin on cholesterol synthesis in type II diabetes patients. Diabetes Care, 18:1362–9.

72. Shapiro, D.J., Imblum, R.L. and Rodwell, V.W. 1969. Thin-layer chromatographic assay for HMG-CoA reductase and mevalomic acid. Anal. Biochem., 31:383–90.

73. Shefer, S., Hauser, S. and Mosbach, E.H. 1968. 7α-Hydroxylation of cholesterol by rat liver microsomes. J. Lipid Res., 9:328–33.

74. Shefer, S., Hauser, S., Lapar, V. et al. 1972. HMG CoA reductase of intestinal mucosa and liver of the rat. J. Lipid Res., 13:402–12.

75. Shefer, S., Nicolau, G. and Mosbach, E.H. 1975. Isotope derivative assay of microsomal cholesterol 7α-hydroxylase. J. Lipid Res., 16:92–6.

76. Shoda, J., Tanaka, N., He, B.F. et al. 1993. Alterations of bile acid composition in gallstones, bile, and liver of patients with hepatolithiasis, and their etiological significance. Dig. Dis. Sci., 38:2130–41.

77. Shoda, J., Miyamoto, J., Kano, M. et al. 1997. Simultaneous determination of plasma mevalonate and 7α-hydroxy-4-cholesten-3-one levels in hyperlipopro-

teinemia: convenient indices for estimating hepatic defects of cholesterol and bile acid syntheses and biliary cholesterol supersaturation. Hepatology, 25:18–26.

78. Siavoshian, S., Simoneau, C., Maugeais, P. et al. 1995. Measurement of mevalonic acid in human urine by bench top gas chromatography-mass spectrometry. Clin. Chim. Acta., 243:129–36.

79. Vlahcevic, Z.R., Pandak, W.M., Heuman, D.M. et al. 1992. Function and regulation of hydroxylases involved in the bile acid biosynthesis pathways. Semin. Liver Dis., 12:403–19.

80. Vlahcevic, Z.R., Stravitz, R.T., Heuman, D.M. et al. 1997. Quantitative estimations of the contribution of different bile acid pathways to total bile acid synthesis in the rat. Gastroenterology, 113:1949–57.

81. Woollen, B.H., Holme, P.C., Northway, W.J. et al. 2001. Determination of mevalonic acid in human urine as mevalonic acid lactone by gas chromatography-mass spectrometry. J. Chromatogr., 760:179–84.

82. Yamashita, H., Kuroki, S. and Nakayama, F. 1989. Assay of cholesterol 7α-hydroxylase utilizing a silica cartridge column and 5α-cholestane3β,7β-diol as an internal standard. J. Chromatogr., 496:255–68.

83. Yoshida, T., Honda, A., Tanaka, N. et al. 1993. Simultaneous determination of mevalonate and 7α-hydroxycholesterol in human plasma by gas chromatography-mass spectrometry as indices of cholesterol and bile acid biosynthesis. J. Chromatogr., 613:185–93.

84. Yoshida, T., Honda, A., Tanaka, N. et al. 1994. Determination of 7α-hydroxy4-cholesten-3-one level in plasma using isotope-dilution mass spectrometry and monitoring its circadian rhythm in human as an index of bile acid biosynthesis. J. Chromatogr. B. Biomed. Appl., 655:179–87.

85. Yoshida, T., Honda, A., Shoda, J. et al. 1997. Short-term effects of 3-hydroxy-3-methylglutaryl-CoA reductase inhibitor on cholesterol and bile acid synthesis in humans. Lipids, 32:873–8.

86. Yoshida, T., Honda, A., Matsuzaki, Y. et al. 1999. Plasma levels of mevalonate and 7α-hydroxy-4-cholesten-3-one in chronic liver disease. J. Gastroenterol. Hepatol., 14:150–5.

CITATION

Originally published under the Creative Commons Attribution License or equivalent. Yoshida T, Honda A, Miyazaki H, Matsuzaki Y. Determination of Key Intermediates in Cholesterol and Bile Acid Biosynthesis by Stable Isotope Dilution Mass Spectrometry. Anal Chem Insights. 2008; 3: 45–60.

A Dynamic Programming Approach for the Alignment of Signal Peaks in Multiple Gas Chromatography-Mass Spectrometry Experiments

Mark D. Robinson, David P. De Souza, Woon Wai Keen,
Eleanor C. Saunders, Malcolm J. McConville,
Terence P. Speed and Vladimir A. Likić

ABSTRACT

Background

Gas chromatography-mass spectrometry (GC-MS) is a robust platform for the profiling of certain classes of small molecules in biological samples. When multiple samples are profiled, including replicates of the same sample and/or different sample states, one needs to account for retention time drifts between experiments. This can be achieved either by the alignment of chromatographic profiles prior to peak detection, or by matching signal peaks after they have been extracted from chromatogram data matrices. Automated retention time correction is particularly important in non-targeted profiling studies.

Results

A new approach for matching signal peaks based on dynamic programming is presented. The proposed approach relies on both peak retention times and mass spectra. The alignment of more than two peak lists involves three steps: (1) all possible pairs of peak lists are aligned, and similarity of each pair of peak lists is estimated; (2) the guide tree is built based on the similarity between the peak lists; (3) peak lists are progressively aligned starting with the two most similar peak lists, following the guide tree until all peak lists are exhausted. When two or more experiments are performed on different sample states and each consisting of multiple replicates, peak lists within each set of replicate experiments are aligned first (within-state alignment), and subsequently the resulting alignments are aligned themselves (between-state alignment). When more than two sets of replicate experiments are present, the between-state alignment also employs the guide tree. We demonstrate the usefulness of this approach on GC-MS metabolic profiling experiments acquired on wild-type and mutant Leishmania mexicana parasites.

Conclusion

We propose a progressive method to match signal peaks across multiple GC-MS experiments based on dynamic programming. A sensitive peak similarity function is proposed to balance peak retention time and peak mass spectra similarities. This approach can produce the optimal alignment between an arbitrary number of peak lists, and models explicitly within-state and between-state peak alignment. The accuracy of the proposed method was close to the accuracy of manually-curated peak matching, which required tens of man-hours for the analyzed data sets. The proposed approach may offer significant advantages for processing of high-throughput metabolomics data, especially when large numbers of experimental replicates and multiple sample states are analyzed.

Background

Metabolomics refers to an all-inclusive profiling of low-molecular weight metabolites with an implicit aim to interpret results in the context of the organism's genome and its global metabolic network [1-4]. The term metabolic profiling is used to denote either targeted or non-targeted profiling of small molecules in biological samples. A targeted analysis focuses on specific, a priori known compounds, and therefore only parts of the chromatogram data matrix generated by the instrument may be considered rele vant. In a non-targeted approach all detectable signals are analyzed, and the aim of such analysis is to achieve as wide coverage of metabolites as permitted by a particular experimental technique. A non-targeted metabolic profiling is an essential component of metabolomics, together with bioinformatics approaches for data analysis, interpretation, and integration [1-4]. In recent years metabolic profiling is increasingly being used in studies of microbial metabolism [5], biomarker discovery [6], toxicology [7,8], nutrition [9,10], and integrated systems biology [4,11]. Hyphenated mass spectrometry approaches, in particular gas chromatography-mass spectrometry (GC-MS) [1,12,13] and liquid chromatography-mass spectrometry (LC-MS) [2,14], are often used for metabolic profiling because of their inherent robustness, sensitivity, and large dynamic range.

GC-MS is particularly well suited for studies of low polarity metabolites with high sensitivity and specificity [15]. The processing of GC-MS data is based on the detection of chromatogram signal peaks, a step performed either with proprietary software or with freely available software such as AMDIS [16]. The net result of peak detection is a set of signal peaks that represent components present in the sample. Metabolic profiling experiments are relatively rapid, and typically multiple replicates per sample state are recorded to facilitate robust statistical analyses [1,2,13].

In order to compare multiple samples, one needs to account for retention time drifts inherent in chromatographic separations [17-23]. The retention time drifts arise because of instrument imperfections (variations in temperature and mobile phase flow rates), column variations (stationary phase saturation and degradation, and stationary phase variations in column-to-column runs), and sample matrix effects (due to variations in sample composition, such as solvent and salts) [17,18]. A review of the literature shows that there is a pressing need for better algorithms for retention time correction in GC-MS small molecule profiling experiments. A summary of approaches used in the past for correcting retention time drifts in hyphenated mass spectrometry experiments is given in Figure 1. An approach often used to correct retention time drifts in practice is a linear correction calculated based on deviations from internal standards [24,25]. This approach has several limitations [21]. For example, addition of internal standards adds new chemicals to the sample, and chromatographic retention time drifts are not linear (Figure 2 and [21]).

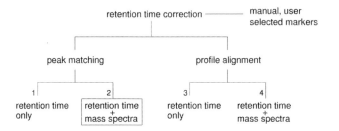

Figure 1. An overview of methods for retention time correction in hyphenated mass spectrometry profiling experiments. A technique often used in retention time correction involves spiking internal standards prior to data acquisition, and then linear time correction is applied manually, based on user selected markers [24, 25]. This approach has significant limitations, as discussed in [21]. The automated approaches for retention time correction in hyphenated mass spectrometry are based on two schools of thought: one is to align the entire chromatographic profiles prior to peak detection (profile alignment), and the other is to perform peak detection first, and then match extracted signal peaks across samples to correct for retention time drifts (peak matching). In either approach one can rely on the time domain data only, or include the information from the m/z data domain (mass spectra). Examples of peak matching algorithms that use retention time only (branch 1) include [18, 19, 22]; examples of peak matching algorithms that use both time domain and m/z data (branch 2) include [20, 21, 23]; algorithms for profile alignment that rely on time domain data only (branch 3) were first proposed in 1979 [39], and include [17, 26, 28, 29]; finally, examples of algorithms for profile alignment that use the entire chromatogram data matrices include [27, 30, 37]. The algorithm proposed here is a peak matching approach, and relies on both time domain data and peak mass spectra (branch 2).

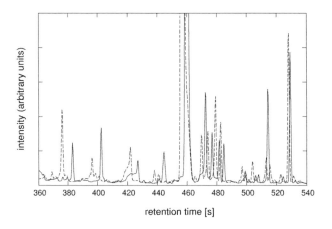

Figure 2. GC-MS total ion chromatograms showing non-linear time shifts. A portion of two GC-MS total ion chromatograms (TICs) of polar extracts of L. mexicana promastigotes harvested in the mid-log growth phase (solid line) and stationary growth phase (dotted lines). The overlay of raw TICs for the segment between 360 s and 540 s shows non-linear shifts in chromatographic separation often observed in practice.

The problem of retention time correction is similar in GC-MS and LC-MS small molecule profiling experiments. There are two schools of thought on how to address this: one is to align entire chromatogram profiles prior to peak detection (profile alignment) [26-30], and the other is to perform peak detection first, and then align extracted signal peaks to correct for retention time drifts (peak alignment or peak matching) [18-23]. In either approach one can rely on the time domain data only, or include highly selective information from the m/z data domain (mass spectra).

Both peak matching and profile alignment methods have been used in the context of GC-MS data. Jonsson and co-authors have developed a strategy for rapid comparison of GC-MS metabolic profiles, based on the division of a chromatogram into time windows [28]. To correct the retention time drifts they relied on finding the maximum covariance between the chromatograms [17]. This is a profile alignment method that uses the time domain data only (Figure 1, branch 3). To be effective this method requires that chromatographic profiles to be aligned are similar to one another, as was the case in the analysis of leaf plant extracts where the main purpose of the alignment was to facilitate setting of time domain windows [28]. A different alignment approach was used in the analysis of GC-MS data obtained from the profiling of tomato volatiles, but the algorithm was not described in detail [31].

The peak matching approach is an alternative to the profile alignment that has received a considerable attention in GC-MS [18,19,22]. Johnson et al. matched GC-MS peaks by retention time within a pre-defined time window [18]. Although this approach was shown to be reasonably effective in the GC-MS analysis of diesel fuels, much remains to be desired for applications in biomedical research. For example, it is unclear how to choose the target-chromatogram to which other chromatograms will be aligned; the choice of the cut-off window is arbitrary; and the proposed method of grouping the peaks depends on the order in which the grouping is performed. Duran and co-workers proposed a clustering procedure which groups peaks starting from the minimum retention time [19]. This approach is symmetrical with respect to all experiments and therefore alleviates the need for an arbitrary choice of a target chromatogram. However, the proposed clustering depends on the peak minimum retention time and the arbitrary retention time window [19]. We have attempted to improve on the peak clustering idea by using hierarchical clustering [22]. However, applications of this method have shown that results are rather sensitive to the quality of input data, especially for large retention time drifts. In summary, the body of work on peak matching suggests that accurate peak alignment is unlikely to be achieved by relying on retention time data only.

The most accurate retention time correction is likely to be achieved when complete chromatogram data matrices are utilized. Such methods have been applied to LC-MS [27] and CE-MS [30] metabolic profiling data. The main drawback of any approach that works with entire chromatogram data matrices is the large amount of (uninformative) noise data that must be handled, which dramatically increases computational costs. Our preliminary calculations showed that progressive alignment of chromatogram matrices aimed to avoid the choice of a "target" data set is computationally prohibitive. Furthermore, it is an open question whether the computational costs can be justified by the potential increase in accuracy compared to peak matching methods, when peak matching is based on both retention times and mass spectra.

The most advanced approaches for correction of retention time drifts have been developed on LC-MS and CE-MS data [20,21,27,30]. To our knowledge the peak matching involving mass spectra has not been applied to GC-MS data, except in an approach proposed for a systematic identification of conserved metabolites [23]. This approach aims specifically at identification of conserved metabolites across experiments, and its peak matching is implicitly serving as input for a motif discovery algorithm [23].

Here we present an approach for peak matching that relies on both peak retention times and mass spectra, capable of accurate alignment of peak lists from a large number of replicate GC-MS experiments. We use dynamic programming to arrive at the optimal solution to the global alignment problem. Dynamic programming was previously used for the alignment of chromatogram data matrices [27,30], but not in peak matching. The proposed approach uses a similarity function which balances effectively similarity in peak retention times with the similarity in mass spectra, mimicking the way an experienced human operator performs peak matching. The proposed algorithm is modelled after the heuristic solution for multiple sequence alignment [32,33], and relies on progressive alignment to match an arbitrary number of data sets. We apply this approach to GC-MS metabolic profiling data acquired on wild-type and mutant Leishmania mexicana parasites. In the absence of a suitable metric and an accepted benchmark for the peak alignment algorithms, we quantify absolute errors relative to manually derived "true" answer. We show that (a) the proposed method performed close to the accuracy of the manually curated alignments by an expert operator; (b) the results are not sensitive to the input parameters, suggesting that the proposed method is robust.

Results

The input for the peak alignment procedure consists of two or more peak lists obtained from hyphenated mass spectrometry experiments (GC-MS data used in this work). These peak lists may be derived from replicate analyses of the same sample (for example, polar extracts from wild-type parasites), or replicate analyses of different samples (for example, polar metabolite extracts of wild-type and mu-tant parasite lines).

Each experiment is represented by a single and unique peak list. In the most general case, a peak list could be understood as a list of peak objects, where each peak object is characterized with one or more attributes. In the work described here, a peak was characterized with a unique peak ID, the retention time at the peak apex, and the mass spectrum at the peak apex. Henceforth, we assume that a peak list consists of peaks ordered by their retention time. Consider a peak list LA that contains peaks p1, p2, …, PnA. In this case the retention time of the peak p2 is greater than that of p1, the retention time of the peak p3 is greater than that of p2, and so on.

Alignment of Two Peak Lists

Consider two peak lists $L_A = [p_1, p_2, …, P_{nA}]$ and $L_B = [q_1, q_2, …, q_{nB}]$ which contain a total number of n_A and n_B peaks, respectively. The alignment of the peak lists L_A and L_B refers to the establishment of a one-to-one correspondence between the peaks from the two lists, with the possibility that any peak from one list has no matching peak in the other list. The alignment between the peak lists L_A and L_B could be represented as a list of peak pairs, where pairing implies peak-to-peak matching. For example,

$$[(p1, q1), (p2, q2), (p3, -), (p4, q3), …] \tag{1}$$

where p_1 is matched with q_1, p_2 is matched with q_2, and so on. The peak p_3 from the list L_A does not have a matching peak in the LB list. The number of elements in the above list will depend on the optimal alignment, but cannot be less than the larger of A and B, and cannot exceed A+B. For brevity, we refer to the alignment between peak lists L_A and L_B as $L_A:L_B$.

It is apparent from the above that the alignment of two peak lists closely resembles the problem of pairwise sequence alignment. The situation for p3 in Equation [1] corresponds to matching a sequence letter to a gap. Furthermore, the analogy can be extended even further if one considers that the peak list is an ordered sequence of peaks. The variations in peak retention times arise from various non-linear effects during the separation stage, such as uneven flow of the carrier phase in the GC or LC column. Such perturbations may affect absolute peak retention times and may shift portions of the chromatogram in a non-linear manner [21], but normally do not change the order of peaks in terms of their retention times. The analogy with pairwise sequence alignment implies that dynamic programming could be applied to find the optimal alignment of two peak lists, provided that a suitable scoring scheme can be devised.

The Scoring Scheme for Peak Alignment

In sequence alignment, the cost function for matching two residue letters is obtained from a pre-computed substitution matrix. In the case of peak lists the cost function should reflect similarity between two peaks (henceforth referred to as the peak similarity function). Since the peak mass-spectrum is a key identifier of a particular peak, the peak similarity function should depend heavily on the similarity in the mass spectra. We propose the following peak similarity function $P(i,j)$, which gives the similarity between the peaks i and j:

$$P(i, j) = S(i, j) . \exp\left(-\frac{(t_i - t_j)^2}{2D^2} \right) \qquad (2)$$

In the above equation $S(i,j)$ is the similarity between the mass spectra of the peaks i and j, ti and tj are retention times of peaks i and j, and D is the retention time tolerance parameter which determines the importance of retention times to the overall peak similarity score. The function $S(i,j)$ can be any function that returns a measure of the similarity in m/z ions detected in mass spectra of the peaks i and j. In our test implementation, $S(i,j)$ was calculated as the cosine of the angle between the two mass spectra vectors (i.e. the normalized dot product). This resulted in values between 0 and 1: when the two mass spectra are identical $S(i,j)$ = 1, and when they are completely dissimilar $S(i,j)$ = 0.

The second term in the equation [2] modulates the similarity in mass spectra. When two peaks have identical retention times (ti = tj) this term equals one, and the peak similarity reduces to the mass spectra similarity ($P(i,j)$ = $S(i,j)$). When the retention times of the two peaks differ, then exp[-(ti - tj)2/2D2] < 1, and $P(i,j)$ < $S(i,j)$. The greater the retention time difference the peak similarity $P(i,j)$ is more reduced relative to the mass spectra similarity $S(i,j)$. The extent of this reduction depends on the relationship between the difference (ti - tj) and the parameter D. When (ti - tj) >> D then exp[-(ti - tj)2/2D2] will be close to zero and peak similarity will be close to zero ($P(i,j)$ = 0), even if mass spectra are identical ($S(i,j)$ = 1).

The peak similarity function gives the "cost" of matching any two peaks. In addition to this, the dynamic programming algorithm requires a gap penalty to be defined. In sequence alignment gaps are treated with the affine gap function, designed to penalize gaps and to favor the extension of existing gaps over the creation of new gaps. In peak alignment favoring gap extensions is not physically justified, therefore, we set the gap penalty to a fixed number (G). With the above definition of $P(i,j)$ (equation [2]) the meaningful range of G is between 0 and 1. A low value of G would favor the insertion of gaps even when peaks are originating from the same chemical compound; a high value of G would favour the alignment of peaks that are actually different.

With the above definitions for the peak similarity function and the choice of gap penalty it is possible to deploy dynamic programming to find the global solution to the problem of optimal alignment of two peak lists. A standard dynamic programming procedure involves the following steps: (1) initialization of a two dimensional score matrix whose rows are indexed with the peaks of one peak list and columns are indexed with the peaks of the other peak list; (2) filling the cells of the score matrix based on the peak similarity function and possibly the gap penalty; (3) the best alignment between two peak lists is deduced from the traceback on the score matrix. We note that the pairwise alignment of two peak lists is analogous to the problem of global sequence alignment [34].

An example pairwise alignment is shown in Figure 3. Figure 3(a) shows TIC segments between 467 s and 534 s for two GC-MS profiling experiments of L. mexicana promastigotes harvested in logarithmic and stationary growth phases, respectively. Peak finding has resulted in fourteen peaks in each segment as shown in Figure 3(b), which also shows the peak matching after the alignment. Figure 4 shows different score matrices and the traceback used to deduce the optimal alignment.

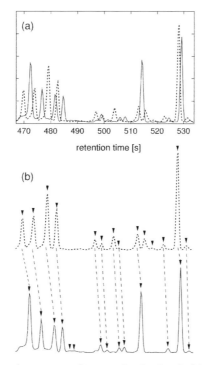

Figure 3. An example total ion chromatogram alignment. Panel a. Detail of the segment between 467 s and 534 s of the TICs showed in Figure 1. Fourteen signal peaks were detected in each segment. Panel b. The peak matching after dynamic programming peak alignment (see also Figure 4).

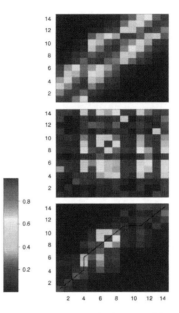

Figure 4. The score matrix for dynamic programming peak alignment. The score matrices for the alignment of two peak lists obtained from the GC-MS segments shown in Figure 3. Three matrices are shown with peak similarity (color coding is according to the scale between 0 and 1). For each score matrix the two axis refer to two peak lists to be aligned, with 14 signal peaks in each list (stationary phase peaks plotted along the x-axis and mid-log phase peaks plotted on the y-axis). The top panel shows the score matrix based on the retention time similarities only (as given the second term in the Equation [2]). The middle panel shows the score matrix based on the similarities in mass spectra taken at the peak apex (as given by the first term in the Equation [2]). The bottom panel is the total peak similarity function, as given by the Equation [2] (this is element-by-element product of the retention time and mass spectra score matrices). The traceback resulting from the application of dynamic programming is also shown in bottom panel. From the traceback the best alignment of peaks was deduced, as shown in Figure 3(b).

Alignment of Two Alignments

Consider an alignment between two internally fixed alignments: one consisting of N peak lists (henceforth referred to as N-alignment) and the other consisting of M peak lists (M-alignment). In order to find the best alignment between two such alignments with two-dimensional dynamic programming the scoring scheme must be extended.

Let the N-alignment contain K peak positions and the M-alignment contain L peak positions. In this case the score matrix may have rows indexed by the peak positions of the N-alignment, and columns indexed by the peak positions of the M-alignment, resulting in a two dimensional table with K rows and L columns. During the dynamic programming procedure the cell (i,j) of this table is filled with the score W(i,j) calculated for the i-th position of the N-alignment and the j-th position of the M-alignment, possibly modified by the gap penalty

G. Alternative definitions of W(i,j) are possible. In our test implementation W(i,j) was calculated as the average similarity between the peaks in the i-th position of the M-alignment and j-th position of the N-alignment. For example, if the position i of the N-alignment contained peaks pi1, pi2,... piN and the position j or the M-alignment contained peaks qj1, qj2,... qjM then:

$$W(i,j) = \frac{\sum\limits_{a=1}^{N} \sum\limits_{b=1}^{M} P(P_{ia}, q_{jb})}{\sum\limits_{a=1}^{N} \sum\limits_{b=1}^{M} I[P(P_{ia}, q_{jb}) > 0]} \tag{3}$$

where I is the indicator function and P is given by Equation [2]. Here we extend the definition of the peak similarity function to include scoring of a peak with a gap, which is always zero (i.e. P(i, -) = P(-, j) = 0). Therefore, it is possible that the denominator is less than NM, since some of the terms in Equation [3] may be zero due to involvement of gaps.

Consider the simple case, an alignment between the alignment LA:LB given by the Equation [1] and a single peak list LC = [r1, r2, ..., rnC]. We note that the alignment LA:LB is a 2-alignment, and the peak list LC could be viewed as 1-alignment. According to Equation [3], the similarity function between the first position in the alignment given by Equation [1] (position (p1, q1)) and the first peak from the list LC (peak r1) will be calculated as:

$$W((p_1, q_1), r_1) = \frac{1}{2}(P(p_1, r_1) + P(q_1, r_1)) \tag{4}$$

The similarity function between the third position of the alignment LA:LB and the first peak from the list LC is W(r1, (p1, -)) = P(r1,p1) since P(r1,-) = 0. In the case of N = 1 and M = 1, the alignment reduces to a simple pairwise alignment discussed above.

It is useful to devise a measure of how good one alignment is relative to another comparable alignment. Given an alignment between an N-alignment and M-alignment we calculate a total alignment score T as follows:

$$T = \sum_{k} Z_k \tag{5}$$

where Z_k is the alignment score for the position k in the alignment, and the summation is over all positions in the alignment. The value of Z_k depends on whether the gap was inserted in the k-th position of the alignment or not. If the gap was inserted either in the N- or M-alignment then Z_k = -G. If the gap was not inserted then Z_k equals W(i,j) given by the Equation [3], where i is the position from the

N-alignment and j is the position from the M-alignment that are aligned to one another to result in the k position of the final (N+M)-alignment.

Alignment of an Arbitrary Number of Peak Lists

Consider the alignment of U peak lists L_A, L_B, L_C,... where the peak list L_A contains a total of n_A peaks, the list L_B contains a total of n_B peaks, the list L_C contains a total of n_C peaks, and so on. The overall goal of the alignment process is to align the peak lists L_A, L_B, L_C,... to obtain the alignment $L_A:L_B:L_C:...$ This alignment can be represented as a table or a matrix with U rows, each corresponding to one peak list. The exact number of columns in the alignment table will depend on the optimal alignment, but must be equal or greater than the maximum of n_A, n_B, n_C,...

Although it is possible to devise an exact dynamic programming solution for a multi-dimensional alignment, this quickly leads to a computationally intractable problem [33,35]. Therefore we modelled our solution after the progressive multiple sequence alignment [32,33], with changes to accommodate the unique requirements of peak alignment. To find the best alignment of peak lists LA, LB, LC,... we first calculate all possible pairwise alignments between the peak lists. From all pairwise alignments the alignment score for each pair of peak lists is calculated (TAB, TAC, TBC,...). In the next step, a dendrogram (guide) tree is built which provides the similarity relationship between the peak lists. A progressive pairwise alignment is performed following the branching order given by the guide tree. In the first step the two most similar peak lists are selected and aligned, resulting in a 2-alignment. The other peak lists are added gradually following the guide tree until all peak lists are exhausted.

Within-State and Between-State Alignment

In the case of experiments performed on different cell states with more than one replicate experiment per cell state it is reasonable to align first replicate experiments performed on each cell state ("within-state alignment"), and then to align the resulting alignments ("between-state alignment") [22]. This is because within each cell state one deals with true experimental replicates, and in the hypothetical case of perfect reproducibility all peaks will be observed in all experiments. In experiments performed on different cell states, such as wild-type and mutant cells, some metabolites may be missing in one state relative to another, and the expectation that all peaks observed in one state will be present in the other state is no longer valid.

Consider an alignment of three cell states (wild-type (WT), mutant-1 (M1), and mutant-2 (M2)), each having eight replicate experiments. In the first step, within-state alignments are performed resulting in three 8-alignments (WT, M1, and M2). To solve the order of between-state alignment all possible pairwise alignments between WT, M1, and M2 alignments are created, the total alignment scores are calculated, and another guide tree is built. The between-state alignment is built progressively from WT, M1, and M2 alignments following the guide tree to result in the final 24-alignment WT:M1:M2.

In practice, the reproducibility in peak retention times of experiments performed on the same cell state is often better compared to experiments performed on different cell states. Therefore, for optimal alignment involving different cell states it may be useful to use different parameters for retention time tolerance (D) and gap penalty (G). We denote these parameters Dw, Gw for within-state alignment and Db, Gb for between-state alignment.

Testing

Metabolite profiling studies were performed on different cultivated stages of the human parasite, Leishmania mexicana. Comparisons were also made between wild type parasites and a mutant cell line lacking three major glucose transporters [36]. Polar metabolites were analyzed by GC-MS and peak lists generated manually or by using the proposed alignment method to assess the accuracy of the alignment and the sensitivity of the latter approach to input parameters such as the gap penalty and the retention time tolerance.

Within-State Alignment of the Wild-Type Cells Profiling Experiments

Eight replicate extracts of L.mexicana wild-type cells were analyzed by GC-MS and peaks were manually pre-processed and aligned. In this way a correct peak alignment table was constructed from 1,337 signal peaks from all eight replicate experiments. The manual analysis showed 173 unique metabolites, and the final alignment table contained 1,384 fields arranged in 8 columns (replicate experiments) and 173 rows (metabolites).

To assess the accuracy of the proposed approach a series of alignments was constructed by applying dynamic programming alignment with different input parameters. The resulting alignment tables were compared to the correct alignment table and analyzed for errors. Two types of errors were observed (Figure 5): peak mixing (type A error) and metabolite splitting (type B error). In peak mixing errors one or more peaks are shifted to a different metabolite row to take a position of a missing peak. Figure 5(a) shows the simplest case of one metabolite being shifted from metabolite-1 to a metabolite-2 row. In practice mixing can involve more than two metabolite rows. In metabolite splitting, one or more peaks are moved to create an extra metabolite row (i.e. metabolite 3 in Figure 5(b)). As a consequence of the removal of spurious metabolites (see Methods) most artificially created metabolite rows due to metabolite splitting will be discarded. The only exception would be metabolites with all eight peaks present, subject to the splitting that has resulted in two metabolite rows with exactly four peaks per row.

(a)

			correct alignment	
metabolite1	6.541	--	6.590	6.554
metabolite2	6.712	**6.690**	6.702	6.710

metabolite1	6.541	**6.690**	6.590	6.554
metabolite2	6.712	--	6.702	6.710

(b)

			correct alignment	
metabolite1	6.541	--	6.590	6.554
metabolite2	6.712	**6.690**	**6.702**	6.710

(1)

metabolite1	6.541	--	6.590	6.554
metabolite2	6.712	--	--	6.710
metabolite3	--	**6.690**	**6.702**	--

(2)

metabolite1		6.541	--	6.590	6.554
metabolite2	✗	6.712	--	--	6.710
metabolite3	✗	--	**6.690**	**6.702**	--

Figure 5. Errors in peak alignment. A portion of the hypothetical alignment table with two types of errors that occur in within-state peak alignment highlighted. The rows of the table represent metabolites and column represent individual experiments. The numbers shown are peak retention times in minutes. Panel (a) shows the type A error (peak mixing) where one or more peaks are shifted to an incorrect metabolite row. Panel (b) shows the type B error (metabolite splitting) where the metabolite row is split to create an artificial metabolite in the alignment table. The condition of minimum peaks is often imposed in practice (see Methods), in which case this type of error results in the deletion of the artificial metabolite (nevertheless the original metabolite is affected as it contains one or more missing peaks). If in the example shown on panel (b) it is assumed that the minimum peak cut-off is three peaks, and therefore both metabolite1 and metabolite2 will be deleted from the alignment table as the net result of the splitting error and the removal of spurious metabolite rows.

Comparison of the obtained peak alignment tables with the ideal alignment table allowed counting of errors. In the case of a simple peak mixing error (such as shown in Figure 5(a)) two metabolites are affected, resulting in an error count of two. If three metabolites were affected by peak mixing three errors were counted. In the case of a metabolite splitting error one or more metabolites may be affected.

Figure 6 shows the number of errors in a peak alignment table as a function of gap penalty in the range $G = 0.10$ to 0.55, for the fixed retention time tolerance of $D = 2.5$ s. For small values of G, metabolite splitting errors (type B) predominated, while for large G values peak mixing errors (type A) predominated. The decrease in the total number of metabolites for low values of G (Figure 6, bottom panel) is counterintuitive, since low values of G would favor the insertion of gaps, which in turn would create more metabolite rows in the alignment table. This effect was however more than countered by the removal of spurious metabolites rows in post-processing (see Methods). In addition, some cases of metabolite splitting resulted in a complete removal of the metabolite from the alignment table, due to the effect shown in Figure 5(b).

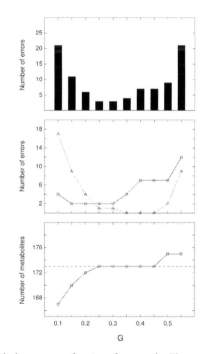

Figure 6. The accuracy of peak alignment as a function of gap penalty. The accuracy of dynamic programming peak alignment as a function of gap penalty G, shown on the x-axis. Eight replicate experiments of L mexicana polar extracts were processed, generating eight peak lists. The peak lists were aligned with the dynamic programming for a range of the gap penalty parameter between 0.10 and 0.55 (the retention time tolerance was fixed D = 2.5 s), and the resulting alignment tables were compared to the correct alignment table built manually. The top panel shows the total number of errors in the alignment. The middle panel shows the number of errors of type A (solid line) and type B (dashed line). The bottom panel shows the total number of metabolites in the resulting alignment. The correct number of metabolites is 173, shown in the dashed line.

A similar picture was observed for a number of errors as a function of the retention time tolerance (Figure 7). Small values of D favor metabolite splitting. This is because for small D any subset of peaks with small but systematically different retention times (due to experimental drift for example) would appear as a different metabolite and would be moved into a different metabolite row. However this effect was countered by the removal of spurious metabolites, and therefore a decrease in the total number of metabolites was observed for low D. Large values of D diminished peak discrimination by retention times resulting in an increase in mixing errors (type A), especially for the parts of the chromatogram which contained compounds with similar mass spectra.

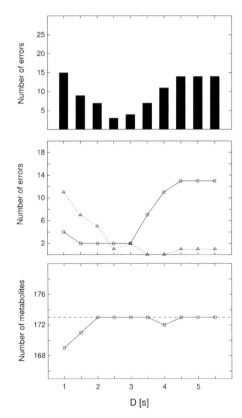

Figure 7. The accuracy of peak alignment as a function of retention time tolerance. The accuracy of dynamic programming peak alignment as a function of retention time tolerance D, shown on the x-axis. Eight replicate analyses of L. mexicana polar extracts were processed, resulting in eight peak lists. The peak lists were aligned with the dynamic programming for a range of retention time tolerances between 1.0 s and 5.5 s (the gap penalty was fixed G = 0.30), and the resulting alignment tables were compared to the correct alignment table built manually. The top panel shows the total number of errors in the alignment. The middle panel shows the number of errors of type A (solid line) and type B (dashed line) as explained in the text. The bottom panel shows the total number of metabolites in the resulting alignment. The correct number of metabolites is 173, shown in the dashed line.

Alignment of Replicate Analyses of Wild Type and Mutant Metabolite Extracts

These analyses were extended to replicate GC-MS chromatograms collected on wild-type and mutant cell states. The input for the alignment consisted of peak lists from 16 experiments (8 independent wild-type and 8 independent mutant extracts; Table 1) with a total of 2,665 signal peaks. Within-state alignment was performed with D_w = 2.5 s and G_w = 0.30, and resulted in 173 unique metabolites for the wild-type cells and 171 unique metabolites for the mutant cells. For the between-states alignment we used the same gap penalty as for within-state alignment (G_b = 0.30), while the retention time penalty was set to D_b = 10.0 s, roughly the value that would allow matching of peaks in the case of large retention time shifts shown in Figure 2. In practice the retention time tolerance can be estimated from the largest retention time shifts between experiments (for data shown in Figure 2 this is 5–7 s). For peaks 10 s apart the peak similarity will equal 0.6 times the similarity in their mass spectra, if D = 10.0 s (Equation [2]).

Table 1. Experiments used in peak alignment

Cell state	Replicates	Aver. No peaks	Metabolites
wild-type, stationary phase	8	167.1 ± 2.3	173
Δgt, stationary phase	8	166.0 ± 4.9	171

A summary of parameters obtained for replicate analyses (8 each) of polar extracts from wild type and mutant (Δgt) parasites, processed using the peak alignment method. Shown is the number of experimental replicates for each cell state, the average number of peak per replicate experiment, and the number of unique metabolites after the within-state peak alignment with D_w = 2.5 s and G_w = 0.30.

The complete alignment of wild-type and glucose transporter mutant experiments resulted in a total of 188 unique metabolites. The correct alignment table involving all 16 experiments was also compiled manually. A total of three errors were found in the within-state alignment of the wild-type experiments, and two errors were found in the within-state alignment of the mutant experiments. A comparison with the correct alignment table showed that incorrect between-state alignment affected a total of ten metabolites. This involved a segment of five misaligned metabolite rows between 10.2 and 10.5 minutes that had a prominent but anomalous feature at m/z = 155 that had created an effect of artificially similar mass spectra. It would be possible to remedy errors of this type with a more stringent retention time tolerance.

Discussion and Conclusion

GC-MS is a robust and sensitive platform for the profiling of certain metabolite classes [1,4,12,13,15]. Due to experimental limitations inherent in all chromatographic separations, slight drifts are observed in GC-MS elution times between experiments [17,18]. These drifts are particularly problematic for non-targeted metabolic profiling studies, which aim to analyze all detectable analytes within multiple experiments [1,12,13,19,22,23]. There are two schools of thought regarding how to address the retention time correction in hyphenated mass spectrometry (Figure 1). Algorithms that rely on time domain data only (Figure 1, branches 1 and 3) ignore highly pertinent information contained in the m/z dimension. Accumulated evidence suggests that effective retention time correction cannot be achieved based on retention time only, regardless of whether the approach involves peak matching or profile alignment [17-19,22,26,28,29].

Profile alignment approaches that use the full chromatogram data matrices are expected to be the most accurate (Figure 1, branch 4). These approaches, however, come at a high cost, both in terms of complexity and computational costs. For example, Bylund and co-authors used correlation optimised warping [26], while Baran and co-authors used dynamic time warping with explicitly specified time shifts [30]. In both cases, an arbitrarily chosen "target" chromatogram was used to align all other chromatograms in a pairwise fashion, with the chromatogram data matrices segmented to facilitate the alignment [27,30]. This raises several difficulties. For example, it is unclear how the choice of the target chromatogram may affect the final alignment. In principle, a more objective alignment of chromatogram data matrices could be achieved by calculating a similarity tree [32,33]. This, however, must involve all possible pairwise alignments, which is likely to be computationally expensive. Furthermore, in the reported examples the chromatogram segmenting was based on strategic or node positions influenced by user chosen parameters [27], or a "representative" set of peaks [30]. This in turn raises the question of how the segmenting method might affect the final alignment. Finally, these approaches handle a large amount of uninformative noise data, since only a small portion of the full chromatogram data matrix is informative signal.

Peak matching algorithms that use both time domain data and mass spectra are particularly promising, and have been applied to both LC-MS [20,21] and more recently to GC-MS data [23]. The input for these algorithms are signal peaks extracted from full chromatogram data matrices. Since the majority of uninformative data is discarded in the peak detection step, these methods operate on a vastly reduced data set while retaining highly selective information contained in the mass spectra. Here we propose an approach that falls into this category, and uses signal peak "objects" which are signals extracted from the chromatogram data matrix. A peak object may be characterized with several attributes including peak retention time (the time taken at the peak apex), peak mass spectrum (the m/z vector taken at the peak apex), experiment/cell state information, a unique peak ID, and so on. The result of an experiment is a list of peak objects, henceforth referred to as a peak list. From this viewpoint the peak matching problem is reduced to the alignment of peaks between multiple peak lists. If we assume that the elution order of peaks is conserved (a reasonable assumption for GC-MS; also an assumption widely used in proteomics based LC-MS [37]), this problem shows resemblance to extensively studied problem of multiple sequence alignment [32,33]. In order to cope with rapidly escalating computational costs of an exact, multidimensional dynamic programming solution, efficient algorithms were developed for multiple sequence alignment [33,35]. We have adapted this approach to the problem of peak alignment in multiple GC-MS experiments.

The method proposed here uses both peak retention times and mass spectra, and relies on dynamic programming to find the optimal solution to the global alignment problem. Any peak matching method that uses both similarity in retention times and mass spectra similarity depends on a balance between the two, and devising an approach that balances this correctly poses a considerable challenge. For example, it is possible to incorporate mass spectra similarity into the total peak similarity score used in progressive hierarchical clustering [22], but it is unclear how to weight the relative contributions of retention time and mass spectra. The problem here is that progressive clustering relies on a single cutoff of the dendrogram tree to delineate peak clusters (essentially, the cutoff is a constant across the entire data set), while the retention times drifts are highly non-linear (Figure 2 and [21]). In the work of Styczynski et al, the peak matching approach used to find conserved metabolites in GC-MS metabolic profiling experiments incorporated both the similarities in mass spectra and peak retention times [23]. However, in the demonstration of this approach the retention time similarity was taken into account only coarsely, with an elution similarity threshold of 1 min [23]. When two metabolites have distinct mass spectra, the retention time information can be neglected altogether and correct peak matching can still be achieved. However, for metabolites that elute in close proximity to one another, and give similar fragmentation patterns, the information provided by retention times is critical for a correct matching. Therefore, we would expect the method of Styczynski and co-authors to have difficulty when metabolites with similar fragmentation patterns elute in close proximity, as demonstrated by their inability to resolve isoleucine and leucine in the test data set [23].

We propose the peak similarity function that incorporates the similarity between peak mass spectra modulated by the similarity in peak retention times (Equation [2]). This function is governed by two parameters: the gap penalty function (G) and the retention time tolerance (D). The first parameter (G) determines how similar mass spectra must be for peaks to be considered to represent the same metabolite; the second parameter (D) is related to expected drifts in retention times between experiments. For example, if retention times in a particular set of experiments are highly reproducible, decreasing the retention time tolerance will enable this information to be leveraged for increased accuracy in the peak alignment.

To achieve the alignment of multiple peak lists, we rely on progressive alignment based on a similarity tree. The similarity tree is calculated from pairwise alignments, and the global alignment is built progressively, starting from the two most similar peak lists, and joining other peak lists in a process guided by the similarity tree [32,33]. When several cell states with multiple replicate experiments per state are analyzed, within each cell state one deals with true experimental replicates, while in experiments performed on different cell states some metabolites may be missing altogether in one state relative to another. To accommodate for this complexity, we first perform a within-state alignment, followed by the between-state alignment built by aligning the within-state alignments. In the case of more than two sets of replicate experiments, a similarity tree is built based on pairwise similarities between fixed within-state alignments, and then the alignments themselves are aligned progressively following the similarity tree. Therefore multiple guide trees are built, one for each set of replicate experiments to facilitate within-state alignment. An additional guide tree may be built to facilitate between-state alignment if more than two states are present.

Several methods for peak matching based on both retention times and mass spectra (Figure 1, branch 2) have been described recently [20,21,23]. Of these only the approach of Styczynski and co-authors was developed on GC-MS data [23], while the software packages MZmine [20] and XCMS [21] were developed on LC-MS data. In our experience the latter software packages tend to overinterpret GC-MS data, assigning a greater number of peaks than expected (probably related to differences in fragmentation patterns between GC-MS and LC-MS). Nevertheless, the alignment algorithms implemented in MZmine and XCMS are relevant and we briefly review them here.

MZmine uses a simple alignment method to build the peak alignment table: one peak at a time is taken and an attempt is made to match it to an existing row of the peak alignment table. If no rows match a new row is created [20]. A secondary peak detection method is used to fill the gaps in the resulting alignment table [20]. We expect this approach to exhibit limitations similar to those observed in hierarchical clustering [22], and discussed above.

XCMS incorporates one of the most advanced peak matching algorithms for metabolite profiling data described to date [21]. In this approach the distribution of peaks along the time domain by using the kernel density estimator is calculated, and regions where many peaks have similar retention times are identified [21]. From this, a fixed time interval that determines each group of peaks is deduced [21]. In practice, the retention time drifts are distributed in a highly irregular fashion, and a fixed time interval is unlikely to be able to capture peak groups correctly. This is suggested by observed peak collisions, where more than one peak from the same experiment is joined into a single group [19]. The same problem was encountered in other peak matching methods [19,22], and requires additional, empirical intervention, such as "collision resolution" [19] or "tie-breaking" in XCMS [21]. The approach we propose inherently prevents peak collisions. Furthermore, it does not rely on any fixed intervals to group peaks, nor does it rely on any assumptions about the distribution of peaks across the samples.

It is of interest to compare the method proposed here to the approach for constructing signal maps, described recently in LC-MS proteomics experiments [37]. Underlying the calculation of signal maps is the method for optimal alignment of chromatogram data matrices, that can be classified as belonging to branch 4 in Figure 1. These authors used the Needleman-Wunsch algorithm to align full chromatogram data matrices obtained from LC-MS proteomics experiments [37]. They also used minimum spanning tree, or progressive merging of pairwise alignments into a consensus run, to produce a global signal map [37]. Apart from the overt difference in that we focus on extracted signals rather than using the full chromatogram data matrices, there are several important yet more subtle differences between the two methods. First, in contrast to the method of [37] the score function proposed here incorporates both similarities in retention times and mass spectra. Second, in constructing signal maps from peptide mass spectra Prakash and co-workers allow for multiple mass spectra in one experiment to correspond to a single mass spectrum in the other experiment [37]. Since we are aligning peak objects rather than the raw signal, our approach inherently allows only one-to-one peak matching, and we consider explicitly the question of gaps (a match-to-nothing). Both of these features are critical for the ability to achieve the alignment of signal peaks in GC-MS. Similarly as in the approach of [37], we rely on the two-dimensional formulation of the global sequence alignment problem. However, during progressive alignment of individual experiments we do not merge individual runs into a "consensus" run, because this has a potential to degrade signal if two unrelated signal peaks are merged. Rather, our approach is based on the generalized solution for the alignment of two alignments, i.e. an N- and M-alignment. This has an additional benefit to be directly extensible to the alignment of pre-computed alignments representing replicates of different cell states (where signal-to-nothing matches may be significant), and allows one to tackle the problem of an arbitrary number of cell states with the same conceptual framework and finite computational resources. We note that the modification proposed for the calculation of mass spectra similarity [37], and results from other proteomics studies [38], could be used in the method proposed here.

It would be useful to compare more directly the performance and accuracy of the approach proposed here to the approaches for peak matching described previously. Unfortunately, this is currently not feasible for two reasons. Firstly, a standard data set which could be used as a benchmark does not exist. Furthermore, it is difficult to create such a data set ad hoc. Most other alignment methods of interest, such as methods implemented in software packages MZmine [20] or XCMS [21], are embedded in the multi-step processing pipelines, and the input to peak matching is generated directly from the output of package-specific peak detection. Secondly, a suitable metric to quantify differences between two methods does not exist. This is particularly problematic, because in even the simplest case the resulting alignment table may involve hundreds of peak entries, and it is not clear how to represent and quantify the differences between two such tables.

The development of a suitable metric to compare two peak matching methods is likely to require a separate research effort. To circumvent this we have performed a detailed analysis of absolute error and compiled error statistics relative to the "correct" alignment table created manually (Figures 6 and 7). The error analysis performed on experimental data consisting of eight replicate GC-MS experiments collected on Leishmania parasites with ~170 peaks per experiment showed that, for near optimal parameters, <5 metabolites were affected by misalignment errors. This approaches the accuracy achieved in a manual alignment, which required tens of man-hours for the tested data set.

The alignment accuracy was not overly sensitive to empirical parameters for a wide range of values (gap penalty and retention time tolerance), suggesting that the method is robust. To address the problem of a benchmark data set, we provide our test data set in the supplementary material. This includes input peak lists, raw chromatogram data matrices, and alignment tables, including those produced for optimal parameters, as well as the "correct" alignment table prepared manually.

The main drawback of the proposed approach is that it operates on a set of signal peaks obtained from peak detection pre-processing. Automated peak detection remains a challenge [30], and any errors introduced during peak detection (such as missing peaks or false peaks) will propagate through to the alignment tables. This is inherent in all peak matching methods, and could be viewed as an advantage as well. Focusing on signal peaks results in a greatly reduced data set which contains the vast majority of interesting signal, and allows one to leverage this information for downstream processing [21], in addition to significantly reducing computational costs required for downstream processing.

Methods

Metabolic Profiling of L. mexicana

L. mexicana wild-type and L. mexicana Δgt promastigotes (lacking three glucose transporters [36] were cultivated in RPMI media, pH 7.4 containing 10 % fetal calf serum (FCS). Parasites were harvested in mid-log phase or stationary phase (24 hr and 96 hr after inoculation, respectively). Prior to harvesting, the culture medium was rapidly chilled to 0°C by immersing the culture flasks in a dry-ice/ethanol bath. The temperature of the medium was monitored with an electronic thermometer and subsequent steps performed at 0°C. Replicate aliquots of the quenched culture medium (containing 4×10^7 cells) were harvested by centrifugation (15,000 g, 30 sec, 0°C) and washed twice with ice-cold phosphate-buffered saline, pH 7.5. The washed cell pellets were suspended in chloroform:methanol:water (containing 1 nmole scyllo-inositol and 1 nmole norleucine internal standards) (1:3:1 v/v/v) and extracted at 60°C for 15 min. Water was added to give a final chloroform:methanol:water ratio of 1:3:3 (v/v) and the aqueous and organic phases separated by centrifugation. The upper aqueous phases were dried in vacuo, suspended in 20 mg/ml methoxyamine in pyridine (20 μl, 16 hr, 25°C) with continuous shaking and then derivatized with MSTFA + 1% TMCS (Pierce; 20 μl, 1 hr, 25°C). Samples (1 μL) were injected onto an Agilent 6890N gas chromatograph interfaced with a 5973 mass selective detector using a 7683 automatic liquid sampler. Gas chromatography was performed on a 30 m DB5-MS column with 0.25 mm inner diameter and 0.25 μm film thickness (J&W Scientific). Injection temperature was 270°C, the interface set at 250°C, and the ion source adjusted to 230°C. The carrier gas was helium (flow rate 1 ml/min). The temperature program was 1 min isothermal heating at 70°C, followed by a 12.5°C/min oven temperature ramp to 295°C, then 25°C/min to 320°C and held for 1 min. Mass spectra were recorded at 3.2 scans/s (m/z 50–500).

Post-Processing and Data Preparation

The total ion-chromatogram was integrated in ChemStation (MSD Chemstation D.01.02.16, Agilent Technologies) by using the default integrator. Resulting peak tables were exported to external files for further processing, together with raw data in ANDI-MS format. Initially the TICs were examined visually for each experiment, and peak lists were edited manually to mark the reference peak and uninformative peaks originating from the derivatizing reagent (TMS). Subsequently the reference peak and uninformative peaks were removed from the peak lists, and peaks in the region 6.5 to 21.0 min of each chromatogram were selected for further processing. Prior to peak alignment the mass spectra at peak apexes were extracted from the ANDI-MS files, and signals at m/z 73 and 147 (largely due to the derivatizing reagent) were suppressed for all peaks. To remove spurious metabolites such as those that arise as artefacts of empirical peak integration, all metabolite rows with less than four peaks were discarded after the alignment (i.e. one half of the original sample size of eight replicate experiments) [22].

Authors' Contributions

MDR was involved in the development of the method and development of the test implementation. DPDS was involved in the development of GC-MS experiments and error analysis. WWK was involved in the development of the test implementation, ECS was involved in GC-MS experiments, MJM was involved in the development of the method and GC-MS experiments, TPS was involved in the development of the method. VAL originally conceived the project and was leading development of the method, development of the test implementation, and error analysis. VAL drafted the manuscript; all authors contributed to the final version. All authors have read and approved the final manuscript.

Acknowledgements

This work was supported by the Bio21 Molecular Science and Biotechnology Institute, University of Melbourne. MJM was supported by the Australian National Health and Medical Research Council (NHMRC) Programme grant and Principal Research Fellowship. DDS was supported by NHMRC "Dora Lush" Biomedical scholarship (Id. No. 359427).

References

1. Fiehn O., Kopka J., Dormann P.., Altmann T, Trethewey RN., Willmitzer L: Metabolite profiling for plant functional genomics. Nat Biotechnol 2000, 18:1157–1161.

2. Allen J., Davey HM., Broadhurst D., Heald JK., Rowland JJ., Oliver SG., Kell DB: High-throughput classification of yeast mutants for functional genomics using metabolic footprinting. Nat Biotechnol 2003, 21:692–696.

3. Sweetlove LJ., Last RL., Fernie AR: Predictive Metabolic Engineering: A Goal for Systems Biology. Plant Physiology 2003, 132:420–425.

4. Fernie AR., Trethewey RN., Krotzky AJ., Willmitzer L: Metabolite profiling: from diagnostics to systems biology. Nat Rev Mol Cell Biol 2004, 5:763–769.

5. Koek MM., Muilwijk B., van der Werf MJ., Hankemeier T: Microbial me-tabolomics with gas chromatography/mass spectrometry. Anal Chem 2006, 78:1272–1281.

6. Sabatine MS., Liu E., Morrow DA., Heller E., McCarroll R., Wiegand R., Berriz GF., Roth FP., Gerszten RE: Metabolomic identification of novel bio-markers of myocardial ischemia. Circulation 2005, 112:3868–3875.

7. Robertson DG: Metabonomics in toxicology: a review. Toxicol Sci 2005, 85:809–822.

8. Keun HC: Metabonomic modeling of drug toxicity. Pharmacol Ther 2006, 109:92–106.

9. Gibney MJ., Walsh M., Brennan L., Roche HM., German B., van Ommen B: Metabolomics in human nutrition: opportunities and challenges. Am J Clin Nutr 2005, 82:497–503.

10. German JB., Watkins SM., Fay LB: Metabolomics in practice: emerging knowl-edge to guide future dietetic advice toward individualized health. J Am Diet Assoc 2005, 105:1425–1432.

11. Nikiforova VJ., Gakiere B., Kempa S., Adamik M., Willmitzer L., Hesse H., Hoefgen R: Towards dissecting nutrient metabolism in plants: a systems biol-ogy case study on sulphur metabolism. J Exp Bot 2004, 55:1861–1870.

12. Roessner U., Luedemann A., Brust D., Fiehn O., Linke T., Willmitzer L., Fernie A: Metabolic profiling allows comprehensive phenotyping of genetically or environmentally modified plant systems. Plant Cell 2001, 13:11–29.

13. Urbanczyk-Wochniak E., Luedemann A., Kopka J., Selbig J., Roessner-Tunali U., Willmitzer L., Fernie AR: Parallel analysis of transcript and metabolic pro-files: a new approach in systems biology. EMBO Rep 2003, 4:989–993.

14. Tolstikov VV., Lommen A., Nakanishi K., Tanaka N., Fiehn O: Monolithic silica-based capillary reversed-phase liquid chromatography/electrospray mass spectrometry for plant metabolomics. Anal Chem 2003, 75:6737–6740.

15. Halket JM., Waterman D., Przyborowska AM., Patel RK., Fraser PD., Bramley PM: Chemical derivatization and mass spectral libraries in metabolic profiling by GC/MS and LC/MS/MS. J Exp Bot 2005, 56:219–243.

16. Stein SE: An Integrated Method for Spectrum Extraction and Compound Iden-tification from Gas Chromatography/Mass Spectrometry Data. J Am Soc Mass Spectrom 1999, 10:770–781.

17. Malmquist G., Danielsson R: Alignment of chromatographic profiles for prin-cipal component analysis: a prerequisite for fingerprinting methods. Journal of Chromatography A 1994, 687:71–88.

18. Johnson KJ., Wright BW., Jarman KH., Synovec RE: High-speed peak matching algorithm for retention time alignment of gas chromatographic data for chemometric analysis. J Chromatogr A 2003, 996:141–155.

19. Duran AL., Yang J., Wang L., Sumner LW: Metabolomics spectral formatting, alignment and conversion tools (MSFACTs). Bioinformatics 2003, 19:2283–2293.

20. Katajamaa M., Oresic M: Processing methods for differential analysis of LC/MS profile data. BMC Bioinformatics 2005, 6:179.

21. Smith CA., Want EJ., O'Maille G., Abagyan R., Siuzdak G: XCMS: processing mass spectrometry data for metabolite profiling using nonlinear peak alignment, matching, and identification. Anal Chem 2006, 78:779–787.

22. De Souza DP., Saunders EC., McConville MJ., Likic VA: Progressive peak clustering in GC-MS Metabolomic experiments applied to Leishmania parasites. Bioinformatics 2006, 22:1391–1396.

23. Styczynski MP., Moxley JF., Tong LV., Walther JL., Jensen KL., Stephanopoulos GN: Systematic identification of conserved metabolites in GC/MS data for metabolomics and biomarker discovery. Anal Chem 2007, 79:966–973.

24. Frenzel T., Miller A., Engel KH: A methodology for automated comparative analysis of metabolite profiling data. European Food Research and Technology 2004, 216:1438–2377.

25. Broeckling CD., Reddy IR., Duran AL., Zhao X., Sumner LW: MET-IDEA: data extraction tool for mass spectrometry-based metabolomics. Anal Chem 2006, 78:4334–4341.

26. Nielsen NPV., Carstensen JM., Smedsgaard J: Aligning of single and multiple wavelength chromatographic profiles for chemometric data analysis using correlation optimised warping. Journal of Chromatography A 1998, 805:17–35.

27. Bylund D., Danielsson R., Malmquist G., Markides KE: Chromatographic alignment by warping and dynamic programming as a pre-processing tool for PARAFAC modelling of liquid chromatography-mass spectrometry data. J Chromatogr A 2002, 961:237–244.

28. Jonsson P., Gullberg J., Nordstrom A., Kusano M, Kowalczyk M., Sjostrom M., Moritz T: A strategy for identifying differences in large series of metabolomic samples analyzed by GC/MS. Anal Chem 2004, 76:1738–1745.

29. Eilers PH: Parametric time warping. Anal Chem 2004, 76:404–411.

30. Baran R., Kochi H., Saito N., Suematsu M., Soga T., Nishioka T., Robert M., Tomita M: MathDAMP: a package for differential analysis of metabolite profiles. BMC Bioinformatics 2006, 7:530.

31. Tikunov Y., Lommen A., de Vos CH., Verhoeven HA., Bino RJ., Hall RD., Bovy AG: A novel approach for nontargeted data analysis for metabolomics. Large-scale profiling of tomato fruit volatiles. Plant Physiol 2005, 139: 1125–1137.

32. Feng DF., Doolittle RF: Progressive sequence alignment as a prerequisite to correct phylogenetic trees. J Mol Evol 1987, 25:351–360.

33. Thompson JD., Higgins DG., Gibson TJ: CLUSTAL W: improving the sensitivity of progressive multiple sequence alignment through sequence weighting, position-specific gap penalties and weight matrix choice. Nucleic Acids Res 1994, 22:4673–4680.

34. Needleman SB., Wunsch CD: A general method applicable to the search for similarities in the amino acid sequence of two proteins. J Mol Biol 1970, 48:443–453.

35. Lipman DJ., Altschul SF., Kececioglu JD: A tool for multiple sequence alignment. Proc Natl Acad Sci U S A 1989, 86:4412–4415.

36. Burchmore RJ., Rodriguez-Contreras D., McBride K., Merkel P., Barrett MP., Modi G., Sacks D., Landfear SM: Genetic characterization of glucose transporter function in Leishmania mexicana. Proc Natl Acad Sci U S A 2003, 100:3901–3906.

37. Prakash A., Mallick P., Whiteaker J., Zhang H., Paulovich A., Flory M., Lee H., Aebersold R., Schwikowski B: Signal maps for mass spectrometry-based comparative proteomics. Mol Cell Proteomics 2006, 5:423–432.

38. Wolski WE., Lalowski M., Martus P., Herwig R., Giavalisco P., Gobom J., Sickmann A., Lehrach H., Reinert K: Transformation and other factors of the peptide mass spectrometry pairwise peak-list comparison process. BMC Bioinformatics 2005, 6:285.

39. Reiner E., Abbey LE., Moran TF., Papamichalis P., Schafer RW: Characterization of normal human cells by pyrolysis gas chromatography mass spectrometry. Biomed Mass Spectrom 1979, 6:491–498.

CITATION

Originally published under the Creative Commons Attribution License or equivalent. Robinson MD, De Souza DP, Keen WW, Saunders EC, McConville MJ, Speed TP, Likić VA. A dynamic programming approach for the alignment of signal peaks in multiple gas chromatography-mass spectrometry experiments. BMC Bioinformatics 2007, 8:419. doi:10.1186/1471-2105-8-419.

Solid-Phase Extraction and Reverse-Phase HPLC: Application to Study the Urinary Excretion Pattern of Benzophenone-3 and its Metabolite 2,4-Dihydroxybenzophenone in Human Urine

Helena Gonzalez, Carl-Eric Jacobson, Ann-Marie Wennberg, Olle Larkö and Anne Farbrot

ABSTRACT

Background

Benzophenone-3 (BZ-3) is a common ultraviolet (UV) absorbing compound in sunscreens. It is the most bioavailable species of all UV-absorbing compounds after topical application and can be found in plasma and urine.

Objectives

The aim of this study was to develop a reverse-phase high performance liquid chromatography (HPLC) method for determining the amounts BZ-3 and its metabolite 2,4-dihydroxybenzophenone (DHB) in human urine. The method had to be suitable for handling a large number of samples. It also had to be rapid and simple, but still sensitive, accurate and reproducible. The assay was applied to study the urinary excretion pattern after repeated whole-body applications of a commercial sunscreen, containing 4% BZ-3, to 25 healthy volunteers.

Methods

Each sample was analyzed with regard to both conjugated/non-conjugated BZ-3 and conjugated/non-conjugated DHB, since both BZ-3 and DHB are extensively conjugated in the body. Solid-phase extraction (SPE) with C8 columns was followed by reverse-phase HPLC. For separation a Genesis C18 column was used with an acethonitrile-water mobile phase and the UV-detector was set at 287 nm.

Results

The assay was linear r2 0.99, with detection limits for BZ-3 and DHB of 0.01 µmol L–1 and 0.16 µmol L–1 respectively. Relative standard deviation (RSD) was less than 10% for BZ-3 and less than 13% for DHB. The excretion pattern varied among the human volunteers; we discerned different patterns among the individuals.

Conclusions

The reverse-phase HPLC assay and extraction procedures developed are suitable for use when a large number of samples need to be analyzed and the method fulfilled our objectives. The differences in excretion pattern may be due to differences in enzyme activity but further studies, especially about genetic polymorphism, need to be performed to verify this finding.

Keywords: benzophenone-3, 2,4-dihydroxybenzophenone, HPLC, reverse-phase HPLC, oxybenzone, sunscreen

Introduction

Ultraviolet (UV) radiation from the sun has adverse effects on human skin; it causes photoaging, sunburn and most seriously, skin cancer. The protective ozone layer has been subject to destruction and the areas where the ozone layer is thin have higher incidence of skin cancer.

Malignant melanoma is one of the most rapidly increasing types of cancer in Sweden and non-melanoma skin cancer is also increasing. UV-radiation from the sun is the most important etiological factor.

Sunscreens are widely used to protect us against harmful radiation. Benzophenone-3 (BZ-3) has been a commonly used filter in sunscreens for the last few decades; it has both UVA and UVB protecting properties. It is an organic chemical absorber with a molecule weight (MW) of 228.25 and CAS number 131-57-7. The structure is shown in Figure 1a (ChemIDplus). The ultimate sunscreen stays on the skin and its protective mechanism functions there. However, BZ-3 is relatively lipophilic (log P 3.64 ± 0.37) and several reports have shown that BZ-3 penetrates the skin and is excreted in urine (Hayden, Roberts et al. 1997; Gustavsson Gonzalez, Farbrot et al. 2002; Janjua, Mogensen et al. 2004; Sarveiya, Risk et al. 2004; Gonzalez, Farbrot et al. 2006). Urine is the major excretion pathway and 2,4-dihydroxybenzophenone (DHB), one of the major metabolites in rats (Okereke, Kadry et al. 1993). DHB has a MW of 214.22 and CAS number 131-56-6. The structure is shown in Figure 1b. BZ-3 is the most bioavailable compound of all UV-absorbing chemicals, (Nash, 2006) and is extensively conjugated in the human body.

We wanted to develop an assay to measure the amounts of BZ-3 and DHB in human urine. The method had to suit our needs. For example, it had to be possible to handle a large number of samples easily; hence it had to be rapid and simple but still sensitive, accurate and reproducible. Several methods using high performance liquid chromatography (HPLC) and BZ-3 have been described (Abdel-Nabi, Kadry et al. 1992; Jiang, Hayden et al. 1996; Vanquerp, Rodriguez et al. 1999; Chisvert, Pascual-Marti et al. 2001) and this work has been developed to some extent on the basis of previously described method (Abdel-Nabi, Kadry et al. 1992). However, there are few methods for extraction of BZ-3 and DHB in human urine after it has been metabolized by the human body (Sarveiya, Risk et al. 2004). Abdel-Nabi et al. used urine from rats (Abdel-Nabi, Kadry et al. 1992). Several other methods are designed for product evaluation and not for extraction from biological media. None of the methods fitted our needs completely and for that reason we developed this method. More than 1000 urine samples were collected and each was analyzed for conjugated/non-conjugated BZ-3 and conjugated/non-conjugated DHB. More than 4000 analyses were performed.

The method we developed was used to study the excretion pattern of BZ-3 and DHB in urine after repeated topical whole-body applications of a sunscreen containing BZ-3 to 25 human volunteers.

Experimental

Reagents

BZ-3 (2-hydroxy-4-methoxybenzophenone), DHB (2,4-dihydroxybenzophenone) and benzophenone (BZ), all purity 99%, (Aldrich Chem).

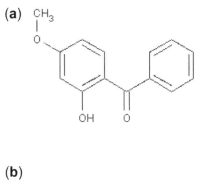

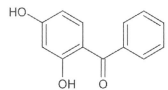

Figure 1. Chemical structure of BZ-3 (a) and DHB (b).

Methanol (HPLC-grade), acethonitrile (HPLC-grade) and trifluoroacetic acid (TFA) (Merck).

β-glucurunidase/arylsulfatase obtained from Helix pomata (Boehringer Mannheim).

Chromatography

A gradient HPLC-system with a pump PU-1580 (Jasco), a 50 μl loop injector and a SPD-10A VP uv-vis detector (Schimadzu) were used. Chromatographic separation was achieved on a Genesis C18 ID (4.6 mm × 150 mm) column. A Genesis C18 (20 mm × 4.0 mm) was used as the precolumn.

The mobile phase was acethonitrile:dionized ultra filtered water (UF) (0.45 μm) 44:66 (v/v) with 1 mL TFA to 10000 mL mobile phase with gradient elution according to Table 1. UV absorption was done at 287 nm. The run time was 31.5 min.

Standard Solutions

One stock solution containing 4.0 mmol L^{-1} BZ-3 and one stock solution containing 3.0 mmol L^{-1} DHB in 70:30 (v/v) methanol: UF water were used. Working standard solutions were prepared in methanol at five concentrations, 0.04 mmol L^{-1} BZ-3, 0.03 mmol L^{-1} DHB, 0.4 mmol L^{-1} BZ-3, 0.3 mmol L^{-1} DHB, 4 mmol L^{-1} BZ-3, 3 mmol L^{-1} DHB, 40 mmol L^{-1} BZ-3, 30 mmol L^{-1} DHB and 100 mmol L^{-1} BZ-3, 75 mmol L^{-1} DHB. All solutions were prepared in volumetric flasks, class A and kept refrigerated at temperature between 2 and 8 °C. Fresh standards were prepared every month.

Table 1. Gradient program

	Gradient program				
Time (min)	0.1	16	18	30	31.5
Flow (ml/min)	1.0	1.0	1.9	1.9	1.0

Sample Pre-Treatment

Pre-Treatment Before Extracting Conjugated Samples

The urine samples were centrifuged at 2100 g for 10 minutes. 1 mL urine was used. 100 μL glucuronidase/arylsulphatase and 25 μL internal standard were added. The sample was incubated at 37 °C 16 hours. The sample was diluted with 2 mL phosphate buffer (50 mmol L–1, pH 6.5). Most samples were diluted 1:20 with 0.9% NaCl-solution.

Pre-Treatment for Non-Conjugated Samples

The urine samples were centrifuged at 2100 g for 10 minutes. 1 mL was used. 25 μL internal standard was added. The sample was diluted with 2 mL phosphate buffer.

Solid-Phase Extraction

Extraction was performed with a vacuum manifold and solid-phase extraction (SPE) columns C8, 100 mg, 6 mL (Isolute Inc., purchased from Sorbent AB).

1. The column was activated with 2 mL methanol. The vacuum was turned off when the methanol reached the top of the sorbent, to prevent the column from drying. The procedure was repeated once.

2. The column was rinsed with 2 mL phosphate buffer. The vacuum was turned off when the buffer reached the top of the sorbent bed, to prevent column from drying. The procedure was repeated twice.

3. The sample was transferred to the column with a pasteur pipette and the sample was then drawn slowly through the column. The column was dried under full vacuum. The procedure was repeated once.

4. The column was rinsed twice with 1 mL phosphate buffer. After each rinse it was dried under full vacuum.

5. The sample was eluted four times with methanol: TFA 99:1 (v/v) 0,25 mL. After each eluation it was dried under full vacuum.

6. The samples were transferred to glass vials.

Laboratory-Made Extraction Controls

One healthy volunteer applied 2 mg cm^{-2} of a sunscreen containing 4% BZ-3, a total of 33 g of sunscreen containing 1.32 g of BZ-3. All urine was collected for 24 hours after the application. The urine was diluted with urine from four healthy volunteers into two different concentrations, 7.0 µmol L^{-1} and 45 µmol L^{-1} for BZ-3 and 0.86 µmol L^{-1} and 6.3 µmol L^{-1} for DHB.

Internal Standard

BZ was used as internal standard at the concentration 5 mmol L^{-1} (0.091 g). BZ was diluted in 100 mL methanol. Volumetric flasks, class A were used. The internal standard was prepared fresh every month and stored in a refrigerator at a temperature between 2 and 8 °C.

Urine Collection and Urine Samples

25 volunteers (16 women and 9 men; mean age 27 years, range 22–42) participated in the study. Height and weight were measured. Their body surface area (BSA) was calculated with the DuBois formula: BSA = 0.007184 × [height (cm)] 0.725 × [weight (kg)] 0.425). The sunscreen used was a commercially available sunscreen, SPF 14 containing 4% BZ-3. Before the first application of sunscreen each volunteer gave a urine sample to confirm that no BZ-3 was found in the

urine prior to this investigation. Each volunteer received 2 mg cm^{-2} according to his or her BSA. The sunscreen was distributed in plastic containers, one for each application. The amount of sunscreen per application varied between the participants from 26 g to 47 g. The total amount of BZ-3 varied between 10.4 g to 18.8 g. BZ-3 was measured in the urine. One volunteer was excluded because the written instructions were not followed accurately.

The volunteers were instructed to apply the sunscreen evenly over the entire body, with the exception of the scalp and genital area, morning and night for 5 days, a total of 10 times. They were allowed one shower/day before the second application. During the five days the sunscreen was applied, all urine was collected, the volume measured and 10 mL from each sample saved and stored at –70 °C. Hence, each volunteer produced a different n umber of urine samples. They collected the urine for 5 days. After the last application they continued to collect urine for another 5 days, making a total of 10 days. The time of day, number and volume for each urine sample were recorded. A total of 1234 urine samples were collected and analyzed.

Results

Minimum Detectable Limits

The minimum detectable limit was defined as three times the baseline noise level. The detection limits for BZ-3 and DHB were 0.01 μmol L^{-1} (0.1 ng per 0.05 mL sample) and 0.16 μmol L^{-1} (2 ng per 0.05 mL sample), respectively.

Chromatography and Selectivity

Figure 2 shows chromatograms of the HPLC-separation of BZ-3, DHB and internal standard.

It was more difficult to achieve separation for DHB since there was more interference in the beginning of the chromatogram.

Calibration

Calibration was done against an external standard at five different concentrations 0.04, 0.4, 4, 40 and 100 μmol L^{-1} BZ-3 and 0.03, 0.3, 3, 30, 75 μmol L^{-1} DHB in a solution made of methanol and UF water.

A calibration curve was made for each sample series, which also consisted of extraction controls at low and high concentrations before each of the 9 urine samples. Each sample series consisted of 60 samples.

Precision and Linearity

Within-day precision showed relative standard deviations (RSD) 10% and 11% for the low concentrations of BZ-3 and DHB and 7.2% and 6.6% for the high concentration of BZ-3 and DHB, respectively. Between-days precision showed RSD 10% and 13% for the low concentrations of BZ-3 and DHB and 8.0% and 8.7% for the high concentration of BZ-3 and DHB respectively. There was no difference regarding precision between BZ-3 and DHB, except for the lowest concentration of DHB, which exhibited greater variation Time (min) owing to interferences in the beginning of the chromatogram (Fig. 2).

BZ-3, DHB and internal standard showed excellent linearity with correlation coeffi cient (r2) 0.99 for all concentrations.

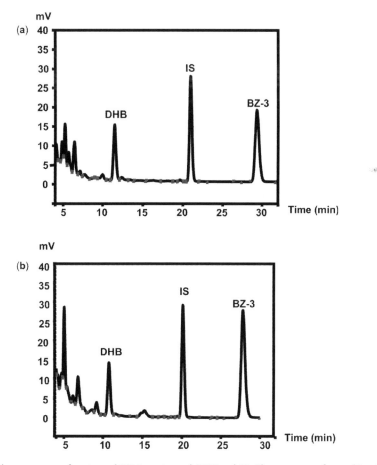

Figure 2. Chromatograms of conjugated BZ-3, conjugated DHB and IS. Chromatograms from subject 8 (a) and high concentration of laboratory-made extraction controls (b).

Excretion Pattern

BZ-3 and DHB were extensively conjugated and only a small proportion was excreted in the non-conjugated form, mean values 5.9% and 8.8% respectively (Fig. 3).

The excretion pattern varied among the individuals. We discerned two patterns and three groups. Nine of the volunteers showed a pattern of rapid excretion of conjugated BZ-3 on the days when the sunscreen was being applied, followed by evenly decreasing excretion on the 5 consecutive days when the sunscreen was not being applied. This is exemplified in Figure 4a. Seven of the volunteers showed a slow and even increase of conjugated BZ-3 during the days the sunscreen was being applied and a slow and even decrease during the days the sunscreen was not applied, resembling a Gaussian curve. This is exemplified in Figure 4b.

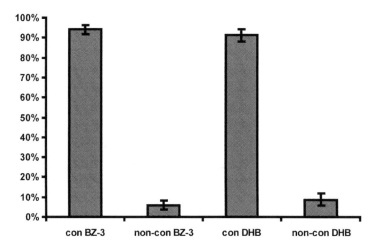

Figure 3. The relationships between conjugated/non-conjugated BZ-3 and conjugated/non-conjugated DHB. Both BZ-3 and DHB are extensively conjugated in the human body.

Eight of the volunteers did not fit into any of the patterns above (Fig. 4c), forming a third group.

The excretion of conjugated DHB followed the excretion of conjugated BZ-3 regardless of pattern.

Discussion

A total of 1234 samples were collected and each sample was analyzed for conjugated/non-conjugated BZ-3 and conjugated/non-conjugated DHB. More than

4000 samples were analyzed and we found this method very suitable for handling large numbers of samples.

Methanol/UF water was used instead of urine for the standard solutions because in high concentrations, the added BZ-3 was precipitated in the urine. For this reason we also included the laboratory-made extraction controls that where extracted with the SPE-columns in the same procedure as the other urine samples. They were also used to control the within-day and between-days variations, which were found to be sufficiently low for the method to be considered stable.

As internal standard BZ was used. It is a closely related structural analogue of BZ-3. The use of an internal standard provides a more correct quantitation. The detection limits were at the same levels as in other methods, (Abdel-Nabi, Kadry et al. 1992; Sarveiya, Risk et al. 2004; Schakel, Kalsbeek et al. 2004) and they were sufficient to analyze the samples on day 10.

Previous studies in rats have shown that BZ-3 is extensively conjugated and excreted in the urine; although a small part is metabolized by cytochromes P-450 to DHB. DHB also undergoes extensive conjugation. A minor route of elimination of BZ-3 is faeces, which were not included in the present study. The metabolite 2,2′-dihydroxy-4-methoxy-benzophenone (DHMB) was not investigated since it was detected only in trace amounts in the urine, but it was the major metabolite found in faeces (Okereke, Kadry et al. 1993).

2,3,4-trihydroxybenzophenone (THB) is also a metabolite of BZ-3. However, Sarvieya et al. have shown that THB is only found in trace amounts in human urine (Sarveiya, Risk et al. 2004). In our experimental set-up THB had unstable properties and it was therefore not analyzed.

Reverse-phase HPLC is a reliable method for detection and quantitation. Compared to other methods such as gas chromatography and mass spectronomy, reverse-phase HPLC is in general more available, most laboratories have access to this method. Many endogenous substrates are also metabolized by conjugation, for example bilirubin and ethinylestradiol (Burchell, Brierley et al. 1995). In 1949 the grey baby syndrome was reported, caused by chloramphenicol toxicity in neonates due to their immature ability to conjugate (Sutherland, 1959). Strasbourg et al. have shown that for children between 13 and 24 months, the hepatic glucuronidation activity was lower for a number of drugs (Strassburg, Strassburg et al. 2002).

Pharmacogenetics is the study of how genes influence an individual's response to drugs. There are several examples of when pharmacogenetics is of importance. In dermatology, hydroxychloroquine is used to treat severe cases of polymorphic light eruption and discoid lupus erythematosus. Inherited deficiency of glucose-6-phosphate dehydrogenase should be excluded before treatment with hydroxychloroquine is started.

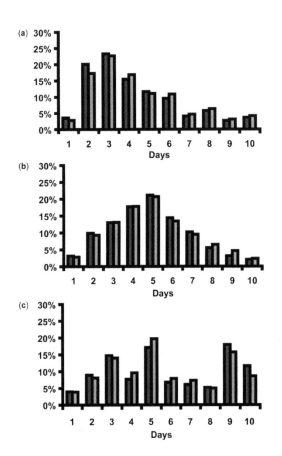

Figure 4. Excretion patterns of conjugated BZ-3 and conjugated DHB during the 10 days urine was collected. Regardless of pattern DHB showed a similar curve to BZ-3. Pattern 1 (9/24) shows rapid excretion on the first days and then evenly decreasing excretion (a). Pattern 2 (7/24) shows a slow and even increase followed by a decrease resembling a Gaussian curve (b). The third group did not fit into patterns 1 or 2 (8/24) (c).

There are also large inter-individual differences in the capacity to conjugate, for example 10% of the North American population is homozygous for the UGT1A1*28 allele. FDA has guidelines about the anti-cancer drug Irinotecan, stating that reduction in the starting dose should be considered for patients known to be homozygous for the UGT1A1*28 allele (FDA, 2005).

This could be one explanation why there are so pronounced differences in the excretion pattern between the subjects.

In the future the genotype might play a role in determining when to decide what sunscreen is best suited for an individual. If the person's ability to conjugate is low, a sunscreen containing BZ-3 may not be the sunscreen of choice. There have been speculations about the estrogenic effects of some sunscreens, including BZ-3. It might not have any potent effect by itself but if a person is medicated with estrogens there might be a more potent effect from an endocrine active sunscreen in some individuals due to saturation of the conjugation enzymes. It seems that sunscreens containing BZ-3 are being phased out. For example, no sunscreens sold at the Swedish pharmacies contain BZ-3.

Future studies about genotype and sunscreens are needed to evaluate the importance of our findings.

Acknowledgements

We thank Göran Oresten, Gunnar Stenhagen and Mats Ohlson for their helpful analytical advice.

References

1. Abdel-Nabi, I.M., Kadry, A.M. et al. 1992. Development and validation of a high-performance liquid chromatographic method for the determination of benzopheno ne-3 in rats. J. Appl. Toxicol., 12(4):255–9.

2. Burchell, B., Brierley, C.H. et al. 1995. Specificity of human UDP-glucurono-syltransferases and xenobiotic glucuronidation. Life Sci., 57(20):1819–31.

3. ChemIDplus http://chem.sis.nlm.nih.gov/chemidplus/.

4. Chisvert, A., Pascual-Marti, M.C. et al. 2001. Determination of the UV filters worldwide authorised in sunscreens by high-performance liquid chromatography. Use of cyclodextrins as mobile phase modifi er. J. Chromatogr. A, 921(2):207–15.

5. FDA, 2005. http://www.fda.gov/cder/foi/label/2005/020571s024,027,028lbl.pdf.

6. Gonzalez, H., Farbrot, A. et al. 2006. Percutaneous absorption of the sunscreen benzophenone-3 after repeated whole-body applications, with and without ultraviolet irradiation. Br. J. Dermatol., 154(2):337–40.

7. Gustavsson Gonzalez, H., Farbrot, A. et al. 2002. Percutaneous absorption of benzophenone-3, a common component of topical sunscreens. Clin. Exp. Dermatol., 27(8):691–4. Analytical Chemistry Insights 2008:3

8. Hayden, C.G., Roberts, M.S. et al. 1997. Systemic absorption of sunscreen after topical application. Lancet, 350(9081):863–4.

9. Janjua, N.R., Mogensen, B. et al. 2004. Systemic absorption of the sunscreens benzophenone-3, octyl-methoxycinnamate and 3-(4-methylbenzylidene) camphor

after whole-body topical application and reproductive hormone levels in humans. J. Invest. Dermatol., 123(1):57–61.

10. Jiang, R., Hayden, C.G. et al. 1996. High-performance liquid chromatographic assay for common sunscreening agents in cosmetic products, bovine serum albumin solution and human plasma. J. Chromatogr B. Biomed. Appl., 682(1):137–45.

11. Nash, J.F. 2006. Human safety and effi cacy of ultraviolet fi lters and sunscreen products. Dermatol. Clin., 24(1):35–51.

12. Okereke, C.S., Kadry, A.M. et al. 1993. Metabolism of benzophenone-3 in rats. Drug. Metab. Dispos., 21(5):788–91.

13. Sarveiya, V., Risk, S. et al. 2004. Liquid chromatographic assay for common sunscreen agents: application to in vivo assessment of skin penetration and systemic absorption in human volunteers. J. Chromatogr. B. Analyt. Technol. Biomed. Life Sci., 803(2):225–31.

14. Schakel, D.J., Kalsbeek, D. et al. 2004. Determination of sixteen UVfilters in suncare formulations by high-performance liquid chromatography. J. Chromatogr. A., 1049(1–2):127–30.

15. Strassburg, C.P., Strassburg, A. et al. 2002. Developmental aspects of human hepatic drug glucuronidation in young children and adultsGut 50(2):259–65.

16. Sutherland, J.M. 1959. Fatal cardiovascular collapse of infants receiving large amounts of chloramphenicol. AMA J. Dis. Child, 97(6):761–7.

17. Vanquerp, V., Rodriguez, C. et al. 1999. High-performance liquid chromatographic method for the comparison of the photostability of five sunscreen agents. J. Chromatogr. A., 832(1-2):273–7.

CITATION

A Simple and Selective Spectrophotometric Method for the Determination of Trace Gold in Real, Environmental, Biological, Geological and Soil Samples Using Bis(Salicylaldehyde) Orthophenylenediamine

Rubina Soomro, M. Jamaluddin Ahmed,
Najma Memon and Humaira Khan

ABSTRACT

A simple high sensitive, selective, and rapid spectrophotometric method for the determination of trace gold based on the rapid reaction of gold(III) with *bis(salicylaldehyde)orthophenylenediamine (BSOPD) in aqueous and micellar media has been developed. BSOPD reacts with gold(III) in slightly acidic solution to form a 1:1 brownish-yellow complex, which has an maximum absorption peak at 490 nm in both aqueous and micellar media. The most remarkable point of this method is that the molar absorptivities of the gold-BSOPD complex form in the presence of the nonionic TritonX-100 surfactant are almost a 10 times higher than the value observed in the aqueous solution, resulting in an increase in the sensitivity and selectivity of the method. The apparent molar absorptivities were found to be 2.3×10^4 L mol^{-1} cm^{-1} and 2.5×10^5 L mol^{-1} cm^{-1} in aqueous and micellar media, respectively. The reaction is instantaneous and the maximum absorbance was obtained after 10 min at 490 nm and remains constant for over 24 h at room temperature. The linear calibration graphs were obtained for 0.1 –30 mg L^{-1} and 0.01 –30 mg L^{-1} of gold(III) in aqueous and surfactant media, respectively. The interference from over 50 cations, anions and complexing agents has been studied at 1 mg L^{-1} of Au(III); most metal ions can be tolerated in considerable amounts in aqueous micellar solutions. The Sandell's sensitivity, the limit of detection and relative standard deviation (n = 9) were found to be 5 ng cm^{-2}, 1 ng mL^{-1} and 2%, respectively in aqueous micellar solutions. Its sensitivity and selectivity are remarkably higher than that of other reagents in the literature. The proposed method was successfully used in the determination of gold in several standard reference materials (alloys and steels), environmental water samples (potable and polluted), and biological samples (blood and urine), geological, soil and complex synthetic mixtures. The results obtained agree well with those samples analyzed by atomic absorption spectrophotometry (AAS).*

Keywords: spectrophotometry, gold determination, BSOPD, aqueous and micellar media, environmental, biological and soil samples

Introduction

The beauty and rarity of gold has led to its use in jewellary and coinage, and like a standard for monetary stems throughout the world. Gold is also one of most important noble metals due to its wide application in industry and economic activity. It has been used in medicine for quite some time.

A simple sensitive and selective method for determination of trace gold has been required. Sophisticated techniques, such as inductively coupled plasma mass spectrometry (ICP-MS) (Juvonen et al. 2002), inductively coupled plasma atomic emission spectrometry (ICP-AES) (Wu et al. 2004), electrochemical (Kavanoz et al. 2004), spectrophotometry (Zaijun et al. 2003), neutran activation analysis (Nat et al. 2004) and atomic absorption spectrophotometry (AAS) (Medved et al. 2004) have widely been applied to the determination of gold in various samples. Some factors such as initial cost of instrument, technical know-how, consumable and costly maintenance of technique restrict the wider applicability of these techniques, particularly in laboratories with limited budget in developing countries and for field work. A wide variety of spectrophotometric methods for determination of gold have been reported, each chromogenic system has its advantages and disadvantages with respect to sensitivity, selectivity and convenience (Alfonso and Gomez Ariza, 1981; Balcerzak et al. 2006; Chen et al. 2006; El-Zawawy et al. 1995; Fujita et al. 1999; Gangadharappa et al. 2004; Gao et al. 2005; Gowda et al. 1984; Hu et al. 2006; Koh et al. 1986; Matouskova et al. 1980; Melwanki et al. 2002; Mirza 1980; Ortuno et al. 1984; Pal 1999; Patell and Lieser, 1986; Zhao et al. 2006). A comparison of few selected procedures; their spectral characteristics and draw backs are enumerated in Table 1. The shiff-base reagents had widely been applied for the determination of noble metal ions, this type of reagent has higher sensitivity and high selectivity (Ahmed and Uddin, 2007). In the search for more sensitive shiff-base reagent, in this work, a new reagent bis(salicylaldehyde) orthophenylenediami-ne(BSOPD) was synthesized according to the method of (Salam and Chowdhury, 1997) and a color reaction of BSOPD with Au(III) in aqueous and micellar media was carefully studied.

The aim of present study is to develop a simpler direct spectrophotometric method for the trace determination of gold with BSOPD in aqueous solutions, and the presence of inexpensive nonionic micelles, such as polyoxyethylene octylphenyl ether (TX-100), in aqueous solutions.

Experimental Section

Instrumentation

A Perkin Elmer (Germany) (Model: Lambda-2) double-beam UV/VIS spectrophotometer with 1 cm matched quartz cells were used for all absorbance measurements. A pH- meter, WTW inolab (Germany) (Model: Level-1) combined electrodes were employed for measuring pH values. A Hitachi Ltd., Model 180–50, S.N.5721–2 atomic absorption spectrophotometer with a deuterium lamp back ground corrector, equipped with graphite furnace GA-3, with Gold hollow cathode lamps of Hitachi, and a Hitachi Model 056 recorder was used for comparison of the results. The experimental conditions were: slit width, 1.3 nm; lamp current, 10.0 mA; wavelength, 242.8 nm; cuvette, cup; carrier gas (argon), 200 mL min^{-1}; sample volume, 10 μL.

Chemicals and Reagents

All chemicals solvents used were of analytical reagent grade or the highest purity available. Doubly distilled de-ionized water, which is nonabsorbent under ultraviolet radiation, was used throughout. Glass vessels were cleaned by soaking in acidified solutions of $KMnO_4$ or $K_2Cr_2O_7$, followed by washing with concentrated HNO_3 and rinsed several times with de–ionized water.

Samples

Stock solutions and environmental water samples (1000 mL each) were kept in polypropylene bottles containing 1 mL of concentrated nitric acid. Biological fluids (blood and urine) were collected in polyethane bottles from affected persons (Jeweler's who suffered from anemia, blood disorder, liver and kidney damage diseases). Immediately after collection, they were stored in a salt-ice mixture and later, at the laboratory, were kept at –20 oC (Ahmed and Mamun, 2001). More rigorous contamination control was used when one gold levels in the specimens were low.

Gold(III) standard solutions (5.08 × 10⁻³ M)

A 1000 mL stock solution (1000 mg mL^{-1}) of gold was prepared by dissolving 1.0 g of gold (purity 99.999%) in aqua regia by warming, evaporating the solution to dryness, dissolving the residue in hydrochloric acid, evaporating the solution to half its volume, cooling and diluting with water to 1000 mL in calibrated fl ask (Zaijun et al. 2003). Working solutions were prepared by appropriate dilution of standard solution.

Bis(salicylaldehyde)orthophenylenediamine(BSOPD) (3.16 × 10⁻³ M)

The reagent was synthesized according to the method of (Salam and Chowdhury, 1997). The solution was prepared by dissolving the requisite amount of BSOPD in a known volume of double distilled ethanol (Merck, Darmstadt). More dilute solution of the reagent was prepared as required.

Polyoxyethylene octylphenyl ether (TX-100) (10%)

A 500 mL T-X100 solution was prepared by dissolving 50 mL of pure polyoxyethylene—octylphenyl ether (E. Merck Darmstadt, Germany) in 250–300 mL in doubly distilled de-ionized water, sonicated for 15 min and diluted up-to the mark with de-ionized water when it became transparent.

Aqueous Ammonia Solution

A 100 mL solution of aqueous ammonia was prepared by diluting 10 mL of concentrated NH3 (28%–30%) ACS grade with de-ionized water. The solution was stored in a polypropylene bottle.

EDTA Solution

A 100 mL stock solution of EDTA (0.1% w/v) was prepared by dissolving 128 mg of ethylenediami-netetraacetic acid, disodium salt dehydrate (Merck, Darmstadt) in 100 mL de-ionized water.

Other Solutions

Solutions of a large number of inorganic ions and complexing agents were prepared from their AnalaR grade, or equivalent grade, water-soluble salts. In the case of insoluble substances, a special dissolution method was adopted (Pal and Chowdhury, 1984).

General Procedure

A series of standard solutions of a neutral aqueous solution containing 0.1–300 µg of gold in a 10 mL calibrated flask was mixed with 10–25-fold molar excess of the BSOPD solution (preferably 1.0 mL of 3.16×10^{-4} M) BSOPD reagent, 1–3.5 mL (preferably 2 mL) of 10% TX-100 solution, 0.5–1.2 mL (preferably 0.5 mL) of 4 M H_2SO_4. The mixture was diluted to the mark with de-ionized water. After standing for 10 min the absorbance was measured at 490 nm against a corresponding reagent blank. The gold content in an unknown samples (e.g. real, environmental, biological and soil samples) were determined using a concurrently prepared calibration graph.

Results and Discussion

Absorption Spectra

The absorption spectra of the Gold (III)-BSOPD system in a 4 M sulfuric acid medium were recorded using a spectrophotometer. The absorption spectra of the Gold(III)-BSOPD is a symmetric curve with the maximum absorbance at 490 nm and an average molar absorption coefficient of 2.3×10^4 L mol^{-1} cm^{-1} and 2.5×10^5 L mol^{-1} cm^{-1} in aqueous and micellar media, respectively Figure 1. The reagent blank exhibited negligible absorbance, despite having a wavelength in the same region. In all instances, measurements were made at 490 nm against a reagent blank.

Composition of the Absorbance

Job's method (Yoe and Jones, 1944) of continuous variation and the molar-ratio method were applied to ascertain the stoichiometric composition of the complex. Au-BSOPD (1:1) complex was indicated by the molar-ratio method (Fig. 2).

It was found that BSOPD has excellent analytical characteristics. In micellar medium (TX-100) BSOPD reacts with gold(III) to form a highly stable brown-yellow complex. Its apparent molar-absorptivity is 2.5 × 105 L mol-1 cm-1, much higher than that of the other reagents in Table 1. Sensitivity can be measured in a very simple and rapid way and selectivity enhancement and low time-consuming as well as environmental and health safety due to avoiding the use of carcinogenic organic solvents are advantages of the method against previous extraction spectrophotometric method mentioned in Table1. The method is far more sensitive, selective, non-extractive, simple and rapid than all of the existing spectrophotometric methods mentioned in Table 1. (Alfonso and Gomez Ariza, 1981; Balcerzak et al. 2006; Chen et al. 2006; El-Zawawy et al. 1995; Fujita et al. 1999; Gangadharappa et al. 2004; Gao et al. 2005; Gowda et al. 1984; Hu et al. 2006; Koh et al. 1986; Matouskova et al. 1980; Melwanki et al. 2002; Mirza, 1980; Ortuno et al. 1984; Pal, 1999; Patell and Lieser, 1986; Zhao et al. 2006). With suitable masking, the reaction can be made highly selective and reagent blank do not show any absorbance. The method is very reliable, and a concentration in the ngg–1 range in aqueous medium at room temperature (25 ± 5ºC).

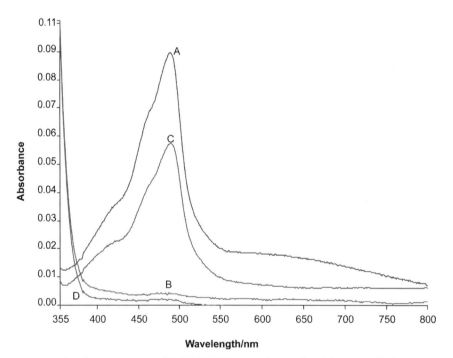

Figure 1. A and B absorption spectra of Au(III)-BSOPD system (1 mg L^{-1}) and the reagent blank (λ_{max} = 490 nm) in micellar media, C and D absorption spectra of Au(III)-BSOPD system (1 mg L^{-1}) and the reagent blank (λ_{max} = 490 nm) in aqueous solutions.

Table 1. Review of reagents for spectrophotometric determination of gold.

Reagent	λ_{max} (nm)	E (L mol^{-1} cm^{-1})	Medium	Beer's Law limits (mg/L)	Comments	Ref
5-(2-Pyridyl) methylene-rhodanine	418	1.1×10^4	HCl	–	1. Less-sensitive,	(Alfonso and Perez-Ruiz, 1981)
5-(6-Methylpyridyl) methylene-rhodanine	420	1.1×10^4	HCl	–	2. Acetate buffer medium. 3. Less-Sensitive 4. pH dependent	
Morin	291	2.02×10^4	HCl	0.2–12	1. Less—sensitive, 2. Indirect method, 3. UV range, 4. Temp: and time dependent.	(Balcerzak et al. 2006)
2-Carboxyl-1-naphthiorhodanine (CNTR)	540	1.35×10^5	Dimethyl forma-mide (DMF) and EmuLsifier-OP (p-Octyl polyethylene glycol phenyl ether)	0.01–2.0	1. Solvent extractive so Lengthy and time consuming. 2. DMF is highly toxic	(Chen et al. 2006)
5 (2,4Dihydroxybenzylidine) rohodamine(DHBR)	558	8.45×10^4	Cetylpyridinium bromide (CPB)	0.16–2.24	1. Less-sensitive 2. Less-slective. 3. pH dependent.	(El-Zawawy et al. 1995)
Thiamine and Phloxine	570	2.1×10^5	Methylcell-ulose	0.02–0.8	1. pH, time and temp: dependent.	(Fujita et al. 1999)
2'-Aminoacetophenone isonicotinoyl hydrazone (2-AAINH)	440	3.50×10^4	Aqueous Dimethyl formamide (DMF)	0.4–5.0	1. Less-sensitive, 2. Organic medium.	(Gangadharappa et al. 2004)
p-Sulfobenzylidene-thiorhodanine (SBDTR)	540	1.05×10^5	HCl and emulsifier-OP	0.1–20	1. Sensitive 2. SoLid Phase Extraction	(Gao et al. 2005)

Table 1. (Continued)

Propericiazine	511	3.85×10^4	H_3PO_4	0.1–7.0	1. Less-sensitive,	(Gowda et al. 1984)
5-(2-hydroxy-5-nitrophenyla zo)thiorhodanine(HNATR)	520	1.37×10^5	Emulsifier-OP (p-Octyl polyethylene glycol phenyl ether)	0.01–3.0	1. Solid Phase Extraction 2. Less-sensitive	(Hu et al. 2006)
Methylene Blue	657	1.08×10^5	$C_2H_2Cl_2$	0.04–1.58	1.Solvent extractive so lengthy and time consuming. 2. pH dependent.	(Koh et al. 1986)
Bromopyrogallol red	400	3.0×10^4	–	0.1–3.0	1. Less-sensitive 2. Strongly affected by ionic strength of solution	(Matouskova et al. 1980)
Ethopazine hydrochloride(EPI)	513	2×10^4	H_3PO_4	0.5–14.1	1. Less-sesitive,	(Melwanki et al. 2002)
Isopendyl hydrochloride(IPH)	512	2.1×10^4	H_3PO_4	0.5–14.5	2. Complex stable only for 45 minutes.	
Tri-iso-octylamine	325	5.8×10^3	CCl_4	–	1. Less-sesitive, 2. Solvent extractive so lengthy and time consuming	(Mirza, 1980)
1,2,4,6-Tetraphenylpyridin-ium perchlorate	313	3.44×10^4	HCl	0.05–0.5	1. Light-sensitive 2. UV-range	(Ortuno et al. 1984)
Photoinitiated gold sol	523	3.06×10^3	Tritron X-100	0–150	1. Less-sensitive, 2. Complex procedure, 3. Time dependent	(Pal, 1999)
Amides and Amidines	320–400	0.06–1.2×10^4	Chloroform or Benzene	–	1. Less-sesitive, 2. Solvent extraction	(Patell and Lieser, 1986)

Table 1. (Continued)

Reagent	λ_{max} (nm)	E ($Lmol^{-1} cm^{-1}$)	Medium	Beer's Law Limits (mg/L)	Comments	Ref.
5-(p-Aminobenzylidene)-thiorhodanine (ABTR)	550	1.23×10^5	Emulsifier-OP(p-Octyl polyethylene glycol phenyl ether) and Dimethyl formamide (DMF)	0.01–3.0	1. Solid Phase Extraction with reversed phase polymer-based. 2. Complex stable for only 5 h.	(Zhao et al. 2006)
Bis (salicylaldehyde) orthophenylenediamine (BSOPD)	490	2.5×10^5	TX-100	0.01–30	1. Ultrasensitive 2. Highly selective 3. Aqueous reaction medium 4. Less toxic surfactant	Present work

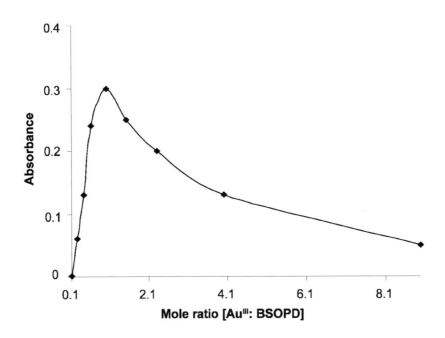

Figure 2. Composition of Au (III)-BSOPD complex by the Mole ratio method in micellar media.

Optimization of Some Parameterson the Absorbance

Effect of Surfactant

Of the various surfactants {nonionic [polyoxyethylenedodecylether (Brij-35), polyoxyethylene sorbitan monopalmitate (Tween-40), polyoxyethylene sorbitan mono-oleate (Tween-80), TritonX-100]; cationic [cetyltrimethylammonium bromide (CTAB), cetylpyridinum chloride (CPC)]; and anionic [sodium dodecyl sulfate (SDS)]} studied, TritronX-100 was found to be the best surfactant for the system. In a 10% TritronX-100 medium, however, the maximum absorbance was observed; hence, a 10% TritronX-100 solution was used in the determination procedure.

Different volumes of 10% TritronX-100 were added to a fixed metal ion concentration, in 10 mL volumetric flask, and the absorbance was measured according to the standard procedure. It was observed that at 5 mg L–1 Au-chelate metal, 1–3.5 mL of 10% TritronX-100 produced a constant absorbance of the Au-chelate. Outside this range of surfactant (i.e. 10%–35%) the absorbance decreased. For all subsequent measurements, 2 mL of 10% TritronX-100 (i.e. 20%) was added.

The basic principle of spectrophotometric determination is Beer-Lambert Law (Beer's) i.e. absorbance (A) is directly proportional to the concentration (C) of the analyte. The absorbance of the analyte is always measured against the corresponding reagent blank. When the sample (analyte) is diluted, in that case absorbance of reagent blank increased and does not obey the Beer's Law.

Effect of Acidity

Of the various acids (nitric, sulfuric, hydrochloric and phosphoric) studied, sulfuric acid was found to be the best acid for the system. The absorbance was at a maximum and constant when a 10 mL of solution (5 mg L–1) contained 0.5–1.2 mL of 4 M (2–4.8 M) sulfuric acid (or pH 0.83–1.2) at room temperature (25 ± 5°C). Outside this range of acidity, 2–4.8 M (or pH 0.83–1.2) the absorbance decreased. For all subsequent measurements, 0.5 mL of 4 M (i.e. 2 M) sulfuric acid (or pH 1.14) was added.

Effect of Time

The reaction is fast. Constant maximum absorbance was obtained just after 10 min of the dilution to volume at room temperature (25 ± 5°C), and remained strictly unaltered for 24 h (Fig. 3).

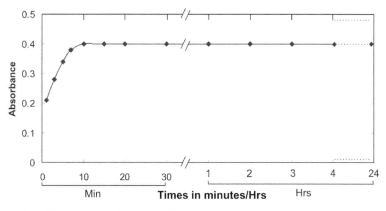

Figure 3. Effect of the time on the absorbance of Au(III)-BSOPD system.

Effect of Temperature

The absorbance at different temperatures, 0–40°C, of a 10 mL solution (5 mg L^{-1}) was measured according to the standard procedure. The absorbance was found to be strictly unaltered throughout the temperature range of 10–40°C. Therefore, all measurements were performed at room temperature (25 ± 5°C).

Effect of the Reagent Concentration

Different molar excesses of BSOPD were added to a fixed metal-ion concentration, and the absorbances were measured according to the general procedure. It was observed that at 0.5 mg L^{-1} Au-metal (optical path length, 1 cm), reagent molar ratios 1:10 and 1:25 produced a constant absorbance of the Au-chelate (Fig. 4). The effect of reagent at different concentration of AuIII (5 mg L^{-1}) was also studied but similar effect was observed. For all subsequent measurements, 1.0 mL of 3.16 × 10^{-3} M BSOPD reagent was added.

Analytical Performance of the Method

Calibration Curve

The effect of metal concentration was studied over 0.01–100 mg L^{-1}, distributed in four different sets (0.01–0.1, 0.1–1, 1–10, 10–100 mg L^{-1}) for convenience of the measurement. The absorbance was linear for 0.1–30 mg L^{-1} and 0.01–30 mg L^{-1} of gold (III) in aqueous and surfactant media, respectively. From the slope of the calibration graph, the average molar absorption coefficient was found to be 2.3 × 10^4 L mol^{-1} cm^{-1} and 2.5 × 10^5 L mol^{-1} cm^{-1} in aqueous and micellar media, respectively. Of the four calibration graphs, the one showing the limit of the linearity range is given in Fig. 5; the next two were straight-line graphs passing through the origin (R2 = 0.9986). The selected analytical parameters obtained with the optimization experiments are summarized in Table 2.

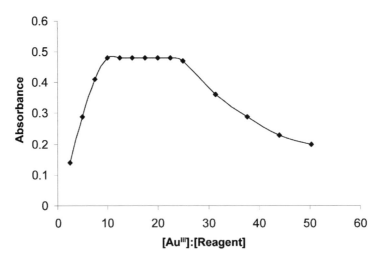

Figure 4. Effect of reagent [BSOPD:AU(III) molar concentration ratio] on the absorbance of Au(III)-BSOPD system in micellar media.

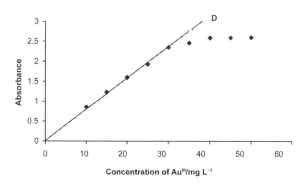

Figure 5. Calibration graph: D, 10–30 mg L–1 of Au (III).

Precision and Accuracy

The precision of the present method was evaluated by determining different concentrations of gold (each analyzed at least five times). The relative standard deviation (n = 5) was 2.5%–0%, for 0.1–300 μg of AuIII in 10 mL, indicating that this method is highly precise and reproducible. The detection limit (Mitra, 2003) (3s of the blank) and Sandell's sensitivity (Sandell, 1965) (concentration for 0.001 absorbance unit) for gold(III) were found to be 1 ng mL^{-1}, 5 ng cm^{-2}, respectively. The method was also tested by analyzing several synthetic mixtures containing gold(III) and diverse ions shown in Table 4. The results of the total gold in a number of real samples were in good agreement with the certified values in Table 5. The reliability of our Au-chelate procedure was tested by recovery studies. The average percentage recovery obtained for the addition of a gold(III) spike to some environmental water samples was quantitative, as shown in Table 6. The results of biological and geological analyses by the spectrophotometric method were in excellent agreement with those obtained by AAS (Table 7). Hence, the precision and accuracy of the method were excellent.

Table 2. Selected analytical parameters obtained with optimization experiments.

Parameter	Selected value in aqueous solutions	Selected value in aqueous micellar solutions
Wavelength, λ/nm	490	490
Acidity/M H$_2$SO$_4$	0.16–0.48 (preferably 0.4)	0.2–0.48 (preferably 0.2 M)
pH	0.48–1.23 (preferably 0.9)	0.83–1.2 (preferably 1.14)
Surfactant/10% TX-100/mL	–	1.0–3.5 (preferably 2)
Time/h	1 min–24 h (preferably 20 min)	1 min–24 h (preferably 5 min)
Temperature/°C	10–40 (preferably 25 ± 5)	10–40 (preferably 25 ± 5)
Reagent (fold molar excess, M:R)	1:10–1:20 (preferably 1:15)	1:10–1:25 (preferably 1:15)
Linear range/mg L^{-1}	0.1–30	0.01–30
Molar absorption coefficient/ L mol^{-1} cm^{-1}	2.31 × 10^4	2.52 × 10^5
Sandell's sensitivity/ngcm^{-2}	60	5
Detection limit/μg L^{-1}	15	1
Reproducibility (% R SD)*	1–5	0–2.5
Correlation coefficient (R^2)	0.9959	0.9986

*Relative Standard Deviation (RSD).

Table 3. Table of tolerance limits of foreign ions[a].

Species x	Tolerance ratio[b] x/Au[III]	Species x	Tolerance ration x/Au[III]
Acetate	1000	Iron (II)	50[c]
Ascorbic Acid	75	Iron (III)	25[d]
Bicarbonate	100	Manganese (II)	100
Carbonate	500	Manganese (VII)	25[e]
Chloride	1000	Magnesium	100
Citrate	50	Molybdenum (VI)	100
Fluorides	100	Mercury (I)	100
EDTA	500	Mercury (II)	100
Nitrate	200	Neodymium(III)	100
Sulfate	1000	Nickel (II)	100
Oxalate	1000	Palladium(II)	25[d]
Phosphate	200	Potassium	100
Tartrate	200	Rhodium(III)	100
Ammonium(I)	100	Ruthenium(III)	100
Aluminum(III)	100	Selenium (IV)	25[d]
Arsenic (III)	100	Silver (I)	25[e]
Barium	25[d]	Sodium	100
Beryllium (II)	100	Strontium	50
Bismuth (III)	100	Thallium (I)	100
Cadmium	100	Tin(IV)	25[d]
Calcium	100	Titanium(IV)	25[f]
Chromium (III)	100	Tungsten (VI)	100
Cobalt (II and III)	100	Vanadium (V)	25[d]
Copper (II)	50[d]	Zirconium(IV)	100
Lanthanum(III)	100	Zinc	100
Lithium	100		

[a]Tolerance limit defined as ratio that causes less than 5% interference.
[b]Tolerance ratio, [Species (x)]/Au[III] (w/w).
[c]With 50 mg L^{-1} EDTA.
[d]With 500 mg L^{-1} EDTA.
[e]With 500 mg L^{-1} Oxalate.
[f]With 50mg L^{-1} Ascorbic acid.

Micellar Mechanism

BSOPD is an organic colorimetric reagent that provides basis of sensitive methods for the determination of a number of metal ions (Ahmed and Uddin, 2007). Gold ion combines with BSOPD to yield non-polar colored complex whose color differ significantly from BSOPD. This non-polar colored complex is generally extracted into organic solvents which is time-consuming and tedious. This problem has been overcome in recent time by introducing a hydrophobic micellar system generated by a surfactant. Micellar systems are convenient to use because they are optically transparent, readily available, and relatively nontoxic and stable (Tani et al. 1998). Nevertheless, the addition of surfactant at concentration above the critical micelle concentration (CMC) to an aqueous medium to form a micellar solution is the most commonly preferred procedure today. Micelles enhance the solubility of organic compounds in water by providing local non-polar environments. This phenomenon of micellar solubilization has been used in the development of many new methods and modification of existing methods of analysis (Paramauro et al. 1992). The use of micelle formation is promising for improving the analytical performance of the spectrophotometric procedure (Khan et al. 2007). Especially, the surfactants have been used to improve UV-visible spectrophotometric determination of metal ions with complexing agents (e.g. BSOPD, BBSOPD, etc). Generally the metal-chelate complexes formed (e.g. with shiff-base reagents) in the surfactant media (e.g. TX-100) are more stable than those formed in the absence of surfactant (e.g. aqueous media) (Ghazy et al. 2006).

Table 4. Determination of gold in some synthetic mixtures.

Sample	Composition of mixture/mg L^{-1}	Gold(III)/mgL^{-1}		Recovery ± s (%)
		Added	Found[a]	
A	Au^{III}	0.5	0.49	98 ± 0.5
		1.00	1.00	100 ± 0.0
B	As in A + Na (25) + Be^{2+} (25)	0.5	0.50	100 ± 0.0
		1.00	0.99	99 ± 0.3
C	As in B + Zn (25) + Co (25) + EDTA(50)	0.50	0.49	98 ± 0.5
		1.00	0.99	99 ± 0.2
D	As in C + Ca(25) + Cr^{3+}(25)	0.5	0.50	102 ± 0.6
		1.00	1.02	102 ± 0.4
E	As in D + Mn^{2+}(25) + Ag^{+}(10)	0.5	0.52	104 ± 1.2
		1.00	1.06	106 ± 1.0
F	As in E + Ni^{2+}(25) + Hg^{2+}(25)	0.5	0.54	108 ± 1.5
		1.00	1.09	109 ± 1.2

[a]Average of five analyses of each sample.

Effect of Foreign Ions

The effect of over 50 cations, anions and complexing agents on the determination of only 1 mg L^{-1} of Au^{III} was studied. The criterion for interference (Ojeda et al. 1987) was an absorbance value varying by more than 5% from the expected value for AuIII alone. There was no interference from the following 1000 fold amount of acetate, chlorides, sulfate or oxalate; a 500-fold amount of EDTA or carbonate; a 200-fold amount of tartrate, phosphate or nitrate. EDTA prevented the interference of 50-fold of iron (II) or copper (II), 25-fold amounts of barium, iron (III), selenium (IV), tin (IV) or vanadium (V). Interferences of 25-fold of manganese (VII) or silver (I) have been completely removed by using oxalate as masking agent. A 25-fold of palladium (II) or Titanium (IV) has been completely removed by using EDTA or ascorbic acid as masking agent. However, for those ions whose tolerance limit has been studied their tolerance ratios are mentioned in Table 3.

Table 5. Determination of gold in certified reference materials.

Certified reference materials* (Composition)	Gold (mg L^{-1})		Relative error (%)
	Certified value	Found (n = 5)	
OXG 60 (SiO_2, Al_2O_3, Na_2O, K_2O, CaO, MgO, TiO_2 , MnO, P_2O_5, Fe_2O_3, LOI)	1.025	1.015	1.0
OXG56 (SiO_2 , Al_2O_3, Na_2O, K_2O, CaO, MgO, TiO_2 , MnO, P_2O_5, Fe_2O_3, LOI)	0.611	0.605	0.6
SH 24 (SiO_2, Al_2O_3, Na_2O, K_2O, CaO, MgO, TiO_2, MnO, P_2O_5, Fe, S)	1.326	1.315	1.1

*These CRMs obtained from Rock labs Ltd., Auckland, New Zealand.

Table 6. Determination of gold in some environmental water samples.

Sample		Gold/µg L^{-1}		Recovery ± s (%)	s_i^b (%)
		Added	Found[a]		
	Tap water	0	5.0		
		100	104.0	99 ± 0.3	0.45
		500	505.0	100 ± 0.0	0.00
	Well water	0	8.5		
		100	109.0	100.4 ± 0.5	0.32
		500	512.0	100.6 ± 0.2	0.35
	Rain water	0	0.0		
		100	100.5	100.5 ± 0.2	0.24
		500	500	500 ± 0.0	0.00
River water	Lake[c] (upper)	0	25.0		
		100	123.0	98 ± 0.3	0.19
		500	520.0	99 ± 0.2	0.25
	Lake[c] (lower)	0	22.0		
		100	120.0	98 ± 0.5	0.30
		500	522.0	100 ± 0.0	0.00
	Indus (upper)	0	12.8		
		100	112.0	99 ± 0.5	0.35
		500	513.0	100.0 ± 0.2	0.19
	Indus (lower)	0	9.5		
		100	110.0	100.5 ± 1.0	0.35
		500	512.0	100.5 ± 0.8	0.21
Sea water	Arabian Sea (upper)	0	4.5		
		100	105	100.5 ± 1.0	0.35
		500	506	100.3 ± 0.7	0.33
	Arabian Sea (lower)	0	5.5		
		100	106	100.5 ± 0.2	0.35
		500	506	100.1 ± 0.3	0.31
Drain water	Jewels wastewater[d]	0	125.0		
		100	225.0	100.0 ± 0.0	0.00
		500	528.0	100.5 ± 0.4	0.28
	Aral wah[e]	0	45.0		
		100	148.0	102 ± 1.0	0.24
		500	545.0	100 ± 0.0	0.00
	Fiber tex[f]	0	25.0		
		100	127.5	102 ± 0.8	0.49
		500	530.0	100.9 ± 0.3	0.35
	Jhampur[g]	0	28.5		
		100	130.0	101 ± 0.9	0.15
		500	525.0	99 ± 0.5	0.26

[a]Average of five replicate determinations.
[b]The measure of precision is the relative deviation(sr).
[c]The Manchar Laka, Dadu. Sindh.
[d]Jewelers shops from Khairpur. Karachi and Hyderabad, Sindh.
[e]Aral wah. Dadu, Sindh.
[f]Kotri, Hyderabad.
[g]Kotri, Hyderabad.

Table 7. Concentration of gold in biological (blood and urine) and geological samples.

Serial No	Sample	Gold/μgL^{-1}				Relative error%	Sample Source
		AAS[a] (n = 5)		Proposed method (n = 5)			
		Found	RSD[b]%	Found	RSD%		
1	Blood	55.2	1.0	56.5	1.0	2.3	Normal adult (Male)
	Urine	13.8	1.2	14.6	1.3	1.4	
2	Blood	423.2	1.5	425.5	1.2	0.5	Jeweler's blood(Male)[c]
	Urine	98.5	1.8	99.8	1.8	1.3	(Hyderabad)
3	Blood	128.8	1.3	127.5	1.5	1.0	Jeweler's blood(Male)[c]
	Urine	32.5	1.8	31.8	2.0	1.6	(Tandojam)
4	Rock$_1$[e]	15.8	1.2	16.3	1.0	3.0	MTL[d]
							Peshawer
5	Rock$_2$[e]	16.7	1.5	17.2	1.2	2.9	Karak Mountain
6	Rock$_3$[e]	15.1	1.3	15.5	1.5	2.5	Baka Khail Mountain

[a]Atomic Absorption Spectrophotometry.
[b]Relative Standard Deviation.
[c]Samples were from Jewelers of Hyderabad.
[d]Mineral Testing Laboratory, 164-C, Industrial Estate, Jamrud Road, Peshawar.
[e]Value in ng g^{-1}.

Applications

The present method was successfully applied to the determination of gold in a series of synthetic mixtures of various compositions (Table 4), and also in a number of real samples such as several standard reference materials (Table 5). The method was also extended to the determination of gold in a number of environmental, geological and soil samples. In view of the unknown composition of environmental water samples, the same equivalent portions of each sample were analyzed for gold content; the recoveries in both the "spiked" (added to the samples before the mineralization or dissolution) and the "unspiked" samples are in good agreement (Table 6). The results of biological and geological analyses by the spectrophotometric method were found to be in excellent agreement with those obtained by AAS (Table 7). The results of soil sample analysis by the spectrophotometric method are shown in Table 8.

Determination of Gold (III) in Synthetic Mixtures

Several synthetic mixtures of varying compositions containing gold (III) and diverse ions of known concentrations were determined by the present method using EDTA as a masking agent; and the results were found to be highly reproducible.

The results are shown in Table 4. Accurate recoveries were achieved in all solutions.

Table 8. Determination of Gold in some surface soil samples.

Serial No	[a]Gold/ng g^{-1}	Sample source
S_1	10.2 ± 0.5[b]	Traffic soil (Hyderabad bus terminal)
S_2	8.5 ± 0.8	Roadside soil (Hyderabad—Karachi highway)
S_3	6.8 ± 0.7	Marine soil (Sand of Arabian Sea)
S_4	20.5 ± 1.0	Industrial Soil (Pharmaceutical Company)
S_5	7.5 ± 0.8	Agricultural soil (Sindh University Campus)

[a]Average of five analyses of each sample.
[b]The measure of precision is the standard deviation, $\pm$ s.

Determination of Gold (III) in Certified Reference Materials

A 0.5 g sample was accurately weighed and placed in a 50 mL Erlenmeyer flask following a method recommended by Parker (Parker, 1983). To this, 10 mL of concentrated HNO_3 and 5 mL of concentrated HCl was added, carefully covering the flask with a watch glass until the brisk reaction subsided. The solution was heated and simmered gently after the addition of 5 mL of concentrated HNO_3, until all carbides were decomposed. The solution was evaporated carefully to dense white fumes to drive off the oxides of nitrogen and then cooled to room temperature 25 ± 5°C). After suitable dilution with de-ionized water, the contents of the Erlenmeyer flask were warmed to dissolve the soluble salts. The solution was then cooled and neutralized with a dilute NH_4OH solution. The resulting solution was filtered, if necessary, through a Whatman No. 42 filter paper into a 25 mL calibrated flask. The residue was washed with a small volume of hot water and the volume was made up to the mark with de-ionized water.

A suitable aliquot of the above solution was taken in to 10 mL calibrated flask and gold content was determined by general procedure using EDTA as masking agent. The results are shown in Table 5. The results for total gold were in good agreement with certified values Table 5.

Determination of Gold (III) in Environmental Water Samples

Each filtered (with Whatman No. 42) environmental water sample (100 mL) evaporated nearly to dryness with 10 mL of concentrated HNO3 in a fume cupboard and was heated with 10 mL of de-ionized water in order to dissolves the salts. The solution was then cooled and neutralized with dilute NH_4OH solution. The resulting solution was then filtered and quantitatively transferred into a 25 mL calibrated fl ask and made up to the mark with de-ionized water.

An aliquot (1–2 mL) of this solution was pipetted into a 10 mL calibrated flask, and the gold content was determined as described under a procedure using EDTA as a masking agent. The analysis of environmental water samples from various sources for gold, and the results are given in Table 6.

Determination of Gold (III) in Biological Samples

Human blood (5–10 mL) or urine (10–20 mL) was collected in polyethane bottles from the affected persons. The samples were taken into a 100 mL micro-Kjeldahl flask. A glass bead and 10 mL of concentrated nitric acid were added and the flask was placed on the digester under gentle heating following a method recommended by Stahr (Stahr, 1991). When the initial brisk reaction was over, the solution was removed and cooled. Five milliliters of concentrated HNO_3 was added carefully, followed by the addition of 0.5 mL of 70% $HClO_4$, and heating was continued to dense white fumes, repeating HNO_3 addition if necessary. Heating was continued for at least ½ h and then cooled. The content of the flask was filtered and neutralized with dilute ammonia. The resultant solution was then transferred quantitatively into a 10 mL calibrated flask and made up to the mark with de-ionized water.

A suitable aliquot (1–2 mL) of the fi nal solution was pipetted out into a 10–mL calibrated fl ask, and the gold content was determine as described under Procedure using EDTA, ascorbic acid, tetra sodium pyrophosphate as a masking agent. The samples were also measured by atomic absorption spectrometry (Pyrzynska, 2005) for comparison of the results. The results of biological (human fluids) analyses by the spectrophotometric method were found to be in excellent agreement with those obtained by AAS. The results are given in Table 7.

Determination of Gold (III) in Geological Samples

An accurately weighed amount of 1 g sample was ground well and placed in 100 mL Erlenmeyer flask and dissolved in 15 mL of the acid mixture (aqua regia) was added, the solution was boiled on hot plate at 120ºC until the dissolution of the sample was completed, the resulting solution was carefully evaporated to small volume (2–3 mL) to remove the NO_3^- ions. After cooling and neutralized with dilute NH_4OH solution. The resulting solution was then filtered and quantitatively transferred into a 20 mL calibrated flask and made up to the mark with de-ionized water.

An aliquot (1–2 mL) of this solution was pipetted into a 10 mL calibrated flask, and the gold content was determined as described under a general procedure using EDTA as a masking agent. The samples were also measured by atomic absorption spectrometry (Pyrzynska, 2005). The results of geological analyses by the spectrophotometric method were found to be in excellent agreement with those obtained by AAS. The results are given in Table 7.

Determination of Gold in Soil Samples

An air-dried homogenized soil sample (100 g) was weighed accurately and placed in a 100 mL micro-Kjeldahl flask. The sample was digested in the presence of oxidizing agent, following the method recommended by Jackson (Jackson, 1965). The content of the fl ask was fi ltered through a What-man No. 42 filter paper into a 25 mL calibrated flask and neutralized with dilute NH_4OH solution. It was then diluted up to the mark with de-ionized water.

Suitable aliquots (1–2 mL) were transferred into a 10 mL calibrated flask. The gold content was then determined, as described under general procedure using EDTA as a masking agent. The results are shown in Table 8.

Conclusions

In the present work, a simple, sensitive, selective and inexpensive micellar method with the Au(III)-BSOPD complex was develop for the determination of gold in industrial, environmental, geological and soil samples. The presence of a micellar system (altered environment) avoids the previous steps of solvent extraction, and reduces the cost and toxicity while enhancing the sensitivity, selectivity and molar absorptivity. The molar absorptivities of the Gold-BSOPD complex formed in presence of the nonionic TX-100 surfactant are almost a 10 times higher than the value observed in the aqueous solution, resulting in an increase in the sensitivity and selectivity of the method. The apparent molar absorptivities were found to be 2.3×10^4 L mol^{-1} cm^{-1} and 2.5×10^5 L mol^{-1} cm^{-1} in aqueous and micellar media, respectively. Compared with other methods in the literature (Table 1), the proposed method has several remarkable analytical characteristics:

(i)   The proposed method is very highly sensitive. The molar absorptivity of the complex is 2.5×10^5 L mol–1 cm–1 and highest of all reagents in the literature (please see Table 1). Thus, amount of ng g 1 of gold can be determined without pre-concentration; its sensitivity corresponds with that of the graphite furnace atomic absorption method.

(ii) The proposed method is highly selective. Because certified reference materials, environmental water and geological samples contain large amounts of metal ions and very low levels of gold, the selectivity of the determination of gold is most important.

(iii) The proposed method is very simple, rapid and stable. The reaction of gold(III) with BSOPD is completed rapidly in micellar medium within 10 min at room temperature.

With suitable masking, the reaction can be made highly selective. The proposed method using BSOPD in the presence of aqueous micellar solutions not only is one of the most sensitive methods for the determination of gold, but also is excellent in terms of selectivity and simplicity. Therefore, this method will be successfully applied to the monitoring of trace amounts of gold, in real, environmental, biological, geological and soil samples.

Disclosure

The authors report no conflicts of interest.

References

1. Ahmed, M.J. and Mamun, M.A. 2001. Spectrophotometric determination of lead in industrial, environmental, biological and soil samples using 2,5-dimercapto-1,3,4-thiadiazole. Talanta, 55:43–54. Ahmed, M.J. and Nasir Uddin, M. 2007. A simple spectrophotometric method for the determination of cobalt in industrial, environmental, biological and soil samples using Bis (Salicylaldehyde) orthophenyl-enediamine. Chemosphere, 67:2020–7.

2. Alfonso, G.G. and Gomez Ariza, J.L. 1981. Derivatives of rhodanine as spectrophotometric analytical reagents I. Condensation at C-5 with aromatic pyridine and nonpyridine aldehydes. Microchem. J., 26:574–85.

3. Balcerzak, M., Kosiorek, A. and Swiecicka, E. 2006. Morin as a spectrophotometric reagent for gold. J. Anal. Chem., 61:119–23.

4. Chen, Z., Huang, Z., Chen, J., Chen, J., Yin, J., Su, Q. and Yang, G. 2006. Spectrophotometric determination of gold in water and ore with 2-Carboxyl-1-Naphthalthiorhodanine. Anal. Lett., 39:579–87.

5. El-Zawawy, F.M., El-Shahat, M.F. and Mohamed, A.A. 1995. Spectrophoto-metric determination of silver and gold with 5-(2,4-dihydroxyben zylidene)rho-danine and cationic surfactants. Analyst., 120:549–54.

6. Fujita, Y., Mori, I. and Matsuo, T. 1999. Spectrophotometric determination of gold(III) by an association complex formation between gold-thiamine and phloxine. Anal. Sci., 15:1009–12.

7. Gangadharappa, M., Reddy, P.R., Reddy, V.R. and Reddy, S. 2004. Direct spec-trophotometric determination of gold(III) using 2'aminoacetophenone isonico-tinoyl hydrazone (2-AAINH). J. Indian Chem. Soc., 81:525–7.

8. Gao, X.Y., Hu, Q.F., Yang, G.Y. and Jia-yuan, Y. 2005. Study on solid phase extraction and spectrophotometric determination of gold with p-sulfoben-zylidene-thiorhodanine. Fenxi. Shiyanshi., 24:67–70.

9. Ghazy, S.E., El-shazly, R.M., El-Shasri, M.S., Al-Hazmi, GAA. and El-Asmy, A.A. 2006. Spectrophotometric determination of cooper(II) in natural waters , vitamins and certified steel scrap samples using acetophenone-chlorophenylthi-osemicarbazone. J. Iranian Chemical Society, 3(2):140–50.

10. Gowda, H.S., Gowda, A.T. and Made Gowda, N.M. 1984. Spectrophotometric determination of microamounts of gold(III) with propericiazine. Microchem. J., 30:259–65.

11. Hu, Q., Chen, X., Yang, X., Huang, Z., Chen, J. and Yang, G. 2006. Sol-id phase extraction and spectrophotometric determination of Au (III) with 5-(2-Hydroxy-5-nitrophenylazo)- thiorhodanine. Anal. Sci., 22:627–30.

12. Jackson, M.L. 1965. Soil Chemical Analysis, Prentice Hall, Englewood Cliffs, NJ. 272.

13. Juvonen, R., Lakomaa, T. and Soikkeli, L. 2002. Determination of gold and the platinum group elements in geological samples by ICP-MS after nickel sulphi-de fire assay: difficulties encountered with different types of geological samples. Talanta, 58:595–603.

14. Kavanoz, M., Gulce, H. and Yildiz, A. 2004. Anodic Stripping voltammetric determination of gold on a polyvinylferrocene coated glassy carbon electrode. Turk J. Chem., 28:287–97.

15. Khan, H., Ahmed, M.J. and Bhanger, M.I. 2007. A rapid spectrophotometric method for the determination of trace level lead using 1,5-Diphenylthiocarba-zone in the presence of aqueous micellar solution. Anal. Sci., 23(2):193–9.

16. Koh, T., Okazaki, T. and Ichikawa, M. 1986. Spectrophotometric determina-tion of gold(III) by formation of dicyanoaurate(I) and its solvent extraction with methylene blue. Anal. Sci., 2:249–53.

17. Matouskova, E., Mcova, I.N. and Suk, V. 1980. The spectrophotometric deter mination of gold with bromopyrogallol red. Microchem. J., 25:403–9.

18. Medved, J., Bujdos, M., Mats, P., Matus, P. and Kubova, J. 2004. Determination of trace amounts of gold in acid-attacked environmental samples by atomic absorption spectrometry with electro thermal atomization after preconcentration. Anal. Bioanal. Chem., 379:60–5.

19. Melwanki, M.B., Masti, S.P. and Seetharamappa, J. 2002. Determination of trace amounts of gold(III) using ethopropazine hydrochloride and isothipendyl hydrochloride: spectrophotometric study. Turk J. Chem., 26:17–22.

20. Mirza, M.Y. 1980. Studies on the extraction of platinum metals with tri-iso-octylamine from hydrochloric and hydrobromic acid separation and determination of gold, palladium and platinum. Talanta, 27:101–6.

21. Mitra, S. 2003. Sample Preparation Techniques in Analytical Chemistry. Wiley-Interscience, New Jersey. 14.

22. Nat., A., Ene, A. and Lupu, R. 2004. Rapid determination of goLd in Romanian auriferous aLLuviaL sands,concentrates and rocks by 14 MeV NAA. J. Radional. Nucl. chem., 261:179–88.

23. Ojeda, C.B., Torres, A.G., Rojas, F.S. and Pavon, JMC. 1987. Fluorimetric determination of trace amounts gallium in biological tissues. Analyst.,112:1499–501.

24. Ortuno, J.A., Perez-Ruiz, T., Sanchez-Pedreno, C. and Buendia, P.M. 1984. 1,2,4,6-Tetraphenylpyridinium perchlorate as a reagent for ion-association complex formation and its use for the spectrophotometric determination of gold. Microchem. J., 30:71–8.

25. Pal, A. 1999. Photoinitiated gold sol generation in aqueous Triton X-100 and its analytical application for spectrophotometric determination of gold. Talanta, 46:583–87.

26. Pal, B. and Chowdhury, B. 1984. Triazene-N.-oxides as new type of fluorimetric reagents. Part I: trace determination of zirconium at the nanogram level, Mikrochim. Acta., 83:121–31.

27. Parker, G.A. 1983. Analytical Chemistry of Molybdenum. Springer-Vergal, Berline.

28. Patell, K.S. and Lieser, K. 1986. Extraction and spectrophotometric determination of gold (III) by use of amides and amidines. Anal. Chem., 58:1547–51.

29. Pramauro, E., Prevot, A.B., Pelizzetti, E., Marchelli, R., Dossena, A. and Biancardi, A. 1992. Quantitative removal of uranyl ions from aqueous solutions using micellar-enhanced ultrafiltration. Anal. Chim. Acta., 264(2):303–10.

30. Pyrzynska, K. 2005. Recent developments in the determination of gold by atomic absorption spectrometry technique. Spectrochemica. Acta., Part(B. 60):1316–22.

31. Salam, M.A. and Chowdhury, D.A. 1997. Preparation and characterization of shiff-base reagents. Bull. Pure Appl. Chem., 16C:45–9.

32. Sandell, E.B. 1965. Colorimetric Determination of Traces of Metals, Inter-science, New York. 269.

33. Stahr, H.M. 1991. Analytical Methods in Toxicology, third ed. John Wiley and sons, New York. 75.

34. Tani, H., Kamidate, T. and Walanabe, H. 1998. Aqueous micellar two-phase system for protein separation. Anal. Sci., 14:875–9.

35. Wu, Y., Jiang, Z., Hu, B. and Duan, J. 2004. Electro thermal vaporization inductively coupled plasma atomic emission spectrometry for the determination of gold, palladium, and platinum using chelating resinYPA4 as both extractant and chemical modifier. Talanta, 63:585–292.

36. Yoe, J.A. and Jones, A.L. 1944. Stability constant of coordination compounds. Ind. Eng. Chem. Anal. Ed., 16:11.

37. Zaijun, L., Jiaomai, P. and Jian, T. 2003. Highly sensitive and selective spectrophotometric method for determination of trace gold in geological samples with 5-(2-hydroxy-5-nitrophenylazo)- rhodanine. Anal. Bioanal. Chem., 375:408–41.

38. Zhao, J., Li, J., Hung, Z., Huang, Z., Wei, Q., Chen, J. and Yang, G. 2006. Study on the solid phase extraction and spectrophotometric determination of gold in water and ore with 5-(p-aminobenzylidene)thiorhodanine. Indian J. Chem., 45A:1651–5.

CITATION

Originally published under the Creative Commons Attribution License or equivalent. Soomro R, Ahmed MJ, Memon N, Khan H. A simple and selective spectrophotometric method for the determination of trace gold in real, environmental, biological, geological and soil samples using bis (salicylaldehyde) orthophenylenediamine. Anal Chem Insights. 2008 Aug 29;3:75-90.

Palm-Based Standard Reference Materials for Iodine Value and Slip Melting Point

Azmil Haizam Ahmad Tarmizi, Siew Wai Lin and Ainie Kuntom

ABSTRACT

This work described study protocols on the production of Palm-Based Standard Reference Materials for iodine value and slip melting point. Thirty-three laboratories collaborated in the inter-laboratory proficiency tests for characterization of iodine value, while thirty-two laboratories for characterization of slip melting point. The iodine value and slip melting point of palm oil, palm olein and palm stearin were determined in accordance to MPOB Test Methods p3.2:2004 and p4.2:2004, respectively. The consensus values and their uncertainties were based on the acceptability of statistical agreement of results obtained from collaborating laboratories. The consensus values and uncertainties for iodine values were 52.63 ± 0.14 Wijs in palm oil, 56.77 ± 0.12 Wijs in palm olein and 33.76 ± 0.18 Wijs in palm stearin. For the slip melting points, the consensus values and uncertainties were 35.6 ± 0.3 °C in palm oil, 22.7 ± 0.4 °C in palm olein and 53.4 ± 0.2 °C in palm stearin. Repeatability and reproducibility relative standard deviations were found to be good and acceptable, with values much lower than that of 10%. Stability of Palm-Based Standard Reference Materials remained stable at temperatures of –20 °C, 0 °C, 6 °C and 24 °C upon storage for one year.

Keywords: palm-based standard reference materials, iodine value, slip melting point, MPOB Test Methods p3.2: 2004 and p4.2:2004, consensus values, uncertainties

Introduction

Palm oil is one of the most important sources of revenue for Malaysia. The total export earnings increased by 41.8% in 2007 compared to that in 2006 (Mohd. Basri, 2008). In fact, palm oil contributes to almost 30% of the vegetable oils production worldwide, in which 60% of the share accounts for the overall global export (Carter et al. 2007). Palm oil has now gained worldwide acceptance due to its unique properties and versatile applications as well as the competitive traded price over other vegetable oils (Choo et al. 2007). As one of the leading countries in the palm oil business, attempts should be taken to ensure that palm oil is of good quality.

Some of the indicators used to characterize palm oil are iodine value (IV) and slip melting point (SMP). The IV measures the degree of unsaturation or double bonds of oils and fats. It also indicates the ease of oxidation of oils and fats (Guided Wave Incorporated, 2008). Meanwhile, the SMP is widely used to characterize the melting and solidification properties of oils and fats. It changes with the chain length of fatty acids, unsaturation ratios, trans fatty acid content and the position of the fatty acids in the glycerol backbone (Karabulut et al. 2004).

Harmonization of the test methods for these characteristics is crucial in producing palm oil within the traded specifications. Thus, a comparison of measurements should be made with the certified values of the standard reference materials to assure the reliability and robustness of the methods used. With the emphasis on the quality aspects, our research group has been engaged in the development of Palm-Based Standard Reference Materials for palm oil analyses due to the unavailability of such standard reference materials in the market. This is to facilitate the use of these standard reference materials for calibration and validation of analytical measurements as well as to assess the capability of analysts in performing the measurements in the palm oil sector.

Currently, Palm-Based Standard Reference Materials for fatty acids composition, solid fat content, IV and SMP have been produced in-house. However, only two characterization works of fatty acids composition and solid fat content have been published. Thus, this paper reports the outcomes of the inter-laboratory proficiency tests of the standard reference materials from palm oil, palm olein and palm stearin for IV and SMP. Stability of the standard reference materials produced upon storage was also evaluated and further discussed in this paper.

Materials and Methods

Production of Palm-Based Standard Reference Materials

Refined, bleached and deodorized palm oil, palm olein and palm stearin were obtained from a local supplier. The antioxidant, tert-butylhydroquinone (97% purity) and 5-mL dark amber glass ampoules were purchased from Sigma-Aldrich (Steinheim, Germany) and Scherf Praezision (Meiningen-Dreissigacker, Germany), respectively.

A batch of palm oil, palm olein and palm stearin, respectively were heated up to 70 °C to liquefy the solid fat prior to the addition of 200 mg kg–1 of tert-butylhydroquinone. These solutions were then mixed thoroughly to ensure homogeneity. Each batch of oil produced about 3000 ampoules of standards. The usage of 5-mL dark amber glass ampoules helps to avoid color changes and photooxidation of oil standards upon storage. Portions of 5 mL of homogenized oils were pipetted into the ampoules prior to flushing with nitrogen and flame sealed. The oil standards were then labelled, packed in fabricated boxes and stored at –20°C until dispatch.

Characterization Exercises

Thirty-three and thirty-two of local and overseas laboratories collaborated in the inter-laboratory proficiency tests for IV and SMP, respectively, which include Analytical and Quality Development Unit, Malaysian Palm Oil Board, Selangor, Malaysia; Kek Seng Berhad, Johor, Malaysia; Research and Development Laboratory, Golden Jomalina Food Industries Sdn. Bhd., Selangor, Malaysia; Aarhuskarlshamn Sweden AB, Analys-Centrum, Karlshamn, Sweden; PT Multimas Nabati Asahan, Sumatera Utara, Indonesia; Southern Edible Oil Industries Sdn. Bhd., Selangor, Malaysia; Wilmar International Ltd., Singapore; School of Sciences and Food Technology, Universiti Kebangsaan Malaysia, Selangor, Malaysia; Quality Assurance Laboratory, Golden Jomalina Food Industries Sdn. Bhd., Selangor, Malaysia; PT Asianagro Agungjaya, West Java, Indonesia; Kempas Edible

Oils Sdn. Bhd., Johor, Malaysia; PGEO Edible Oils Sdn. Bhd., Johor, Malaysia; IOI Edible Oils Sdn. Bhd., Sabah, Malaysia; Oleochemical Products Services Unit, Malaysian Palm Oil Board, Selangor, Malaysia; Pan-Century Edible Oils Sdn. Bhd., Johor, Malaysia; Edtech Associates Sdn. Bhd., Penang, Malaysia; Chemara Laboratory Sdn. Bhd., Negeri Sembilan, Malaysia; ITS Testing Services Sdn. Bhd., Selangor, Malaysia; Kuala Lumpur-Kepong Berhad, Selangor, Malaysia; Biochem Laboratories Sdn. Bhd., Penang, Malaysia; Allied Chemists Sdn. Bhd., Johor, Malaysia; Felda-Johore Bulkers Sdn. Bhd., Johor, Malaysia; Chemsain Konsultant Sdn. Bhd., Sarawak, Malaysia; Lotus Laboratory Services Sdn. Bhd., Johor, Malaysia; SGS Laboratory Services Sdn. Bhd., Johor, Malaysia; Alami Technological Services Sdn. Bhd., Selangor, Malaysia; Chemical Laboratory Sdn. Bhd., Johor, Malaysia; Kuala Lumpur-Kepong Edible Oils Sdn. Bhd. (East Malaysia), Sabah, Malaysia; Testing Services (Sabah) Sdn. Bhd., Sabah, Malaysia; MM Vitaoils Sdn. Bhd., Selangor, Malaysia; PGEO Edible Oils Sdn. Bhd. (Prai Division), Penang, Malaysia; KL-Kepong Edible Oils Sdn. Bhd., Johor, Malaysia; Lam Soon Edible Oils Sdn. Bhd., Johor, Malaysia.

Four ampoules of each oil standards (palm oil, palm olein and palm stearin), which correspond to four replications, were sent to the collaborating laboratories. The IV and SMP measurements should be conducted using MPOB Test Methods p3.2:2004 and p4.2:2004 (Ainie et al. 2004), respectively. The MPOB Test Methods p3.2:2004 was technically equivalent to ISO 3961:1996. The MPOB Test Methods p4.2:2004 was originated from AOCS Official Method Cc 3-25 (Firestone, 1998).

In the determination of IV, about 0.2 g of oil sample, in a 20 mL mixture of cyclohexane (Merck, Darmstadt, Germany) and glacial acetic acid (Systerm, Shah Alam, Malaysia) was reacted with 25 mL of Wijs reagent (Merck, Darmstadt, Germany) followed by addition of 20 mL of 100 g L–1 potassium iodide (Systerm, Shah Alam, Malaysia) and distilled water after storage in the dark for 1 h. The liberated iodine was titrated with 0.1 M sodium thiosulfate (Univar, Seven Hills, Australia) until the yellowish iodine color disappeared. A small amount of starch solution (Merck, Darmstadt, Germany) was then added to the solution as an indicator and the titration continued until the blue color has also disappeared. The IV is calculated by the following equation:

$$IV(g/100\,g) = \frac{12.69C(V_1 - V_2)}{m}$$

where,

C is the concentration of sodium thiosulfate solution (mole L–1);

V1 is the volume (mL) of sodium thiosulfate solution used for the blank test;

V2 is the volume (mL) of sodium thiosulfate solution used for the sample; and m is the mass (g) of the sample

Wijs reagent, which contains iodine monochloride in acetic acid, can be also prepared manually in the laboratory. The ratio between iodine and chlorine of the Wijs reagent shall be within the limits of 1.10 ± 0.1. However, the preparation of the reagent is time consuming and hence is highly recommended to use the commercially available Wijs reagent.

SMP was measured using the following procedure. At least three clean capillary tubes were initially dipped into a completely melted oil sample to a depth of 10 mm. The tubes were then chilled until the oil sample was solidified prior to placing them in a test tube and held in a beaker of water equilibrated at 10 °C for 16 h in a thermostat water bath (Huber, Offenburg, Germany). The capillary tubes were subsequently removed from the test tube and attached to a thermometer with a rubber band such that the lower ends of the tubes were at the same level as the bottom of the mercury bulb of the thermometer. The thermometer was suspended in a beaker containing 400 mL of boiled distilled water. The thermometer should be immersed in the water to a depth of 30-mm. The initial temperature of the thermostat water bath was adjusted between 8 to 10 °C below the expected SMP of the oil sample. The water bath was agitated using a magnetic stirrer and heat was supplied at the rate of 1 °C min–1 and reduced to 0.5 °C min–1. The temperature at which the sample in the tubes started to melt and become clear is defined as the SMP. The difference between values of the measurement carried out by the same analyst on the same test sample shall not exceed 0.8 °C for palm oil and 0.5 °C for palm olein and palm stearin.

Apart from the oil standards, each collaborator was also supplied with detailed instructions of the study protocol and reporting cards to compute their analysis results. The IV and SMP analyses should be carried out within two month's time before sending out the results to the Malaysian Palm Oil Board.

Statistical Evaluation

Inter-laboratory proficiency tests results were assessed using the SoftCRM Version 1.2.0 software (Bonas, 1997). The software is particularly applied to evaluate the standard reference materials data as well as to document the quality of the standard reference materials produced. Consensus values for IV and SMP of each oil standard were generated at 95% confidence interval (CI). Outlying data or extreme values were discarded based on Grubb and Cochran tests. Grubb test determines outlying mean values (variability) between laboratories, whereas Cochran test identifies data variability within laboratories (Pocklington and Wagstaffe, 1987; ISO 5725-2 1994; Azmil Haizam et al. 2008).

Repeatability relative standard deviation, RSDr (relative standard deviation within laboratory) and reproducibility relative standard deviation, RSDR (relative standard deviation between laboratories) were also calculated for IV and SMP (ISO 5725-2 1994). The RSDr is determined from the test results generated under repeatability conditions of the same method on identical test items in the same laboratory by the same analyst using the same instrument within short intervals of time. Meanwhile RSDR is identified from the test results generated under reproducibility conditions in which the results are obtained using the same method on identical test items in different laboratories and analysts using different instruments (ISO 3541-1 1993).

Stability Test

Stability of the Palm-Based Standard Reference Materials was monitored for 12 months at four storage conditions of –20°C, 0°C, 4°C and 24°C. The oil standards were randomly analyzed for their IV and SMP at predetermined storage intervals.

Table 1. Statistical evaluation of iodine value in Palm-Based Standard Reference Materials.

Standards	P^a	N^b	Consensus value (Uncertainty) ^{c}	$s_r^{\,d}$	$RSD_r^{\,e}$	R^f	$S_R^{\,g}$	$RSD_R^{\,h}$	R^i
Palm Oil	30	120	52.63 (0.14)	0.21	0.40	0.58	0.42	0.79	1.16
Palm Olein	30	120	56.77 (0.12)	0.17	0.31	0.49	0.37	0.64	1.02
Palm Stearin	30	120	33.76 (0.18)	0.22	0.64	0.61	0.51	1.50	1.41

[a]Number of laboratories retained after eliminating outliers.
[b]Number of accepted test results (replicates).
[c]Consensus value and uncertainty (Wijs) generated as 95% confidence interval.
[d]Repeatability standard deviation.
[e]Repeatability relative standard deviation.
[f]Repeatability limit.
[g]Reproducibility standard deviation.
[h]Reproducibility relative standard deviation.
[i]Reproducibility limit.

Results and Discussion

Characterization of Iodine Value

Statistical evaluations for IV in palm oil, palm olein and palm stearin are tabulated in Table 1. The consensus values (IV) and their uncertainties were calculated at 95% CI using the SoftCRM 1.2.0 software. Data of thirty laboratories were retained for IV characterization in all palm products, while only three laboratories were discarded. The RSDr of the characterized IV in the three oils ranged from 0.40 to 0.64%, while RSDR of 0.79%, 0.64% and 1.50% was achieved in palm oil, palm olein and palm stearin, respectively. Both RSDr and RSDR were observed to be more than 5-fold lower than that of 10%, which signified that the IV analysis performed was within the acceptable variability. Example of intra- and inter-laboratory variability of IV in palm oil is illustrated in Figure 1. The bar graph consists of laboratory codes with their individual means and standard deviations after elimination of outliers through Cochran and Grubb tests at 95% CI. From that, the overall mean of means and its standard deviation was then determined from the accepted values. All the values were obtained according to ISO Guide 35 (1989).

Table 2. Statistical evaluation of slip melting point in Palm-Based Standard Reference Materials.

Standards	P[a]	N[b]	Consensus value (Uncertainty)[c]	s_r[d]	RSD$_r$[e]	R[f]	S_R[g]	RSD$_R$[h]	R[i]
Palm Oil	29	116	35.6 (0.3)	0.27	0.76	0.76	0.91	2.57	2.56
Palm Olein	25	100	22.7 (0.4)	0.13	0.59	0.38	0.93	4.08	2.60
Palm Stearin	26	104	53.4 (0.2)	0.14	0.27	0.40	0.46	0.87	1.30

[a]Number of laboratories retained after eliminating outliers.
[b]Number of accepted test results (replicates).
[c]Consensus value and uncertainty (°C) generated as 95% confidence interval.
[d]Repeatability standard deviation.
[e]Repeatability relative standard deviation.
[f]Repeatability limit.
[g]Reproducibility standard deviation.
[h]Reproducibility relative standard deviation.
[i]Reproducibility limit.

Table 3. Changes of iodine value in Palm-Based Standard Reference Materials.

Standards	T[a] (°C)	Storage period (month)[b]								
		0	2	4	6	8	10	12	Mean	SD[c]
Palm Oil	−20	52.25 ± 0.19	52.65	52.55	52.15	52.55	52.40	52.50	52.47	0.18
	0		52.55	52.60	51.90	52.20	52.40	52.70	52.39	0.30
	6		52.60	52.60	51.70	52.45	52.50	52.55	52.40	0.35
	24		52.70	52.60	52.05	52.45	53.30	52.60	52.62	0.41
Palm Olein	−20	57.20 ± 0.08	56.80	56.65	56.15	56.20	56.25	55.40	56.24	0.49
	0		57.00	56.70	56.25	56.55	56.00	56.00	56.42	0.40
	6		56.80	56.80	56.15	56.60	56.20	56.15	56.45	0.32
	24		56.05	56.45	56.20	56.65	56.10	55.90	56.23	0.28
Palm Stearin	−20	34.18 ± 0.32	33.90	34.70	33.70	34.10	34.20	34.20	34.13	0.34
	0		33.95	34.70	33.40	34.20	34.30	34.20	34.13	0.43
	6		34.10	34.70	33.65	34.20	34.15	34.25	34.18	0.34
	24		33.90	34.70	33.90	34.20	34.30	34.30	34.22	0.30

[a]Storage temperature.
[b]Mean of duplicate (Wijs).
[c]Standard deviation.

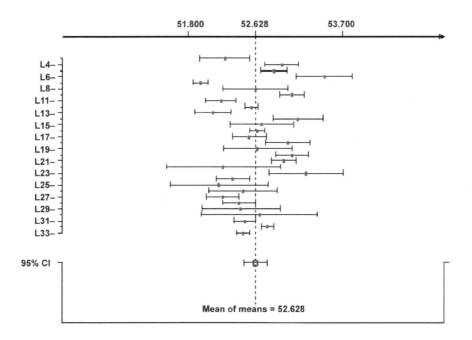

Figure 1. Example of bar graph for characterising iodine value in palm oil. The results plotted correspond to four replications. Mean of means designates the average result of total individual laboratory means at 95% confidence interval (CI).

Characterization of Slip Melting Point

Table 2 summarizes the statistical evaluations of SMP in palm oil, palm olein and palm stearin. The RSDr and RSDR were also identified for the characterized SMP. Three out of 32 laboratories were eliminated for characterization of SMP in palm oil while 25 and 26 laboratories were retained in palm olein and palm stearin, respectively. In general, both RSDr and RSDR were found to be less than 10%. The RSDr of palm oil, palm olein and palm stearin was 0.76%, 0.59% and 0.27%, respectively. These values were almost comparable with that of RSDr for IV characterization. The RSDR, however, exhibited slightly higher values for palm oil (2.57%) and palm olein (4.08%) in comparison to that of IV characterization. The higher RSDR is generally due to run-to-run variations, which may be caused by the variation in the instrument sensitivity, change of environmental conditions or even uncontrolled change of instrument parameters (Azmil Haizam et al. 2007; Azmil Haizam et al. 2008). Such effects are important, which could influence the RSDR when the samples were analysed by different laboratories. Only palm stearin produced a lower RSDR of 0.87%. Furthermore, both RSDr and RSDR were still well below 10% and hence the SMP results were acceptable. Similar to IV, the SMP assessment can also be expressed in the form of bar graphs. Figure 2 shows an example of within- and between- laboratory variability of SMP in palm oil.

Stability of Palm-Based Standard Reference Materials

Results of storage stability tests for IV and SMP are showed in Table 3 and Table 4, respectively. The test results of the four storage conditions (–20 °C, 0 °C, 4 °C and 24 °C) were compared with oil standards that were initially prepared and stored at –20 °C (t = 0). No detectable changes of IV and SMP were perceived even at 24 °C. As the Palm-Based Standard Reference Materials of these characteristics were observed to have good stability, their shelf-life could be extended for more than one year.

Conclusion

From the characterization exercises, it can be concluded that the production of Palm-Based Standard Reference Materials (palm oil, palm olein and palm stearin) for iodine value and slip melting point is achievable through inter-laboratory proficiency tests. The establishment of the consensus values (certified values and their uncertainties) at an acceptable level of 95% confidence interval has been attained using the SoftCRM 1.2.0 software. Consensus values of these Standard Reference Materials remained stable after one year of storage and hence their shelf-life could be prolonged for another one year.

Acknowledgements

The authors are grateful to the Director-General of the Malaysian Palm Oil Board for approval to publish this research work; Director of Product Development and Advisory Services Division and Head of the Analytical and Quality Development Unit for their valuable suggestions; collaborators of the inter-laboratory proficiency tests for their contributions; and laboratory staff of the Analytical and Quality Development Unit for their technical assistance.

Table 4. Changes of slip melting point in Palm-Based Standard Reference Materials.

Standards	T^a (°C)	Storage period (month)[b]								
		0	2	4	6	8	10	12	Mean	SD[c]
Palm Oil	−20	36.4 ± 0.1	36.4	36.3	35.7	36.0	35.8	36.1	36.05	0.27
	0		36.1	35.8	36.1	35.8	36.0	35.9	35.95	0.14
	6		36.0	36.3	35.9	36.0	35.9	36.3	36.07	0.19
	24		36.2	35.8	36.0	35.7	36.2	36.0	35.98	0.20
Palm Olein	−20	22.4 ± 0.2	22.4	22.4	22.9	22.6	23.1	22.3	22.62	0.32
	0		22.7	22.5	22.4	23.0	22.6	22.8	22.67	0.22
	6		22.5	22.9	23.0	23.2	22.5	22.6	22.78	0.29
	24		22.4	22.2	22.8	23.0	22.8	23.1	22.72	0.35
Palm Stearin	−20	53.2 ± 0.1	53.4	53.1	53.0	53.4	53.0	53.2	53.18	0.18
	0		53.0	52.9	53.4	53.5	53.3	54.0	53.35	0.39
	6		53.2	53.4	53.7	53.1	53.0	52.9	53.22	0.29
	24		53.4	53.3	53.2	53.1	53.5	53.0	53.25	0.19

[a]Storage temperature.
[b]Mean of duplicate (°C).
[c]Standard deviation.

Disclosure

The authors report no conflicts of interest.

References

1. Ainie, K., Siew, W.L., Tan, Y.A. et al. 2004. MPOB. Test Methods— A Compendium of Test on Palm Oil Products, Palm Kernel Products, Fatty Acids, Food Related Products and Others. Malaysian Palm Oil Board, Selangor.

2. Azmil Haizam, A.T., Elina, H., Siew, W.L. et al. 2007. Commercialization of Standard Reference Materials from Palm Oil Products, MPOB. 117th Viva Committee Meeting, Viva Report No. 391/2007 (07). Malaysian Palm Oil Board, Selangor.

3. Azmil Haizam, A.T., Siew, W.L. and Ainie, K. 2008. Development of palm-based reference materials for the quantification of fatty acids composition. J. Oleo. Sci., 57:275–85.

4. Bonas, G. SoftCRM version 1.2.0. 1996. Funded by Standards, Measurements and Testing Programmes.

5. Carter, C., Finley, W., Fry, J. et al. 2007. Palm oil markets and future supply. Eur. J. Lipid Sci. Technol., 109:307–14.

6. Choo, Y.M., Nik Mohd Aznizan, N.I. and Norihan, A.M. 2007. Introduction to MPOB. and Malaysian palm oil industry in Selected Readings of the 27th

Palm Oil Familiarization Programme. Malaysian Palm Oil Board, Selangor, 1–23.

7. Firestone, D. 1998. Offical Methods and Recommended Practices of the AOCS. 5th ed. American Oil Chemists' Society, Champaign.

8. Guided Wave Incorporated. Application Note—Iodine value with Model 412 [online]. Accessed 7 May 2008. URL: http://www.guided-wave.com/ support/ notes/.

9. ISO Guide 35 1989. Certification of reference materials—General and statistical principle. 2nd ed. International Organization for Standardization, Switzerland.

10. ISO 3541 1993. Statistics, vocabulary and symbols—Part 1: Probability and general statistical terms. 1st ed. International Organization for Standardization, Switzerland.

11. ISO 5725-2 1994. Accuracy (trueness and precision) of measurement methods and results—Part 1: Basic method for the determination of repeatability and reproducibility of a standard measurement method. 1st ed. International Organization for Standardization, Switzerland.

12. ISO 3961 1996. Animal and vegetable fats and oils—Determination of iodine value. 3rd ed. International Organization for Standardization, Switzerland.

13. Karabulut, I., Turan, S. and Ergin, G. 2004. Effects of chemical interesterification on solid fat content and slip melting point of fat/oil blends. Eur. Food Res. Technol., 218:224–9.

14. Mohd Basri, W. Overview of the Malaysian palm oil industry 2007 [online]. Accessed 7 May 2008. URL: http://econ.mpob.gov.my/economy/ overview07. htm.

15. Pocklington, W.D. and Wagstaffe, P.J. 1987. The certification of the fatty acid profile of two edible oil and fat materials, BCR. Information Reference Materials, Report EUR. 11002 EN. Community of Bureau of Reference, Commission of the European Communities, Brussels.

CITATION

Originally published under the Creative Commons Attribution License or equivalent. Tarmizi AH1, Lin SW, Kuntom A. Palm-based standard reference materials for iodine value and slip melting point. Anal Chem Insights. 2008 Sep 22;3:127-33.

Biomedical and Forensic Applications of Combined Catalytic Hydrogenation-Stable Isotope Ratio Analysis

Mark A. Sephton, Will Meredith,
Cheng-Gong Sun and Colin E. Snape

ABSTRACT

Studies of biological molecules such as fatty acids and the steroid hormones have the potential to benefit enormously from stable carbon isotope ratio measurements of individual molecules. In their natural form, however, the body's molecules interact too readily with laboratory equipment designed to separate them for accurate measurements to be made. Some methods overcome this problem by adding carbon to the target molecule, but this can irreversibly overprint the carbon source 'signal'. Hydropyrolysis is a newly-applied catalytic technique that delicately strips molecules of their functional groups but retains their carbon skeletons and stereochemistries intact, allowing precise determination of the carbon source. By solving analytical problems, the new technique is increasing the ability of scientists to pinpoint molecular indicators of disease, elucidate metabolic pathways and recognise administered substances in forensic investigations.

Keywords: fatty acids, steroids, stable isotopes, hydropyrolysis

Introduction

Stable isotope ratio measurements are becoming important for methods aimed at determining the origin of organic molecules found in biological fluids. In particular the relative abundance of stable isotopes of carbon (carbon-12 and carbon-13) is very useful and can be determined by modern instruments at high sensitivities and high precision.

During analysis materials are usually converted by combustion to CO_2 and the ratios of 13C to 12C in the gas are determined by mass spectrometry (combustion-MS). Variations in values are quoted differentially (compared to an internationally agreed standard, PDB) where: $\delta 13C = [(13C/12C) \text{ sample} /(13C/12C) \text{ standard} -1] \times 1000$ in per mil (parts per thousand; symbol ‰).

In recent times, methods have been developed where stable isotope ratios can be determined for individual molecules, a procedure termed compound specific isotope analysis (abbreviated to CSIA). This is accomplished using a relatively new development of the carbon isotope ratio technique that combines a separation procedure (gas chromatography, GC) with mass spectrometry (MS) via a combustion interface (Matthews and Hayes, 1978). The integrated methodology, termed GC-C-IRMS, can provide high precision (± 0.2‰) measurements on small samples (nanogram quantities) of individual compounds separated from mixtures.

Yet despite the continued development of mass spectrometers that achieve new and unprecedented levels of sensitivity and precision for individual molecules, sample preparation methods for biological molecules represent a bottleneck in analytical advancement. A recent reviewer stated "despite their importance for high-precision compound specific isotope analysis, dedicated studies addressing the issues of sample preparation are few and far between" (MeierAugenstein, 1999).

The problem arises because, in their natural form (Fig. 1a) biological molecules interact too readily with the laboratory equipment designed to separate them prior to isotope ratio analysis. Existing strategies to avoid separation problems involve attaching small molecules to the functional groups (MeierAugenstein, 1999) that reduce the ability of functional groups to interact with the separation equipment (Fig. 1b). This derivatization approach solves the separation problem but the addition of extra carbon atoms can corrupt the original carbon isotope signal of the target molecules. Stable isotopic changes occur when derivatization takes place in a nonquantitive manner. Commonly applied derivatization methods include esterification, acetlyation and silylation (MeierAugenstein, 2004). Esterification and acetlyation involve modification of carbon atoms and can induce kinetic isotope effects. Silylation can produce derivatives that interfere with the conversion of the target molecule to CO_2, a necessary step for CSIA by GC-C-IRMS. Other potentially useful methods remove the functional groups of biological molecules by chemical treatments such as catalytic hydrogenation to convert starting compounds to the parent hydrocarbon and/or the next lower homolog (Fig. 1c). Regrettably, in the past, this approach has removed part of the carbon skeleton of the molecule and information about the structure of the molecule is lost making the products difficult to identify (Beroza, 1962). Furthermore, following the loss of carbon, the carbon isotope ratios of the products may become dissimilar to that of the starting material.

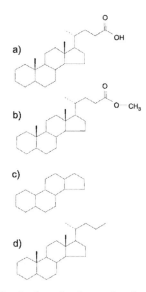

Figure 1. Strategies to make biological molecules such as fatty acids and steroids amenable to stable isotope ratio analysis by GC-C-IRMS. a) A common steroid molecule (cholanoic acid) with a functional group that causes it to interact too readily with the separation equipment. A successful isotope ratio analysis must measure the carbon isotope ratio of the carbon skeleton. b) One approach is to modify the functional groups by attaching small molecules but this adds carbon. c) Another approach is to remove functional groups by catalytic reactions but this has previously removed parts of the carbon skeleton. d) The hydropyrolysis approach removes functional groups but retains the carbon skeleton intact for stable isotope analysis.

A new analytical method has been developed to replace the functional groups of biological molecules with hydrogen but retain the carbon skeleton and stereochemistry of the molecule intact (Fig. 1d). Hydropyrolysis involves the catalytic addition of hydrogen to the carbon skeleton at relatively high pressures using a dispersed molybdenum catalyst. Indeed, during hydropyrolysis, all that is lost from the molecule is that which analysts would want to remove prior to carbon isotopic analysis. During the procedure a catalyst, ammonium dioxydithiomolybdate [(NH4)2MoO2S2], decomposes in situ above 250°C to form a catalytically-active molybdenum sulphide phase. Hydropyrolysis experiments are performed in a continuous flow temperature-programmed reactor configuration (Fig. 2), which has been described in the literature (Love et al. 1995, 1997; Russell et al. 2004; Sephton et al. 2005a), with products rapidly swept from the reactor to a silica trap.

Two of the most important classes of molecule starting to benefit from stable isotope studies are the fatty acids and steroid hormones. In their natural form both types of molecule contain functional groups that hinder chromatographic separation during GC-C-IRMS. This paper summarizes and extends previously published work on these molecules (Sephton et al. 2005a; Sephton et al. 2005b).

Fatty Acids

Fatty acids are one of the most important classes of molecules in metabolism. Fatty acids serve as precursors in the synthesis of other compounds, act as a high density source of calories and facilitate the transport of essential nutrients such as fat-soluble vitamins. Furthermore, some diseases involve disturbances in fatty acid metabolism including diabetes mellitus, sudden infant death syndrome and Reye's syndrome.

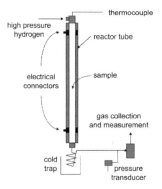

Figure 2. Schematic of the hydropyrolysis equipment.

To examine the efficacy of hydropyrolysis for transforming functionalised molecules to their hydrocarbon counterparts Sephton et al. (2005a) tested the procedure on a simple fatty acid (n-octadecanoic or "stearic" acid). The isotopic composition of this starting material was analysed by combustion-IRMS (Table 1). The n-octadecanoic acid was prepared for GC-C-IRMS by subjecting it to hydropyrolysis at 520°C. n-Octadecanoic acid was adsorbed onto silica and mixed with quartz sand before the catalyst was added. Samples were then heated resistively from 50 °C to 250 °C at 30 °C min–1, and then from 250 °C to 520 °C at 8 °C min–1, under a hydrogen pressure of 15 MPa. A hydrogen sweep gas flow of 10 dm3 min–1, measured at ambient temperature ensured that the products were quickly removed from the reactor vessel. Fig. 3 displays GC-C-IRMS analyses of the untreated n-octadecanoic acid and its converted counterpart, n-octadecane (Sephton et al. 2005a).

Table 1. Carbon isotope ratios of starting materials and hydropyrolysis products. (Data from Sephton et al. 2005a and Sephton et al. 2005b).

Compound	$\delta^{13}C$ (‰)	±1σ	Technique
Octadecanoic acid			
n-Octadecanoic acid	−28.3	<0.1	Combustion-MS
n-Octadecane	−28.0	<0.1	GC-C-IRMS
Cholanic acid			
5β-Cholanic acid	−14.6	0.1	Combustion-MS
5β-Cholane	−14.6	0.2	GC-C-IRMS
Cholestanol			
5α-Cholestanol	−25.0	0.1	Combustion-MS
5α-Cholestane	−24.6	0.2	GC-C-IRMS
Cholesterol			
Cholesterol	−25.1	0.1	Combustion-MS
5β-Cholestane	−24.7	0.3	GC-C-IRMS
5α-Cholestane	−24.2	0.3	GC-C-IRMS

Fig. 3 reveals that n-octadecanoic acid displayed significant peak tailing. It follows that when analysed within a complex mixture, this peak tailing would lead to peak overlap precluding accurate stable isotope ratio measurements. Broader peaks also increase the proportion of background measured with the analyte, degrading the ultimate stable isotope ratio determination. In contrast, the hydropyrolysis product (n-octadecane) gave a much sharper peak that would be less likely to overlap with those for other compounds when present within biological fluids and would produce a measurement with relatively a low contribution from background. In summary, the benefits of the procedure for fatty acid anayoco appear to be (i) more precise measurements and (ii) smaller amounts of sample required, both owing to the increased signal to noise ratio associated with hydro-pyrolysis products relative to their starting materials.

Comparison of carbon isotopic determinations for the untreated n-octade-canoic acid by combus-tion-IRMS and the products from hydropyrolysis by GC-C-IRMS (Fig. 3, Table 1), indicated that the isotopic composition of the processed sample was representative of the starting material. It appears that no isotopic effects are associated with the conversion from acid to alkane. Thus, the technique allows the effective determination of the carbon isotopic composition of individual fatty acids without the use of derivatizing agents.

Steroid Hormones

Steroid hormones perform vital biochemical functions including regulation of sexual development and function, suppression of inflammation, stress control, and maintenance of salt and water balance. Understanding the origin and fate of individual steroids within the human is essential for both endocrine studies and forensic investigations.

In biochemical investigations of the endocrine system, "labelled" steroids are deliberately enriched in the heavy stable isotope of carbon and are tracked as they pass along their metabolic pathways (e.g. Wolthers and Kraan, 1999). Stable isotope tracers have significant advantages over more conventional radioactive counterparts because they have no negative physiological effects (Koletzko et al. 1997).

Forensic investigations in athletics use natural abundances of stable carbon isotopes to determine the origin of androgenic and anabolic steroids and their me-tabolites in biological fluids. For example, androgenic-anabolic steroids are part of the World Anti-Doping Agency's prohibited list (WADA, 2006). The abuse of these substances can be difficult to detect because steroids such as testosterone are found naturally in the body and exogenous analogues may be used to top up nor-mal levels. Moreover, some steroids have uncertain origins and controversy exists about whether they are produced endogenously. For instance, there is increasing evidence that nandrolone metabolites can appear in the urine of people without the administration of exogenous nandrolone (Kohler and Lambert, 2002). Other steroids may be ingested in contaminated meat products or dietary supplements (Debruyckere inet al. 1992). Fortunately for the forensic scientist, steroids des-tined for pharmaceutical applications are produced by modifying steroid mol-ecules from plants (Coppen, 1979). Plants and humans are isotopically distinct and stable isotope studies can discriminate between exogenous and endogenous sources. GC-C-IRMS has been applied with some success to individual steroids in urine samples from humans (e.g. Becchi et al. 1994).

Hydropyrolysis of steroids represents a significant analytical challenge owing to their structural complexity, both in terms of oxygen functionality and the presence of carbon double bonds. To assess the efficacy of hydropyrolysis for steroids Sephon et al. (2005b) subjected 5β cholanic acid, 5α cholestanol and cholesterol to the hydropyrolysis procedure previously described for fatty acids.

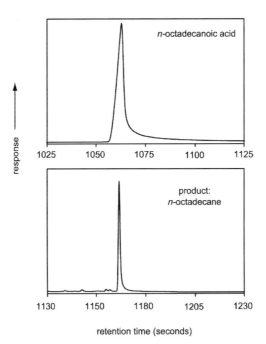

Figure 3. GC-C-IRMS traces of n-octadecanoic acid and its hydro-pyrolysis product, n-octadecane, displaying a marked increase in chromatographic performance (Sephton et al. 2005a).

GC-C-IRMS analyses indicated that hydropyrolysis of 5β cholanic acid produced the single molecular product 5β cholane (Fig. 4). The efficient conversion of 5β cholanic acid to its hydrocarbon counterpart was expected owing to the position of the oxygen functionality on an aliphatic side chain. Hence the conversion is similar to that observed for n-octadecanoic acid.

GC-C-IRMS analysis of the hydropyrolysis products of cholestanol also displayed largely a single molecular product in 5αcholestane (Fig. 5). The results illustrated that, in addition to defunctionalizing aliphatic side chains, hydropyrolysis also efficiently eliminates exocyclic oxygen-containing functional groups.

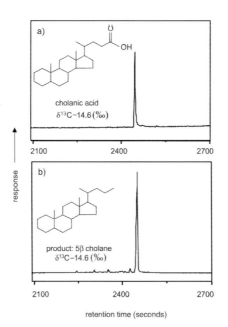

Figure 4. GC-C-IRMS traces of (a) cholanic acid and (b) its hydropyrolysis product 5b cholane displaying an increase, chromatographic performance.

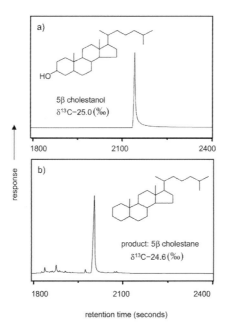

Figure 5. GC-C-IRMS traces of (a) 5α cholestanol and (b) its hydropyrolysis product 5α cholestane displaying an increase in chromatographic performance (Sephton et al. 2005b).

For cholesterol, effective hydrogenation would be expected to produce two cholestane isomers (5α and 5β) owing to the non-selective nature of the hydrogenation reaction for the carbon double bond adjacent to ring-joining positions. However GC-C-IRMS analysis of the products (Fig. 6) indicated that extensive rearrangement occurs with the hydropyrolysis procedure giving, in addition to the two expected cholestane isomers, four diasteranes, and an unresolved complex mixture comprising other diasteranes, cholestenes and diasterenes. Multiple products are less suitable for effective carbon isotope analysis and the cholesterol data suggested that a catalyst system and temperature regime that minimises such rearrangements should be a high priority for future development. Comparison of carbon isotopic determin ations for the untreated cholanic acid, cholestanol and cholesterol by combustion-IRMS and for the products from hydropyrolysis by GC-C-IRMS (Table 1), indicated that the isotopic compositions of the processed samples were faithful expressions of the starting materials. It appears that no isotopic effects were associated with the conversion from functionalized steroid to the hydrocarbon counterpart. The technique seems to enable carbon isotopic compositions of individual steroids to be obtained without the corrupting effects associated with derivatization.

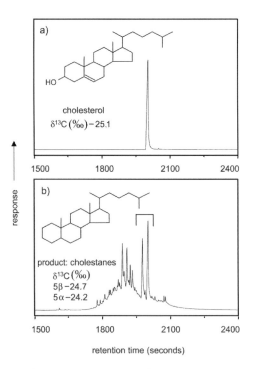

Figure 6. GC-C-IRMS traces of (a) cholesterol and (b) its hydropyrolysis products (mainly 5b and 5a cholestane) (Sephton et al. 2005b).

Conversion Results

With the possibility of structural rearrangements occurring during hydropyrolysis (Fig. 6) Sephton et al. (2005b) attempted to constrain the isotopic effects of partial conversion of a functionalised molecule to its hydrocarbon skeleton. This was achieved by operating the procedure at a range of temperatures (400, 500 and 550 °C) with the lower temperatures representing suboptimal conditions. Partial conversion may result in the product exhibiting enrichment in the more reactive ^{12}C isotope while the residual starting material becomes enriched in the less reactive ^{13}C. Such incomplete transformation could give hydropyrolysis products with carbon isotopic compositions not fully representative of the molecules of interest.

To examine the effect of incomplete conversion on stable isotopic compositions, Sephton et al. (2005b) compared the hydropyrolysis products of n-octadecane and 5β cholanic acid produced at various temperatures and degrees of conversion. Figure 7 indicates that limited variation occurred between yields of 50 and 80‰ conversion probably reflecting the lack of carbon-carbon bond disruption associated with the procedure. Therefore, even under less than optimal conditions and with only partial conversion, hydropyrolysis products appear to be effective indicators of the carbon isotopic composition of the starting material, suggesting significant analytical contingency in the process.

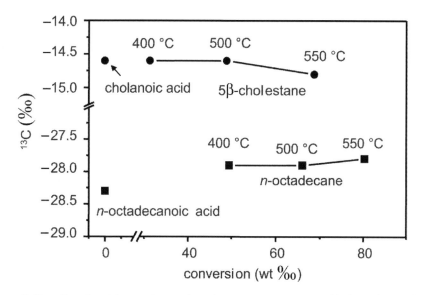

Figure 7. The effect of partial conversion on the carbon isotopic composition of hydropyrolysis products (Sephton et al. 2005b).

Conclusions

Hydropyrolysis shows great potential as an effective preparative technique to facilitate GC-C-IRMS analysis of fatty aids and steroids. The absence of reactions which add or remove carbon prior to analysis make the carbon isotopic compositions of hydropyrolysis products faithful representatives of the starting materials, and the much improved chromatographic performance of the products should allow GC-C-IRMS to be applied to increasingly complex mixtures of fatty acids, steroid hormones and their metabolites. This analytical advance introduces the future possibility of detecting more subtle cases of diseases and endogenous steroid abuse than possible by current methods.

Acknowledgements

We are grateful to two anonymous reviewers for their constructive comments.

References

1. Becchi, M., Aguilera, R., Farizon, Y. et al. 1994. Gas-chromatography combustion isotope ratio mass-spectrometry analysis of urinary steroids to detect misuse of testosterone in sport. Rapid Communications in Mass Spectrometry, 8:304–308.

2. Beroza, M. 1962. Determination of the chemical structure of microgram amounts of organic compounds by gas chromatography. Analytical Chemistry, 34:1801–1811.

3. Coppen, J.J.W. 1979. Steroids: from plants to pills—the changing picture. Tropical Science 21:125–141.

4. Debruyckere, G., de, Sagher, R. and Van Peteghem, C. 1992. Clostebolpositive urine after consumption of contaminated meat. Clinical Chemistry, 38:1869–1873.

5. Kohler, R.M.N. and Lambert, M.I. 2002. Urine nandrolone metabolites: false positive doping test? British Journal of Sports Medicine, 36:325–329. Koletzko, B., Sauerwald, T. and Demmelmair, H. 1997. Safety of stable isotope use. European Journal of Pediatrics, 156:S12–S17.

6. Matthews, D.E. and Hayes, J.M. 1978. Isotope-ratio-monitoring gas chromatography-mass spectrometry. Analytical Chemistry, 50:1465–1473.

7. MeierAugenstein, W. 1999. Applied gas chromatography coupled to isotope ratio mass spectrometry. Journal of Chromatography, A. 842:351–371.

8. MeierAugenstein, W. 2004. GC and IRMS technology for 13C and 15N analysis on organic compounds and related gases. In Handbook of Stable Isotope Analytical Techniques, (ed de Groot PA) Vol. 1, 153–176.

9. Sephton, M.A., Meredith, W., Sun, C-G, and Snape, C.E. 2005a. Hydropyrolysis as a preparative method for the compound-specifi c carbon isotope analysis of fatty acids. Rapid Communications in Mass Spectrometry, 19:323–325.

10. Sephton, M.A., Meredith, W., Sun, C.G. and Snape, C.E. 2005b. Hydropyrolysis of steroids: a preparative step for compound-specific carbon isotope ratio analysis. Rapid Communications in Mass Spectrometry 19:3339–3342.

11. WADA. 2006. World Anti-Doping Code, The 2006 Prohibited List, International Standard. [online] Accessed 21 September 2006. URL: http://www.wada-ama.org/rtecontent/document/2006_LIST.pdf.

12. Weykamp, C.W., Penders, T.J., Schmidt, N.A. et al. 1989. Steroid profile for urine: reference values. Clinical Chemistry, 35:2281–2284.

13. Wolthers, B.G. and Kraan, G.P.B. 1999. Clinical applications of gas chromatography and gas chromatography- mass spectrometry of steroids. Journal of Chromatography, A. 843:247–274. Analytical Chemistry Insights 2007: 2.

CITATION
Originally published under the Creative Commons Attribution License or equivalent. Sephton MA1, Meredith W, Sun CG, Snape CE. Biomedical and forensic applications of combined catalytic hydrogenation-stable isotope ratio analysis. Anal Chem Insights. 2007 Sep 6;2:37-42.

Identification and Quantitation of Asparagine and Citrulline Using High-Performance Liquid Chromatography (HPLC)

Cheng Bai, Charles C. Reilly and Bruce W. Wood

ABSTRACT

High-performance liquid chromatography (HPLC) analysis was used for identification of two problematic ureides, asparagine and citrulline. We report here a technique that takes advantage of the predictable delay in retention time of the co-asparagine/citrulline peak to enable both qualitative and quantitative analysis of asparagine and citrulline using the Platinum EPS reverse-phase C18 column (Alltech Associates). Asparagine alone is eluted earlier than citrulline alone, but when both of them are present in biological samples they may co-elute. HPLC retention times for asparagine and citrul line were influenced by other ureides in the mixture. We found that at various asparagines and citrulline ratios [= 3:1, 1:1, and 1:3; corresponding to 75:25, 50:50, and 25:75 (µMol ml–1/µMol ml–1)], the resulting peak exhibited different retention times. Adjustment of ureide ratios as internal standards enables peak identification and quantification. Both chemicals were quantified in xylem sap samples of pecan [Carya illinoinensis (Wangenh.) K. Koch] trees. Analysis revealed that tree nickel nutrition status affects relative concentrations of Urea Cycle intermediates, asparagine and citrulline, present in sap. Consequently, we concluded that the HPLC methods are presented to enable qualitative and quantitative analysis of these metabolically important ureides.

Keywords: HPLC identification, pecan, ureides, asparagine, citrulline.

Introduction

High-performance liquid chromatography (HPLC) is a potentially powerful tool for quantitative and qualitative analysis of substances possessing similar chemical characteristics. While varying column types, column conditions, and eluate characteristics generally enable peak resolution sufficient for fractionation and analysis of similar chemical constituents in samples, such manipulations are often unsuccessful for ureide mixtures containing both asparagine and citrulline. Ureides are cyclic or acyclic acyl derivatives of urea and play key role in nitrogen metabolism and cycling in certain species of higher plants. Qualitative and quantitative analysis of these two ureides is critical for studies assessing Urea Cycle activity, a metabolic cycle involving nitrogen cycling in organisms, and factors affecting cycle functionality.

Most reversed-phase silica-based media are covalently bonded. The Platinum EPS phase is unique for HPLC analysis based on the utilization of the interactions of the underlying base silica to enhance the selectivity of polar organic analysis (Alltech Associate, Inc. 2002). The Platinum EPS reversed-phase C18 column (Alltech Associates) has been used for ureide analysis (Bai et al. 2006), although its utility is limited in that analysis of asparagine and citrulline, both key Urea Cycle ureides, is problematic. Ureides can also be analyzed via the ZORBAX Rx-SIL(Agilent Technologies) column (made from porous silica microspheres). It is used for normal-phase chromatography and is designed for high stability at low pH (Agilent Technologies, 2006), but was found by the authors to be inferior to the Platinum EPS column for separation of asparagine and citrulline.

Both asparagine and citrulline are considered as nonessential amino acids. Asparagine is one of the 20 most common natural amino acids in biology. It has carboxamide as the side chain's functional group, thus allowing for efficient hydrogen bond interactions with peptide backbones. The precursor to asparagine is oxaloacetate, which is converted to aspartate by a transaminase (Nelson and Cox, 2000). Citrulline is an amino acid critical to the detoxification and elimination of unwanted ammonia within cells. During metabolism, citrulline is a precursor of arginine (Nelson and Cox, 2000). Chemical structures for asparagine and citrulline are as follows:

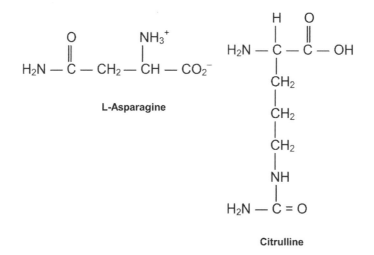

We present here an important yet simple technique that overcomes co-elution problems often encountered with HPLC elution of asparagine and citrulline, thus enabling quantitative and qualitative analysis of L-asparagine and citrulline by HPLC using a Platinum EPS C18 column.

Experimental Section

Biological Materials and Standards

Xylem sap of pecan [Carya illinoinensis (Wangenh.) K. Koch] trees were used for ureide analysis. Two classes of trees were assessed, those that were nickel sufficient (Ni-S) and those nickel deficient (Ni-D). Samples were from multiple 3-year-old greenhouse-grown seedling trees exhibiting either of these two Ni nutritional classes. Classification of Ni deficiency was based on presence/absence of morphological symptoms of Ni deficiency and by tissue Ni concentration (Bai et al.

2006). The ureide standards were acquired from commercial sources. L asparagine anhydrous (99.5%) was purchased from Fluka Biochemika (St. Louis, Mo), whereas citrulline (98%) and other ureides (e.g. allantoic acid and allantoin) were from Sigma (St. Louis, Mo).

HPLC Analysis of Ureides

The methods for ureide analysis were similar to those described previously (Bai et al. 2006). Analysis was performed on a Spectra System SCM 1000 HPLC linked with a Spectra System UV 1000 detector (Thermo Electron Corporation, San Jose, CA) using a Platinum EPS C18 column, 5 μm, 250 mm × 4.6 mm (Alltech Associates, Deerfield, IL). Mobile phase was acetonitrile: 0.03 M potassium phosphate, pH 3.2 (20:80). Flow rate was 0.5 ml/min at a column temperature of 30°C. Sample injection volume was 20 μl. Ureides were detected at 190 nm. Tentative ureide identification was based on identical retention times (RT) compared to ureide standards. The standard ureides (asparagine, citrulline, allantoic acid and allantoin) and internal standard ureides (the same as mentioned as above) were used at a range of 10–100 μMoles ml^{-1} [in acetonitrile: 0.03 M potassium phosphate monobasic pH 3.2, (20:80)]. Plant samples were analyzed for fi ve replicates of each Ni treatment class. A ZORBAX Rx-SIL (Agilent Technologies) column (5 μm, 4.6 mm × 150 mm) with a mobile phase consisting of acetonitrile, 20 mM potassium phosphate, pH 7.2 (90:10) was used to verify the ureide analysis using the Platinum EPC column.

Influence of Different Ratios of Asparagine and Citrulline

The influence of asparagine and citrulline ratio on retention time was evaluated using either 100 μMol ml^{-1} total concentration. The concentrations of asparagine and citrulline were 75:25, 50:50, and 25:75 (μMol ml^{-1}:μMol ml^{-1}), corresponding to the ratios of 3:1, 1:1, and 1:3. Analysis was performed using the Platinum EPS C18 column. The protocol typically resulted in very small differences in retention times for asparagine and citrulline (5.84 min and 5.87 min, respectively) when analyzed as a mixture at equivalent concentrations of allantoic acid, asparagine, citrulline and allantoin. A mixture of these four ureides was then analyzed as above but at different ureide ratios of 1:7:1:1 and 1:1:7:1 of allantoic acid: asparagine: citrulline: allantoin solutions (total concentrations: 100 μMol ml^{-1}). In the present study, we added a relatively high concentration of either asparagine or citrulline as a carrier to the plant sample; thus, the retention time for either citrulline or asparagine in the mixed sample is essentially the same as that for the standard alone. It is noteworthy that the influence on the retention times for asparagine and citrulline diminishes, and resulting peaks are resolved if several ureides are mixed for HPLC analysis. The amount of either asparagine or citrulline in the biological sample is then determined by subtracting the peak area of "a single standard of another run" from the relatively large peak area obtained from a run containing both a "biological sample plus a single standard". The calculation is as follows:

$$S = \text{the peak area of ureide (i.e. asparagine or citrulline) standard} \quad (1)$$

Note that the value of S is determined after a standard is run alone.

$$A = M - S2 \quad (2)$$

where

A = the peak area of the ureide in the biological sample;

M = the peak area of combined mixture of sample and ureide standard (S_1); and

S_2 = the peak area of the ureide standard alone, determined in a separate HPLC run.

Note that the combined mixture of the respective ureide standard at a relatively high concentration (S1) and the respective ureide from biological sample results in a relatively large peak area (M) as compared with other peak area, which are relatively small. S1 and S2 are analyzed at equal concentrations of the standard ureide in both runs.

Preparation of Xylem Sap Fractions

Xylem sap samples were collected from several Ni-D and Ni-S trees at bud break in late March. Sap was collected via vacuum extracted from stems severed about 2 cm above the root collar and again just below the apical tip. Phloem and bark were removed for the distance of 1 cm from the base, and the base of severed stems placed in a vacuum chamber, the exuding xylem sap collected into 2 ml vials, and the sap samples immediately frozen.

Xylem sap samples from twelve representative trees of each of the two Ni treatments were randomly selected and then processed in triplicate for reduced-N and urease analysis. Equivalent volumes of xylem sap for each sample were thawed and twice centrifuged (20,000g) for 30 min. The supernatant was further purified by removing molecules ≥10-kD by twice filtering through a Centricon-10 filter (Millipore filter units, Millipore, Bedford, MA) after centrifugation (5,000g) for 75 min. The partially purified samples were then analyzed for ureides using UV spectroscopy and HPLC. For identification purposes, samples were processed on the two above described HPLC columns with different eluants.

Statistical Analyses

Concentrations of organic molecules in Ni-D and Ni-S samples were square root transformed, prior to analysis. Differences in concentrations in two classes were compared by t-test (P ≤ 0.05; SAS 2001). Mean values were analyzed with four tissue replicates.

Results and Discussion

Initial Analysis of Asparagineand Citrulline Mixture

HPLC analysis, using the above described Platinum EPS C18 column and elute conditions, of a mixture of asparagine and citrulline yields a single peak reflecting co-elution of asparagine and citrulline (5.85 min) (Figure 1). The ZORBAX Rx-SIL column enabled slight separation of these two ureides, giving only slight differences in retention times (5.41 min for asparagine and 5.44 min for citrulline). Also, one other ureide, allantoic acid, elutes at essentially the same time (5.40 min) as asparagine and citrulline. Thus, asparagine and citrulline analysis using the ZORBAX Rx-SIL column is problematic, especially in the presence of allantoic acid. Thus, the ZORBAX column is unsuitable for quantitative analysis using the above described conditions.

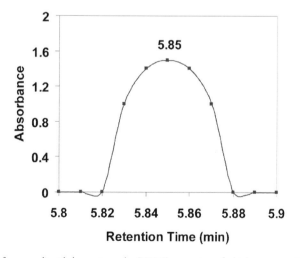

Figure 1. High-performance liquid chromatography (HPLC) separation of a high concentration of 1:1 mixture (100 μMol ml⁻¹:100 μMol ml⁻¹) of asparagine and citrulline using a Platinum EPS C18 column (5 μm, 250 mm × 4.6 mm) at attenuation 16 and detected at 190 nm. Mobile phase was acetonitrile: 0.03 M potassium phosphate, pH 3.2 (20:80); flow rate was 0.5 ml/min; column temperature was 30°C; and sample injection volume was 20 μl.

Analysis of Asparagineand Citrulline Mixture

Retention times of eluting ureides generally change with concentration of the target molecule, eluting pH, temperature, and mobile phase composition. While the most common ureides likely to be found in biological samples were marginally resolvable using the two HPLC protocols described above, asparagine and citrulline analysis remained problematic, as shown in the chromatograms (Figure 2). Ureide separation using the Platinum EPS C18 column was such that the retention time for asparagine alone 2 is earlier (Figure 2A) than the mixtures of asparagine and citrulline (Figure 2B, 2C and 2D) and citrulline alone (Figure 2E). Asparagine alone is eluted at 5.82 min (Figure 2A), whereas citrulline alone eluted at 5.89 min (Figure 2E). Different retention times indicate that they satisfactorily separate, thus A giving the initial appearance of being an efficacious means of analyzing the two problematic ureides while enabling resolution of other ureides, such as allantoic acid (5.44 min), allantoin (5.93 min), and uric acid (6.06 min) (Bai et al. 2006). However, when asparagine and citrulline are mixed at different ratios the peaks fail to resolve and retention time of the single co-eluting peak slows as citrulline concentration increases (Figure 2). For example: retention times progressively increase to 5.84, 5.85 and 5.87 min, respectively, for a 3:1, 1:1 and 1:3 asparagine/citrulline mixture (Figure 2B, 2C, and 2D). The co-elution is obviously due to an interaction between asparagine, citrulline, and the column, 5.8 5.82 5.84 5.86 5.88 5.9 making it difficult to assess results.

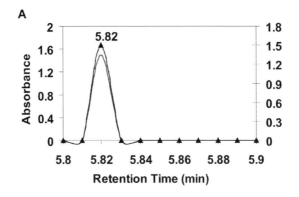

Figure 2. HPLC separation of mixed asparagine and citrulline at various concentration ratios using a Platinum EPS C18 column. Ratios are (A) 4:0 (100 μMol ml⁻¹:0; asparagine alone); (B) 3:1 (75 μMol ml⁻¹:25 μMol ml⁻¹); (C) 1:1 (50 μMol ml⁻¹:50 μMol ml⁻¹); (D) 1:3 (25 μMol ml⁻¹:75 μMol ml⁻¹); and (E) 0:4 (0:100 μMol ml⁻¹; citrulline alone) at attenuation 32 and detected at 190 nm. The mobile phase and conditions are as described in Figure 1.

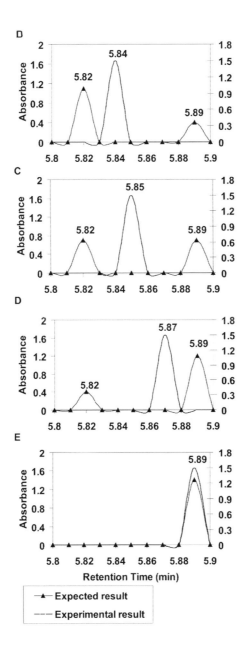

Figure 2. (Continued)

Peak symmetry was maintained for all ureide ratio mixtures. It is noteworthy that retention times on the Platinum EPS C18 column increase as the asparagine and citrulline ratios decrease; thus reflecting a complex interaction between these two ureides, the column mediums, and eluting conditions. One possible explanation is that the ureide aggregates arising from the various ratio mixes (e.g. 3:1, 1:1 and 1:3 ratios) differ in the nature of weak chemical interactions associated with hydrogen bond interaction (Nelson and Cox, 2000). Ureide separation using the Platinum EPS C18 column was improved with ureide ratios of 1:1:7:1 and 1:7:1:1 (allantoic acid: asparagine: citrulline: allantoin), with asparagine clearly separating from citrulline (Figure 3). This indicates a reduction in the physiochemical interaction between asparagine and citrulline that otherwise caused co-elution problems for asparagine and citrulline.

The lengthening of asparagine retention time and shortening of citrulline retention time, respectably, within mixtures of ureides is proportional to the relative concentration of these two ureides. However, the influence of citrulline in a sample on asparagine retention time is minor if an additional amount of asparagine is added to the test sample at a relatively high concentration. Similarly, the influence of asparagine on citrulline retention time is minor if an additional amount of citrulline is added to the test sample at a relatively high concentration. When a plant sample is analyzed, analysis is performed using either individual asparagine or citrulline as the major reference for identification with both the Platinum EPS C18 column and the ZORBAX Rx-SIL column. This influence on retention time diminishes if the sample contains several ureides (allantoic acid, allantoin, asparagine, citrulline and uric acid).

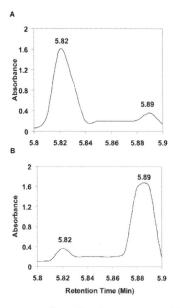

Figure 3. HPLC separation of asparagine and citrulline from a mixture of four ureides at two concentration ratios using a Platinum EPS C18 column. Asparagine was separated from citrulline at ratios of (A) 1:7:1:1 and (B) 1:1:7:1 (allantoic acid: asparagine: citrulline: allantoin; total concentrations were 100 μMol ml⁻¹) at attenuation 16 and detected at 190 nm. The mobile phase and conditions are as described in Figure 1.

Table 1. High-performance liquid chromatography (HPLC) separation of asparagine and citrulline from xylem sap of young pecan trees differing in nickel nutritional status.[a]

Sample type/Ureide	(Min)	X ± Std[b] (μMol ml^{-1})	
		Ni-deficient	Ni-sufficient
Asparagine	5.82	≤0.5b	39.2 ± 4.7a
Citrulline	5.89	28.7 ± 0.3a	≤0.5b

[a]Separation was obtained using an Alltech Associates Platinum EPS C18 column. Ni-S = nickel sufficient. Ni-D = nickel deficient.
[b]Means (± Std) followed by different letters are significantly different (t-test, $P \leq 0.05$).

Analysis of Asparagine and Citrullinein Xylem Sap of Trees with Different Ni Nutritional Status

Pecan trees experience a nickel deficiency in certain growth environments (Wood et al. 2004). Comparative study of Urea Cycle intermediates was made of xylem sap extracts from Ni-S vs. Ni-D trees using the two HPLC columns and methods described above. The identification of these two ureides has been carried out with the Platinum EPS C18 column and the ZORBAX Rx-SIL column. But the concentration of these two ureides listed in Table 1 is derived from the Platinum EPS C18 column. Analysis of the partially purified ≤10 kDa fractions found that asparagine concentration was the highest in spring xylem sap of Ni-S trees, a 78-fold difference in Ni-D trees and that citrulline level was the highest in Ni-D trees, a 57-fold difference in concentration (Table 1). These results identified substantial differences in asparagine and citrulline concentrations in pecan xylem sap. Similar differences were previously noted in pecan foliage (Bai et al. 2006), where asparagine is at a relatively low concentration and citrulline is at relatively high concentration for both Ni-D and Ni-S classes. A substantially greater level (about 1.4-fold) of citrulline was found in foliage of Ni-D trees than that of Ni-S trees (Bai et al. 2006).

In conclusion, we provide a useful means of analyzing asparagine and citrulline in biological samples using HPLC analysis, and present an example of two disrupting ureides interacting with HPLC columns to potentially cause problems with analysis of asparagine or citrulline. We also show that nickel nutritional status can quantitatively affect Urea Cycle intermediates in pecan trees.

References

1. Agilent Technologies. 2006. Agilent Zorbax RX-Sil. Data Sheet. URL: www.chem.agilent.com.

2. Alltech Associate, Inc. 2002. Platinum HPLC column for extended polar selectivity. www.discoverysciences.com/productinfo/technical/datasheets/u32001.pdf.

3. Bai, C., Reilly, C.C. and Wood, B.W. 2006. Nickel defi ciency disrupts metabolism of ureides, amino acids, and organic acids of young pecan foliage. Plant Physiol. 140:433–43.

4. Nelson DL, M. and Cox, M.J. 2000. Lehninger Principles of Biochemistry, 3rd edition. Worth Publishers.

5. Wood, B.W., Reilly, C.C. and Nyczepir, A.P. 2004. Mouse-ear of pecan: a nickel defi ciency. HortScience, 39:1238–42. Analytical Chemistry Insights 2007: 2.

CITATION

Searching for New Clues about the Molecular Cause of Endomyocardial Fibrosis by Way of In Silico Proteomics and Analytical Chemistry

Misaki Wayengera

ABSTRACT

Background

Endomyocardial Fibrosis (EMF) is a chronic inflammatory disease of the heart with related pathology to that of late stage Chaga's disease. Indeed, both diseases are thought to result from auto-immune responses against myocardial tissue. As is the case that molecular mimicry between the acidic termini of Trypanosoma cruzi ribosomal P0, P1 and P2β (or simply TcP0, TcP1, and TcP2β) proteins and myocardial tissue causes Chaga's disease, excessive exposure to certain infections, toxins including cassava ones, allergy and malnutrition has been suggested as the possible cause for EMF. Recent studies have defined the proteomic characteristics of the T. cruzi ribosomal P protein-C-termini involved in mediating auto-immunity against Beta1-adrenergic receptors of the heart in Chaga's disease. This study aimed to investigate the similarity of C-termini of TcP0/TcP2β to sequences and molecules of several plants, microbial, viral and chemical elements- most prior thought to be possible causative agents for EMF.

Methods and Principal Findings

Comparative Sequence alignments and phylogeny using the BLAST-P tool at the Swiss Institute of Biotechnology (SIB) revealed homologs of C-termini of TcP0 and TcP2β among related proteins from several eukaryotes including the animals (Homo sapiens, C. elegans, D. melanogaster), plants (Arabidopsis thaliana, Zea mays, Glycina Max, Oryza sativa, Rhizopus oryzae) and protozoa (P. falciparum, T. gondii, Leishmania spp). The chemical formulae of the two T.cruzi ribosomal protein C-terminal peptides were found to be $C_{61}H_{83}N_{13}O_{26}S_1$ and $C_{64}H_{87}N_{13}O_{28}S_1$ respectively by Protparam. Both peptides are heavily negatively charged. Constitutively, both auto-antigens predominantly contain Asparagine (D), Glycine (G) and Phenylamine (F), with a balanced Leucine (L) and Methionine (M) percent composition of 7.7%. The afore going composition, found to be non-homologous to all molecules of chemical species in the databases searched, suggests the possible role of a metabolic pathway in the pathogenesis of EMF if aligned with our "molecular mimicry" hypothesis.

Conclusions

Our findings provide a "window" to suggest that cross reactivity of antibodies against C-terminal sequences of several animal, plant and protozoal ribosomal P proteins with heart tissue may mediate EMF in a similar manner as C- termini of T. cruzi do for Chaga's disease.

Introduction

Endomyocardial fibrosis or simply EMF is a restrictive cardiomyopathy known to affect persons of defined geographical locales and socioeconomic status [1], [2]. First described at the Department of Pathology-Makerere University, Uganda by the Pathologist J.N.P Davies in 1948[3], the important features of this disease—namely, geographical distribution, cardiac specificity and preference for the socio-economically poor, have evaded a complete scientific explanation despite the intense scientific scrutiny to which the disease has been subjected[4], [5]. Although the pathological lesions in EMF have been clearly found to comprise fibrosis and calcification, possibly resulting from long standing inflammatory responses, no natural insult is evidenced to cause such pathology [5], [6]. Specifically, in as much as several potential insults have been proposed as the primary cause for EMF, including Infection (Toxoplasmosis, Rheumatic fever, Malaria, Myocarditis and Helminthes [7]), allergy (Autoimmunity and Eosinophilia [8]), malnutrition (Protein or Magnesium deficiency[5], [7]) and toxic agents(Cassava, other plant toxins, Arsenic[9], Cerium, Thorium, Serotonin, or Vitamin D[5]); no single one is proven[5], [10]. Existing evidence for an ethnic predisposition points to a possible genetic idiosyncrasy [11], [12]. Largely because of the above lack of evidence for a particular causative insult, the disease remains unpreventable [5]. Recent studies indicate that there might indeed be a decline in the incidence of EMF paralleled to improvement in the socioeconomic welfare of high risk populations [4]. Until now, the only evidenced benefit for drug use in EMF-deterring progression of the inflammatory pathology, has revolved around steroids [13], with the list of trial drugs expanding to include, more lately, serotonin receptor inhibitors [14], [15]. Surgery, mainly that involving cardiomyoectomy of pathological lesions (plus reconstruction of the heart architecture), has a role despite its infrequent use due to poor state of heart surgery available in regions where EMF is similarly prevalent [15]. Ideally, all EMF patients with stage III and IV heart failure would benefit from a heart transplant [15]. The foregoing picture underlines the need to devise novel, cheap and yet still effective medical interventions against EMF.

In the past, the pathophysiology of EMF has been closely related to that of several other cardiomyopathies, including the hypereosinophilic syndrome (Loffler's disease)[15], and Chaga's disease[16]. Specifically, all diseases are known to possess a spectrum of pathology that encompasses hypereosinophilia, fibrosis and or, in long standing cases, calcification [15], [16]. Recent studies have established molecular mimicry as the mechanisms for pathology in some of the above EMF related (particularly Chaga's) cardiomyopathies [16]. Specifically, auto antibodies to the acidic C- termini of two Trypanosoma cruzi (or simply T.cruzi) ribosomal proteins (TcP0 & TcP2β, respectively: EDDDDDFGMGALF and EE-EDDDMGFGLFD) have been associated with the chronic cardiac pathology of Chaga's disease in humans [16]. Martin et al. [17] have recently described, using 3-dimensional modeling and docking experiments, a more clear interaction of the structural elements involved in the autoimmune mechanism of anti-P auto-antibodies cross-reaction and stimulation of the β1-adrenoreceptor, results that may lead eventually to the development of treatments to abolish receptor mediated symptoms in Chaga's disease (see Figure 1 for illustration [17]). Given the prior observed related pathology in both diseases, we hypothesized, that molecular mimicry may explain the pathology seen in both diseases too. By so doing, we also sub-hypothesized that the molecular insult in Chaga's disease may bear similarity (used interchangeably with resemblance here to imply analogy and not necessary homology) to the insult responsible for EMF. This study was conducted to examine the specific-hypothesis that resemblance (analogy) between the C-termini of the two T. cruzi ribosomal proteins TcP0/TcP2β and prior suspected causative insults for EMF explains the commonality of gross pathology. Initially designed to comprise an initial exploratory In Silico phase exploiting comparative sequence alignments [18], [19] and subsequent In Situ hybridization proof of concept phase, the herein presented non-specific results of the exploratory phase made it difficult to conduct parallel confirmatory In Situ inquiry due to a wide spectrum of test candidates and limited resources. Specifically, contrarily to prior data pointing to an architectural conservation of ribosomal P protein-structure across some life domains, no sequence similarity was found between the acidic termini of T.cruzi ribosomal P proteins TcP0/TcP2β and sequences of all searched plant, microbial and viral databases by initial NCBI microbial BLAST-P at default. Repeat BLAST at SIB, however, revealed that both C-termini of T. cruzi ribosomal P protein TcP0 and TcP2β exhibit homology to acidic termini of respective eukaryotic proteins. Further, the C-termini of TcP0 and TcP2β are noted to possess characteristic amino acid composition that confer unto them acidity and negative charge. Overall, we provide evidence to suggest that cross reactivity of antibodies against C-terminal sequences of several animal, plant and protozoal ribosomal P proteins with heart tissue may mediate EMF in a similar manner as C-termini of T. cruzi do for Chaga's disease. It is, never the less, still possible that the mechanisms of molecular mimicry between the suspected EMF-insults and myocardial tissue are mediated via different myocardial antigens altogether- thereby, making the specified protein-portions in our study not the likely cause of EMF.

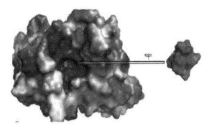

Figure 1. Showing the binding site of the acidic C-termini of the T. cruzi ribosomal P proteins on the specific human antibody. The Figure illustrates the unique binding feature conferred by the positive charge within the binding site of the acidic C-termini of the T. cruzi ribosomal P proteins (themselves negatively charged) on the specific human antibody. The van der Waals surface is coloured according to the electrostatic potential calculated with the program Poisson-Boltzman electrostatics calculated employing using APBS as implemented in PyMol with default charge settings and dielectric constant 80 (Receptor coloured by calculated charge from light gray −1 to dark gray +1). Note that the binding site of the peptides is a positive charged cavity. This work is reminiscent of the recent findings towards a better understanding of the molecular pathology of Chaga's disease, citation [17] Martin OA, Villegas ME, Aguilar CF (2009) Three-dimensional studies of pathogenic peptides from the c-terminal of Trypanosoma cruzi ribosomal P proteins and their interaction with a monoclonal antibody structural model.

Results

Similarity of C-Termini of TcP0 and TcP2β to Analyzed Pathogen, Plant, Viral Proteins and the Human Proteome

NCBI BLAST-P at Default and Other Settings

But for the hits on the source organism's ribosomal proteins (corresponding to the queries: acidic termini of TcP0 and TcP2β [20]–[24]), no similarity was found to proteins of all searched organismal, plant and viral protein genome wide databases (PGDB; for details see Table 1), findings that were considered ambiguous in light of prior studies [20]–[24]; and SIB BLAST-P tool[18], [19], [25] generated results discussed below. Specifically, noticeable was that, despite the presence of completed Toxoplasma genomes among the microbes and several other protozoa including the pathogens of malaria and leishmaniasis (that have previously been suspected to be potential causative insults for EMF [5]); and particularly ones whose whole length ribosomal proteins P0, P1, and P2 have been phylogenetically related to those of T.cruzi[26], [28], no sequence homology or analogy was found with C-termini of P0, P1, and P2 proteins of over 2,000 searched pathogens, plants, and viruses species. We found this to be an error of default (discussed further below) prompting us to repeat the alignments using the BLAST-P tool at the Swiss Institute of Biotechnology (SIB).

Table 1. Showing the hits in the T.cruzi Proteome as the native source of the TcP0 C-terminus queried.

Seq. ID	Sequence Producing significant alignment	Sore (Bits)	E value
gb|EAN99267.1|	60S acidic ribosomal protein P0 [Trypanosoma c...	29.3	2.2
gb|EAN99266.1|	60S acidic ribosomal protein P0 [Trypanosoma c	29.3	2.2

Note that the low score values above do not indicate a lower similarity, but rather the fact that the query only made up a small portion of the entire 60S acidic ribosomal protein P0 of T.cruzi. Overall, both hits are 100% homologous to the respective match segments of 60S acidic terminus of the T.cruzi ribosomal protein P0.
doi:10.1371/journal.pone.0007420.t001

SIB BLAST-P at Default Setting.

Repeat alignments of acidic termini of TcP0 and TcP2β sequences with sequences of proteins available in the Swiss Prot database using BLAST-P at SIB, contrarily to findings of the NCBI BLAST-P tool above, revealed homologs of both TcP0 and TcP2β acidic termini of eukaryotic origin; including from animals (Homo sapiens, C. elegans, D. melanogaster), plants (Arabidopsis thaliana, Zea mays, Glycina Max, Oryza sativa, Rhizopus oryzae) and protozoal (P. falciparum, T. gondii, Leishmania spp.) (See Table 2). The other species that possess homologous C-termini of P ribosomal proteins to those of T cruzi are listed here along with their "common names": Toxoptera citricida aka citrus aphid, Acyrthosiphon pisum aka pea aphid, Argas monolakensis aka mano lake bird tick, Diaphorina citri aka Asian Citrus Phyllid, Ixodes scapularis aka Black-legged tick; Haemaph-ysalis longicornis aka Bush tick, Artemia salina and Artemia franciscana aka Brine shrimp, Blomia tropicalis aka Mite, Ceratitis capitata aka Mediterranean fruit fly; Pichia pastorilis aka yeast, Saccharomyces cerevisiae aka Baker's yeast; Leishmania spp; Bombyx mori aka Silk moth; Suberites domuncula aka Sponge, Asterina pectinifera a aka Starfish; and Candida sphaerica. Figure 2 shows the schematics of these BLAST hits obtained by querying TcP2B on the Swiss Prot database.

Figure 2. Showing the schematics of the BLAST hits obtained by querying the C-termini of TcP2β across a Swiss Prot database using the BLAST tool at SIB. The Figure illustrates the schematics of the Scores and E values of hits obtained by querying the 13 amino acid sequences of the C-termini of TcP2β (EEEDDDMGFGLFD) across a Swiss Prot database using the BLAST tool at SIB. Note the presence of a key at the bottom to annotate meaning to the colors. Interpretation of this schematics may be best done with Table 2. Briefly, % identity of hits declines as the green color (signifying 100% identity) fades from green to finally red (signifying 0% identity). The data was generated by the BLAST tool at the following URL: SIB availablehttp://www.expasy.ch/cgi-bin/blast.pl

Table 2. List of Eukaryota spp. with homologous sequences to TcP2B.

Species	spp. Sub-type/Protein where known	% Identity
Trypanosoma spp.		**100**
	T.cruzi/P2B	
	T.cruzi/P2A-P-JLS/L12E	
	T.cruzi/P1	
	T.cruzi/P2	
	T.brucei/? P2	
Drosophila spp.		**100**
	D.Melanogaster/P0 or DNA-APE	
	D. pseudoobscura pseudoobscura/?	
	D. simulans	
	D. yakuba	
	D. willistoni	
	D. virilis	
	D. mojavensis	
	D. grimshawi	
	D. sechellia	
	D. erecta	
	D. ananassae/P0	
Leishmania spp		**92**
	L. infantum/P21-LIP	
	L. Donovani/P2	
	L. braziliensis/P2B	
	L. peruviana/P1	
	L. major/P2	
Others spp		**>90; ≤100**
	Sarcophaga crassipalpis/P0	100
	Toxoptera citricida/P0	100
	Acyrthosiphon pisum/P0	100
	Diaphorina citri/P0	100
	Ixodes scapularis/P0	100
	Haemaphysalis longicornis/P0	100
	Kluyveromyces polysporus/?	100
	Vanderwaltozyma polysporus/?	100
	Artemia salina/P2	100
	Blomia tropicalis/Blot alt-a6 allergen	100
	Ceratitis capitata/P0	100
	Babesia rodhaini/P0	92
	Pichia pastoris/P2A	92
	Melampsora medusae f. sp. Deltoidis/P1	92
	Branchiostoma floridae-Amphioxus/P2	92
	Saccharomyces cerevisiae/P0	92
	Branchiostoma belcheri-Amphioxus/P1	92
	Bombyx mori/P0	92
	Suberites domuncula-Sponge/P2	92
	Suberites domuncula-Spongeb/L10e/P0	92
	Pectinaria gouldii –Trumpet worm/P1	92
	Lepeophtheirus salmonis–louse/P2, P0	92

Note that some of the hits in *Trypanosoma* spp., *Drosophila* spp., and *Leishmania* spp., although possessing the indicated identities, have score and E-values that places them at a lower place(for details, see supporting File 2).
doi:10.1371/journal.pone.0007420.t002

Primary Structure and Chemical Composition of the C-Terminus of TcP0 and TcP2β

The chemical formulae of the two T.cruzi ribosomal P proteins' acid termini were found to be $C_{61}H_{83}N_{13}O_{26}S_1$ and $C_{64}H_{87}N_{13}O_{28}S_1$ respectively. Chemically, both peptides may be classified as polycarbonated. Surprisingly, although both

C-terminal peptides were noted to contain closely similar amino acid compositions, their respective predicted Instability Indices (II) are differing, with the terminus of TcP0 being stable (at an II value of 11.7) and that of TcP2β unstable (II value of 83.69). Generally, any peptide with an II above 40 is denoted as unstable [25]. These findings-discussed further below, serve to emphasize, how, proteins with closely related amino acid composition may possess differing In Vivo biophysical profiles and thus possibly functions as well.

Similarity of the C-Terminus of TcP0 and TcP2β to Known Chemical Species

Constitutively, both auto-antigens (acidic termini of TcP0 and TcP2β) were found to predominantly contain Asparagine (D), Glycine(G) and Phenylamine(F), with a balanced leucine(L) and methionine(M) percent composition of 7.7%; composition that was found to be non-homologous to any known chemical species to date. Note that this composition is associated with extensive negativity of the peptides- overall charge, uniquely imposed by the amino acids Asparagine and glutamine that are most present within the constitution of both peptides. This is not surprising, since both these terminal peptides are found localized within the acidic C-termini of the T. cruzi ribosomal proteins [16], [17]. Specifically, these findings tally with the findings of Martin et al.[17]'s 3-dimentional structure of TcP0/TcP2β acidic termini specific human antibody model that exhibits a most remarkable feature in the active site, the positively charged, narrow and deep cavity where these acidic residues 3 to 6 are accommodated, further emphasizing the fact that the most important elements in the molecular peptide recognition by the antibody may be the shape of the loop and the presence of negative charges in positions 3–5 of the acidic peptides P0, P2β [17]. Although rather improbable(and apparently non-evidenced); the notable absence of the positively charged amino acid residues Argenine and Lysine in both the C-terminal sequences of TcP0 and TcP2β prompted us to speculate, whether or not, a deficient diet on Arginine and Lysine could lead to EMF. Several pathways are possible here fore, including: either eliciting a plethora of "systemic effects"(because of their role as precursor or intermediates on several metabolic networks) or having a direct pathological impact on protein biosynthesis (as direct precursor of proteins) [5], [7].

Discussion

Our findings offer the first ever evidence to support the postulate that cross reactivity of antibodies against C-terminal sequences of ribosomal P proteins from several animals, plant and protozoal with heart tissue may mediate EMF in a similar manner as C-termini of T. cruzi do for Chaga's disease. Overall, despite previ ous studies implicating several factors in the etiology of EMF[5], including the evidenced role of ethnicity [6] and suspicions around Infections (Toxoplasmosis, Rheumatic fever, Malaria, Myocarditis and Helminthes [7]), allergy (Autoimmunity and Eosinophilia [8]), malnutrition (Protein or Magnesium deficiency[5], [7]) and toxic agents(Cassava, other plant toxins, Arsenic[9], Cerium, Thorium, Serotonin, or Vitamin D[5]) as the primary EMF insult; none is yet proven[5], [10]. Collectively, the pathology seen in EMF has been suspected to be mediated via similar molecular mimicry mechanisms as is seen in Loffler's and Chaga's disease [16]. In light of the recent advances towards understanding the mechanism of molecular mimicry seen in Chaga's disease resulting from resemblance of the C-terminal peptides of T.cruzi ribosomal P (TcP0 and TcP2β) proteins to cardiac tissue (see Figure 1) [16], [17], we felt it responsive to investigate the potential resemblance (analogy) of the above various suspected insults in EMF to the same (C-termini of T. cruzi ribosomal proteins TcP0 and TcP2β). Both P0 and P2 are a major component of the GTPase center of the large ribosomal subunit. The GTPase center, which is located at the N-termini, and functions as a landing platform for translation factors- is regarded as one of the oldest structures in the ribosome and is, presumably, one universally conserved structure in all domains of life [26]–[28]. It has been hypothesized that this structure could indeed be responsible for the major breakthrough on the way to the RNA/protein world, since its appearance would have dramatically increased the rate and accuracy of protein synthesis. Notably, one of the most characteristic ribosomal structures is the stalk: a highly flexible and universal lateral protuberance on the large subunit which is directly involved in the interaction of elongation factors, participating in the translocation mechanism [28]. In eukaryotes the stalk is formed by the pentameric complex P0–(P1)2 (P2)2 that is reminiscent of the bacterial complex L10–(L7/L12)4. In particular, the P0 protein is the eukaryotic L10 equivalent and has a key role in the stalk structure[z]. Interestingly, prior studies have actually pointed to conservation of ribosomal proteins from species within the related life Domain [26], [27]. Functionally, these proteins bind to the highly conserved 26S/28S rRNA GTPase center through the N-terminal domain [28] at sites that are equivalent to those found in bacteria [26], [28]. The P0 C-terminal domain, in particular, is known to interact with the acidic phosphoproteins P1 and P2 (the L7/L12 equivalents) through their N-terminal domains, forming the tip of the stalk [28]. The main functional part of the stalk in all domains of life is composed of small L12/P proteins- and these have, until now, been believed to form an evolutionarily conserved group in all species. We show in Tables 1 and 2 that although no sequence similarity was found between the acidic termini of T.cruzi ribosomal P proteins TcP0/TcP2β and sequences of all searched plant, microbial and viral databases by initial NCBI microbial BLAST-P at default settings, in line with prior data pointing to an architectural conservation of ribosomal P protein-structure across some life domains[26]–[28], repeat alignments using the BLAST-P Software and algorithms at the Swiss Institute of Biotechnology (SIB), revealed homologs of both studied C-termini of TcP0 and TcP2β with ribosomal P proteins (and in one incidence-D. melanogaster: DNA Apurinic apyramidinic endonucleases-APE) of several eukaryotes including the animals (Homo sapiens, C. elegans, D. melanogaster), plants (Arabidopsis thaliana, Zea mays, Glycina Max, Oryza sativa, Rhizopus oryzae) and protozoa (P. falciparum, T. gondii, Leishmania spp. The schematics of those BLAST hits obtained by querying TcP2B on the Swiss Prot database are shown in Figure 2. Grela and colleagues [28] recently performed a comprehensive comparative analysis of the L12/P proteins from the three domains of life and found that bacterial and archaeo/eukaryal L12/P-proteins are not structurally related and, therefore, might not be linked evolutionarily either. Consequently, it has been suggested that proteins be regarded as analogous rather than homologous systems and probably appeared on the ribosomal particle in two independent events in the course of evolution [28]. Therefore, in as much as prior insights into the structure of the ribosomes and their components at high resolution leaves no question that the overall architecture of the translational machinery of the cell has been strongly conserved in all kingdoms, it is worth noting that inter-kingdom differences among ribosomal components may inevitably exist, even though the functional significance of these structural variations has not been clarified yet [26]–[28]. Overall, our findings of several eukaryotic homologs of T.cruzi ribosomal P protein acidic termini provide a "window" to suggest that cross reaction of antibodies against C-terminal sequences of several animal, plant and protozoal ribosomal P proteins with heart tissue possibly mediates EMF in a similar manner as C- termini of T. cruzi do for Chaga's disease. It is, nevertheless, equally still likely that the mechanisms of molecular mimicry between prior suspected EMF-insults and myocardial tissue are mediated via different myocardial antigens- thereby, making the specified protein-portions in our study not the likely cause of EMF.

Considering that our "molecular mimicry" hypothesis is affirmed in animal models for EMF as has been for Chaga's[17], one of the major challenges to explore in future will be determining the likely mechanism of exposure to the these, now observed as, eukaryota conserved ribosomal P protein C-termini. We postulate that metabolic uptake may be such one candidate route of exposure to consider. The possibility that metabolic uptake of the still undefined insult plays a role in the aetiology of EMF arises in light of earlier work that had speculated that metabolites of plantain ingestion (specifically 5-hydroxyindolylacetic) may be the cause of EMF [29], [30]. In a latter controlled study of thirty Nigerians with established endomyocardial fibrosis, however, Ojo [31] found that no significant increase in serum 5-hydroxytryptamine levels occurred in these patients after plantain ingestion. This finding underlined the difference between endomyocardial fibrosis and carcinoid heart disease by proving that no correlation exists between the incidence of endomyocardial fibrosis and the high content of 5-hydroxytryptamine in the local dietary staples [31]. Note, however, that it did not rule out the possibility that another nutritional metabolite other than 5-hydroxytryptamine may be the primary insult, a fact that is further underscored here by the characteristic amino acid composition and negative charge orientation of both studied T. cruzi ribosomal proteins- TcP0 and TcP2β C-termini (notably, high content of Asparagine(D), Glycine(G) and Phenylamine(F), with balanced leucine(L) and methionine(M) percent composition of 7.7%; and large negative (-6) charged. The latter, charge orientation and amino acid content, have lately been determined to be the major determinants for interaction between acidic termini of TcP0 and TcP2β and specific auto- antibodies (see Figure. 1 for illustration) [17]. Our search for alternate such possible related chemical species, nevertheless, failed to yield any matching chemical species. There is hence need to conduct metabolome wide association studies (MWAS) to ascertain what metabolites are common among persons with EMF or EMF preceded hypereosinophilia.

We note a number of specific short-comings of using the BLAST-P method for this study. First, while it is clear that protein and peptide composition are important for their biological function, the same is not always straightforward in their primary structure formats. Rather, the 3D structure and the dynamics of peptides and proteins in solution are much more important, in terms of biological function. Against this background, perhaps this study would better have been conducted using 3-D structural searches. All in all, it is widely established that only those proteins with a sequence identity of at least ~30% are highly probable to share an evolutionary ancestor and to share the same overall fold (with the probability increasing as the identity sequence increase)[26], [28]. Second, in as much as the afore-going account is a useful rule of thumb for proteins in general, it is likely that, for peptides, the panorama is much more complex. For instance, as described by Martin et al. [17], the C-terminal portion of P2B protein from Leishmania brazilensis, Trypanosoma cruzi and Homo sapiens, respectively differ in only one amino acid. However, while R13 and H13 are capable of binding to a specific monoclonal antibody; A13 is not capable of such action. On the other hand, the second extracellular loop of rhodopsin is able to bind to the same antibody (less tightly) in spite of the poor sequence similarity to R13 or H13. Lastly, even with our positive findings, there is a slight possibility that no shared mechanism of pathology in Davies and Chaga's diseases actually exists. In assuming the latter position (one that is likely for some as our study coins no particular EMF causative insult), those of this line of philosophy may need to note that it is equally agreeable at the moment to further postulate that the fundamental mechanisms of aetiology are similar (in that, they involve autoimmunity), but EMF responsible auto-antigens are completely different from the acidic termini of the T.cruzi ribosomal proteins TcP0 and TcP2β that cause disease in Chaga's [16], [17]. Again, such differences, we sustain can only but be speculative; given that our search for related chemical species in known databases by way of NIST and SureChem yielded no currently known species with the primary structures C61H83N13O26S1 and C64H87N13O28S1 consistent with the T.cruzi ribosomal acidic C-termini peptides. This also rules-out the possibility that previous chemical toxins such as Cerium, Thorium [5], and Arsenic [9] may mediate EMF through similar antigens/charge as do the acidic termini of TcP0 and TcP2β, for Chaga's diseases.

Lastly, besides the above noted potential shortcomings of our methodologies; two outstanding ambiguities in our study findings warrant further discussion and explanation. First, that the NCBI BLAST-P tool at default setting yielded results contradictory of existing data on the phylogenetic relationship of ribosomal P protein C-terminal repeats among eukaryotes [26]–[29] requires explanation. Need for such an explanation was, objectively, made further necessary in light of the fact that in a recent study, we successfully used related approaches to evidence the species specific conservation of a sub-group of mosquito non-Long Terminal Repeat (non-LTR) small coding RNAs of the Long Interspersed Nuclear Elements (LINE) class-retroposons [32]; with insignificant differences observed from results obtained by other tools including a protein clock and genome cross-referencing or XREFdb [33]–[35]. However, in order to reduce errors in alignment searches, the default setting of the NCBI BLAST-P tool are designed not to permit database searches employing short-repetitive sequence queries [18], [19]. Therefore, unless the filter is turned off, no result (hit) can be obtained-the case observed in our report. Second, despite both peptides sharing related negative charge orientation, closely similar structures and amino acids content (see Figure 1), their respective predicted Instability Indices (II) were differing, with TcP0 being stable (at an II value of 11.7) while TcP2β is unstable (II value of 83.69). Generally, any peptide with an II above 40 is denoted as unstable [25]. This was interpreted-likely prematurely; as indicative of the possibility that variability in stability of both peptides per se does not influence their biological half life, both of which are shown to be 1 hour within mammalian reticulocytes. Third and last, while previous studies of EMF clearly show the possible role of genetic variants, geographical locales and socioeconomic status in the etiology of EMF [5], [6], our work is limited in that, although studying the insults possibly common in these groups, it never took consideration of those other factors including age and genetic variations. Specifically, Freer et al.[6], in an unmatched case control study in Mulago Hospital, Kampala of 61 EMF patients and 120 controls, show that EMF patients were significantly more likely than controls to have Rwanda/Burundi ethnic origins

(P = 0.008), be peasants (P <0.001), and to come from defined geographical locations (P = 0.003). Elsewhere, Mocumbi et al. [11] not only highlighted the familial and endemic nature of the disease in tropics but also identified early disease and asymptomatology to occur among such subjects. Therefore, amidst the emerging role of genomics in infectious and neglected tropical diseases [36], it may equally be necessary to conduct genome wide association studies (GWAS) to establish the particular small nuclear polymorphisms (SNPs) among persons from these established ethnic and geographical locales that make them highly susceptible to EMF. The fusion protein FIP1L1-PDGFRa, a constitutively activated tyrosine kinase found in as many as half of those with the idiopathic hypereosinophilic syndrome, has emerged as a therapeutic target for imatinib [5], [37]. The recent finding that serotonin acts as a chemotactic factor for eosinophils also underlines the need for inquiries into the role of this pathway in EMF [38]. Zanettini and colleagues [39] have found that some anti- Parkinson medications induce valvular fibrosis via their action on 5HT2B receptors. GWAS studies are called for to determine whether or not, polymorphisms in these and other receptors influence susceptibility to EMF in the presence of intermittent Eosinophilia, in which case, existing drugs may be tried in EMF.

In conclusion, our findings provide a "window" to suggest that cross reactivity of antibodies against ribosomal P protein C-termini of several animal, plant and protozoal with heart tissue may mediate EMF in a similar manner as C- termini of T. cruzi do for Chaga's disease. It is, however, equally possible that the mechanisms of molecular mimicry between the suspected EMF-insults and myocardial tissue are mediated via different myocardial antigens- thereby, denoting these-our study alluded species-protein portions, not the likely cause of EMF.

Materials and Methods

Comparative Alignment of TcP0 and TcP2β Acidic Terminal Sequences With Over 1, 789 Organismal Proteomes

NCBI BLAST-P.

Sequence alignments were conducted by querying sequences of the acidic termini of the two T.cruzi ribosomal P proteins TcP0 and TcP2β [the latters' Swiss Prot Accessions numbers are P26796 and Q26957, respectively]'s known to mediate autoimmune responses in Chaga's diseases: EDDDDDFGMGALF and EEEDDDMGFGLFD across an over 1,789 organismal proteome database, 20 plant (including cassava) proteome database by way of BLAST-P Software and Algorithms[18], [19] at default setting[and hence after, repeated with altered algorithms E = 1–10, A & D = 10–100]. On the other hand, similarity to viral genomes was determined by querying entire peptides' cross referenced DNA/mRNA sequences (derived from mRNA sequences of mRNA of the TcP0 [20], [21], [22] -NCBI Accession X65066; and the whole genomic short gun of T.cruzi [23], [24]- NCBI Accession NW_001849569) across a database of ssRNA, ss-DNA, dsRNA and dsDNA viral genomes employing NCBI's viral genotyping and dsDNA BLAST-N respectively[18], [19]. The latter approach (use of whole reference gene nucleotide sequences rather than amino acids sequences] was employed because the currently available BLAST algorithm and software linked to the viral databases only support nucleotide (N) searches

SIB BLAST-P Tool.

Repeat alignment of acidic termini of TcP0 and TcP2β with sequences of proteins available in the Swiss Prot database using BLAST-P at SIB were conducted as prior described by Altschul et al., [18], [19].

Computational Derivation of TcP0 and TcP2β Acidic Terminal Chemical Composition

To be able to search across chemical species databases (see method 3 below), we needed to derive the chemical formulae of the acidic termini of both TcP0 and TcP2β. The Expasy software Protparam [25] was used to achieve this. Briefly, Protparam [25] is a proprietary computational software available at Swiss Prot/UniProt that can be used to derive series biophysical profiles as well as determine the amino acid composition of any protein or polypeptide of interest, provided its primary structure (linear alignment of amino acids) is known. The Software has a user interface that allows one to feed the respective peptide or polypeptide primary structure into it, hence fore computing the parameters.

Searching for Homologous Chemical Species to TcP0 and TcP2β Acidic Termini

The chemical formulae of the acidic termini of TcP0 and TcP2β obtained above: $C_{61}H_{83}N_{13}O_{26}S_1$ and $C_{64}H_{87}N_{13}O_{28}S_1$ respectively, were used to search for relationship to chemical toxins like Cerium, Thorium, Arsenic, Serotonin, or Vitamin D that are suspected causative insults for EMF[5] as well as all other known chemical species. The National Institute of Standard and Technology (NIST) and SureChem Chemical searches (for details, see URL link below) were finally employed to relate the former formulae to chemical formulae and structure of all known compounds in the respective databases.

Databases, Software and Algorithms

The Databases, Software and algorithms used in this study can all freely be accessed by the reader at the following world-wide web sites:

1. All NCBI BLAST Assembled Organismal genomes are available at the NCBI URL: http://www.ncbi.nlm.nih.gov/mapview/

2. SIB BLAST Tool is available at the following Expasy URL: http://www.expasy.ch/cgi-bin/blast.pl)

3. The NCBI Viral BLAST against dsDNA viruses is available at the NCBI URL: http://www.ncbi.nlm.nih.gov/genomes/VIRUSES/Bitor.cgi?db=VOG&data =vog&gdata=dsdna.defl

4. The NCBI Viral Genotyping BLAST tool is available at the NCBI URL: http://www.ncbi.nlm.nih.gov/projects/genotyping/formpage.cgi

5. The Expasy Software Protparam used to determine chemical composition is available at the Swiss Prot/Uniprot URL: http://www.expasy.ch/tools/protparam.html

6. The SureChem software and algorithm is available at the URL: http://www.freepatentsonline.com/

7. The National Institute of Standards and Technology(NIST) Chemical Formula Search Tool is available at the NIST URL: http://webbook.nist.gov/chemistry/form-ser.html

Author Contributions

Conceived and designed the experiments: MW. Performed the experiments: MW. Analyzed the data: MW. Contributed reagents/materials/analysis tools: MW. Wrote the paper: MW.

References

1. Parry EH., Abrahams DG (1965) The natural history of endomyocardial fibrosis. Q J Med 34: 383–408.

2. Connor DH., Somers K., Hutt MS., Manion WC., D'Arbela PG (1967) Endomyocardial fibrosis in Uganda (Davies' disease): An epidemiologic, clinical, and pathologic study. Am Heart J 74: 687–709.

3. Davies JNP (1948) Endomyocardial fibrosis in Uganda. East Afr Med J 25: 10–16.

4. Sivasankaran S (2009) Restrictive cardiomyopathy in India: the story of a vanishing mystery. Heart 95(1): 9–14.

5. Bukhman G., Ziegler J., Parry E (2008) Endomyocardial Fibrosis: Still a Mystery after 60 Years. PLoS Negl Trop 2(2): e97.

6. Freers J., Mayanja-Kizza H., Rutakingirwa M, Gerwing E (1996) Endomyocardial fibrosis: why is there striking ascites with little or no peripheral oedema? Lancet 347: 197.

7. Andy JJ., Ogunowo PO., Akpan NA., Odigwe CO., Ekanem IA., et al. (1998) Helminth associated hypereosinophilia and tropical endomyocardial fibrosis (EMF) in Nigeria. Acta Trop 69: 127–140.

8. Beisel WR (1995) Herman award lecture, 1995: infection-induced malnutrition-from cholera to cytokines. Am J Clin Nutr 62: 813–9.

9. Edge J (1946) Myocardial fibrosis following arsenical therapy: report of a case. Lancet 248: 675–677.

10. Iglezias SD., Benvenuti LA., Calabrese F., Salemi VM., Silva AM., et al. (2008) Endomyocardial fibrosis: pathological and molecular findings of surgically resected ventricular endomyocardium. Virchows Arch 453(3): 233–41.

11. Mocumbi AO., Ferriera MB., Sidi D., et al. (2008) A population study of endomycardial fibrosis in a rural area of Mozambique. N Engl J Med 359: 43–9.

12. Rutakingirwa M., Ziegler JL., Newton R., Freers J (1999) Poverty and eosinophilia are risk factors for endomyocardial fibrosis (EMF) in Uganda. Trop Med Int Health 4: 229–235.

13. Spry CJ., Take M., Tai PC (1985) Eosinophilic disorders affecting the myocardium and endocardium: a review. Heart Vessels Suppl 1240–2.

14. Ntusi NB., Mayosi BM (2009) Epidemiology of heart failure in sub-Saharan Africa. Expert Rev Cardiovasc Ther 7(2): 169–80.

15. Turan AA., Karayel F., Akyildiz EU., Ozdes T., Yilmaz E., et al. (2008) Sudden death due to eosinophilic endomyocardial diseases: three case reports. Am J Forensic Med Pathol 29(4): 354–7.

16. Sepulveda P., Liegeard P., Wallukat G., Levin MJ., Hontebeyrie M (2000) Modulation of Cardiocyte Functional Activity by Antibodies against Trypanosoma cruzi Ribosomal P2 Protein C Terminus. Infection and Immunity 68(9): 5114–5119.

17. Martin OA., Villegas ME., Aguilar CF (2009) Three-dimensional studies of pathogenic peptides from the c-terminal of Trypanosoma cruzi ribosomal P proteins and their interaction with a monoclonal antibody structural model. PMC Biophys 2(1): 4.

18. Altschul SF., Madden TL., Schäffer AA., Zhang J., Zhang Z., et al. (1997) Gapped BLAST and PSI-BLAST: a new generation of protein database search programs. Nucleic Acids Res 25: 3389–3402.

19. Altschul SF., Wootton JC., Gertz EM., Agarwala R., Morgulis A, et al. (2005) Protein database searches using compositionally adjusted substitution matrices. FEBS J 272: 5101–5109.

20. Skeiky YA.., Benson R., Parsons M., Elkon KB., Reed SG (1992) Cloning and expression of Trypanosoma cruzi ribosomal protein P0 and epitope analysis of anti-P0 autoantibodies in Chagas' disease patients. J Exp Med 176: 201–211.

21. Schijman AG., Levin MJ (1992) Nucleotide sequence of a cDNA encoding a Trypanosoma cruzi acidic ribosomal PO protein: a novel C-terminal domain in T. cruzi ribosomal P proteins. Nucleic Acids Res 20: 2894–2894.

22. Schijman AG., Levitus G., Levin MJ (1992) Characterization of the C-terminal region of a Trypanosoma cruzi 38-kDa ribosomal P0 protein that does not react with lupus anti-P autoantibodies. Immunol Lett 33: 15–20.

23. Schijman AG., Dusetti NJ., Vazquez MP., Lafon S., Levy-Yeyati P, et al. (1990) Nucleotide cDNA and complete deduced amino acid sequence of a Trypanosoma cruzi ribosomal P protein (P-JL5). Nucleic Acids Res 18: 3399–3399.

24. El-Sayed NM., Myler PJ., Bartholomeu DC., Nilsson D., Aggarwal G., et al. (2005) The genome sequence of Trypanosoma cruzi, etiologic agent of Chagas disease. Science 309: 409–415.

25. Gasteiger E., Hoogland C., Gattiker., Duvaud S., Wilkins MR., et al. (2005) Protein Identification and Analysis Tools on the ExPASy Server. In: Walker JM, editor. The Proteomics Protocols Handbook. pp. 571–607. Humana Press.

26. Grela P., Bernadó P., Svergun D., Kwiatowski J., Abramczyk D., et al. (2008) Structural Relationships Among the Ribosomal Stalk Proteins from the Three Domains of Life. J Mol Evol 67: 154–167.

27. Hasler P., Brot N., Weissbach H., Parnassa AP., Elkon KB (1991) Ribosomal proteins P0, P1, and P2 are phosphorylated by casein kinase II at their conserved carboxyl termini. J Biol Chem 266: 13815–13820.

28. Gagou ME., Gabriel MAR., Ballesta JPG., Kouyanou S (2000) The ribosomal P-proteins of the medfly Ceratitis capitata form a heterogeneous stalk structure interacting with the endogenous P-proteins, in conditional P0-null strains of the yeast Saccharomyces cerevisiae. Nucl Acids Res 28(3): 736–743.

29. Lewis CE (1958) Timed excretion of 5-hydroxy indoleacetic acid after oral administration of bananas and 5-hydroxytryptamine. Proc Soc Exp Biol Med 99(2): 523–5.

30. Crawford MA (1962) Excretion of 5-hydroxyindolylacetic acid in East Africans. Lancet 1(7225): 352–3.

31. Ojo GO (1970) The pathogenesis of endomyocardial fibrosis: the question of 5-hydroxytryptamine. Br Heart J 32: 671–674.

32. Byarugaba W., Kajumbula H., Wayengera M (2009) In Silico evidence for the species-specific conservation of mosquito retroposons: implications as a molecular biomarker. Theor Biol Med Model 6: 14.

33. Feng DF., Cho G., Doolittle RF (1997) Determining divergence times with a protein clock: Update and reevaluation. Proc Natl Acad Sci USA 94: 13028–13033.

34. Bassett DE., Boguski MS Jr., Spencer F., Reeves R., Kim S., et al. (1997) Genome cross-referencing and XREFdb: Implications for the identification and analysis of genes mutated in human disease. Nature Genet 15: 339–344.

35. Osamu G (2008) A space-efficient and accurate method for mapping and aligning cDNA sequences onto genomic sequence. Nucl Acids Res 36(8): 2630–2638.

36. Wayengera M., Byarugaba W (2008) Emphasizing the vitality of genomics related research in the area of Infectious diseases. Sci Res Essay 3(4): 125–131.

37. Cools J., DeAngelo DJ., Gotlib J., Stover EH., Legare RD., et al. (2003) A tyrosine kinase created by fusion of the PDGFRA and FIP1L1 genes as a therapeutic target of imatinib in idiopathic hypereosinophilic syndrome. N Engl J Med 348: 1201–1214.

38. Boehme SA., Lio FM., Sikora L., Pandit TS., Lavrador K., et al. (2004) Cutting edge: serotonin is a chemotactic factor for eosinophils and functions additively with eotaxin. J Immunol 173: 3599–3603.

39. Zanettini R., Antonini A., Gatto G., Gentile R., Tesei S., et al. (2007) Valvular heart disease and the use of dopamine agonists for Parkinson's disease. N Engl J Med 356: 39–46.

CITATION

Originally published under the Creative Commons Attribution License or equivalent. Wayengera M. Searching for New Clues about the Molecular Cause of Endomyocardial Fibrosis by Way of In Silico Proteomics and Analytical Chemistry. PLoS One. 2009; 4(10): e7420. doi:10.1371/journal.pone.0007420.

Evaluation of Fructooligosaccharides and Inulins as Potentially Health Benefiting Food Ingredients by HPAEC-PED and MALDI-TOF MS

Chiara Borromei, Maria Careri, Antonella Cavazza,
Claudio Corradini, Lisa Elviri, Alessandro Mangia
and Cristiana Merusi

ABSTRACT

This paper describes the complementarity of high-performance anion exchange chromatography coupled with pulsed electrochemical detection (HPAEC-PED) and matrix-assisted laser desorption/ionization mass spectrometry

(MALDI-TOF-MS) to evaluate commercial available fructans (fructooligosaccharides (FOS) and inulins), having different degrees of polymerization (DP) which are usually employed by food industry as functional ingredients either for their prebiotic properties or as a fat replacer, giving a fat-like mouth feel and texture. The developed HPAEC-PED methods are able to analyze FOS (fructans with DP 3–10) and inulins (DP ranging from 3 to 80) with a good resolution and relatively short retention times to evaluate structural differences between fructooligosaccharide and inulins and the possible presence of inulooligosaccharides as well as of branching. To characterize FOS and inulin at different degrees of polymerization and to assure correct molecular assignment, MALDI-TOF MS analysis was also investigated. The 2,5-dihydroxy benzoic acid (2,5-DHB) was found to be the best matrix for FOS analysis as Actilight and Raftilose P95 products, while 3-aminoquinoline (3-AQ) seems to be the best matrix for inulin with higher DP. The applicability of the optimized methods to the identification and determination of FOS contained in a symbiotic milk as well as a type of inulin added as functional ingredient to a cooked ham is demonstrated.

Introduction

Fructans are carbohydrate polymers consisting of a sucrose molecule that is elongated by a chain of fructosyl units connected through β-(2→1) or β-(2→6) linkages [1], depending on the linkage type they are called inulin and levans, respectively.

Inulin has been defined as a polydisperse carbohydrate material consisting mainly, if not exclusively, of β-(2→1) fructosyl-fructose linkes, containing one terminal glucose as in sucrose and having the generic chemical structure GFn (with G as glucose, F as fructose, and n indicating DP). When referring to the definition of inulin, both GFn and Fn compounds, consisting exclusively, of β-(2→1) fructosyl-fructose linkes, are considered to be included under this same nomenclature. Several inulin types occur in nature and they differ for the degree of polymerization and molecular weight, depending on the source, the harvest time and processing conditions [2].

Fructooligosaccharides (FOS) with DP 3–9 (average DP 4.5) are produced during the process of chemical degradation or controlled enzymatic hydrolysis of inulin by endoglycosidases [3, 4]. Furthermore, FOS can be produced on a commercial scale, from sucrose, using a fungal enzyme from either Aureobasidium sp. [5] or Aspergillus niger [6].

FOS and inulin are recognized as health-promoting food ingredients. The variety of chemical and structural conformations that characterize FOS and inulins makes them flexible and appealing ingredients for different food applications. Inulin has been reported to develop a gel-like structure when thoroughly mixed with water or other aqueous liquid, forming a gel with a white creamy appearance, which can be easily incorporated into foods to replace fats making inulin an interesting ingredient to deliver structure in low or zero fat-food products [7]. Correlation between inulin gel properties and its chemical structure (oligo- and polysaccharides) has been evaluated [8]. Furthermore, FOS and inulin exhibit prebiotic function stimulating the growth and/or activity of one or a limited number of bacteria in the colon that can improve host health [9] and, therefore, they can be employed in functional food formulations [10]. In a previous work, prebiotic effectiveness of FOS and inulin of different degrees of polymerization was reported [11], and the response of bifidobacteria to differently lengthened fructans was analyzed in pure and fecal cultures, confirming that fermentation of FOS and inulins in the colon can be correlated to different metabolic activities carried out by several intestinal microorganisms [12]. Carbohydrate analyses conducted at the end of batch fermentations by high-performance anion-exchange chromatography (HPAEC) with pulsed electrochemical detection (PED) technique demonstrated a very heterogeneous strain-dependent capability to degrade FOS or inulins. It was demonstrated that during batch fermentations, the short fructans were fermented first then gradually the longer ones were consumed. However, regarding the investigated carbohydrates, only qualitative indications were given with respect to the chain polymerization degree.

Within the panel of analytical techniques available for the characterization of fructooligosaccharides (FOS) and inulin, HPAEC-PED can be a useful and sensitive tool for the qualitative chain-length analysis of oligo- and polysaccharides from polydisperse preparations such as FOS and inulin at different degrees of polymerization [8–11]. Besides, HPAEC-PED is routinely used to separate neutral and charged oligosaccharides differing by branch, linkage, and positional isomerism [13]; from the chromatograms generated by HPAEC it is not possible to identify each observed component, without access to reference material.

The lack of standards is an obvious problem when investigating FOS and inulin contain linear homologous series of fructan polysaccharides, of which there are no commercial standards available.

Matrix-assisted laser desorption/ionization time-of-flight mass spectrometry (MALDI-TOF-MS) has been used in molecular sizing of carbohydrates and the technique can conveniently be combined with other methods such as HPLC, demonstrating to be a powerful tool for the characterization of carbohydrates

[14–16]. Furthermore, MALDI-TOF-MS has been used to determine chain length distribution of FOS and inulin [17, 18] as well as for both qualitative and quantitative analyses in selected food samples [19, 20].

This paper deals with the development of HPAEC-PED methods to characterize and compare FOS and inulins having different degrees of polymerization. The qualitative HPAEC-PED profiles were then compared with molecular weight distribution evaluated by MALDI-TOF-MS. Our results could be important for labeling or supporting a prebiotic claim in food as they give indications regarding DP distribution and the amount of FOS or inulin contained. Furthermore, this work describes the validation and application of HPAEC-PED methods for the quantitative determination of short-chain FOS (sc-FOS) added to a symbiotic milk and inulin added to a cooked ham as a functional ingredient.

Materials and Methods

Chemicals

The deionized water (18 MΩ cm resistivity) was obtained from a Milli-Q element water purification system (Millipore, Bedford, Mass, USA). Acetonitrile, methanol (both of HPLC purity), trifluoroacetic acid, and formic acid (analytical reagent grade) were purchased from Carlo Erba (Milan, Italy). Sodium hydroxide and sodium nitrate were from J. T. Baker (Deventer, The Netherlands). Carrez reagent I (potassium hexacyanoferrate(II) trihydrate), carrez reagent II (zinc acetate), glucose, fructose, and sucrose and lactose were purchased from Sigma-Aldrich (Milan, Italy). 1-kestose was from Fluka (Milan, Italy). All sample solutions were filtered through a Type 0.45 μm single-use membrane filter (Millipore, Bedford, Mass, USA).

HPAEC-PED Apparatus

All HPAEC-PED experiments were performed using a DX 500 system equipped with a GP40 pump, using a CarboPac PA-200 column (Dionex, 3×250 mm) and a CarboPac PA-100 (Dionex, 4×250 mm), connected to the associated guard column. Carbohydrates were detected by a model ED40 electrochemical detector in its integrated pulsed amperometric detection mode, applying the following potentials and durations: E_1=0.10 V (t_1=0.40 second), E_2=−2.00 V(t_2=0.01 second), E_3=0.60 V (t_3=0.01 second), E_4=−0.10 V (t_4=0.06 second). Integration is between 0.20 and 0.40 seconds. The chromatographic system was interfaced, via proprietary network chromatographic software (PeakNet TM) to a personal computer, for instrumentation control, data acquisition, and processing. All were from Dionex Corporation (Sunnyvale, Calif, USA).

Chromatographic Conditions

To separate FOS and inulin at different degrees of polymerization (DP), on a CarboPac PA-200 column, the mobile phase consisted of deionized water (eluent A), 600 mM aqueous sodium hydroxide (eluent B), and 250 mM aqueous sodium nitrate solution (eluent C), employing a gradient program as reported in method 1, Table 1. A similar procedure was developed to characterize the oligosaccharide distribution present in a commercial available product containing short chain fructooligosaccharides (scFOS) (Actilight 950P), in which carbohydrates were separated on the CarboPac PA100 column, eluting by the gradient reported in method 2, Table 1, in which eluent A was water, eluent B 600 mM aqueous sodium hydroxide solution, and eluent C 500 mM aqueous sodium acetate solution. The gradient elution program reported in Table 2 was applied to elute oligosaccharide fraction present in inulin added to cooked ham.

Table 1. Gradient elution program to elute (method 1) FOS and inulin and (method 2) scFOS.

HPAEC-PED method 1					HPAEC-PED method 2				
Elution time (min)	A(%)	B(%)	C[1](%)	Comment	Elution time (min)	A(%)	B(%)	C[2](%)	Comment
−40[a]	0	100	0	Start cleaning step	−45[a]	0	100	0	Start cleaning step
−30[a]	0	100	0	End cleaning step	−30[a]	0	100	0	End cleaning step
−29.9[a]	89	10	1	Start conditioning step	−29.9[a]	89	10	1	Start conditioning step
0	89	10	1	End conditioning step	0	89	10	1	End conditioning step
0.1	89	10	1	Injection, acquisition start	0.1	89	10	1	Injection, acquisition start
4	84	15	1	End first gradient step	5	89	10	1	End isocratic elution
					45	60	20	20	End first gradient step
80	79	15	40	End third gradient step	55	50	20	30	End second gradient step

A: deionized water; B: sodium hydroxide (600 mM); C[1]: sodium nitrate (250 mM); C[2]: sodium acetate (500 mM).
[a] Negative time indicates time prior to sample injection.

Table 2. Gradient elution program to elute oligosaccharide fraction present in inulin added to cooked ham.

Elution time (min)	A(%)	B(%)	C[1](%)	Comment
−35[a]	0	100	0	Start cleaning step
−25[a]	0	100	0	End cleaning step
−24.9[a]	79	16	5	Start conditioning step
0	79	16	5	End conditioning step
0.1	74	16	10	Injection, acquisition start
20	62	16	22	End first gradient step

A: deionized water; B: sodium hydroxide (600 mM); C: sodium acetate (500 mM).
[a] Negative time indicates time prior to sample injection.

All mobile phases were sparged and pressurized with helium to prevent adsorption of atmospheric carbon dioxide and subsequent production of carbonate, which would act as displacing ion and shorten retention time.

MALDI-TOF-MS Analysis of FOS

MALDI-MS measurements were performed using an MALDI-LR time-of-flight mass spectrometer (Micromass, Manchester, UK) operating in the positive linear ion mode. Ions formed by a pulsed UV laser beam (λ=337 nm) were accelerated at 15 keV. Laser strength was varied from sample to sample to obtain the best signal.

For all samples, different matrices were tested: 2,5-dihydroxy benzoic acid (2,5-DHB) (Sigma-Aldrich), trihydroxyacetophenone (THAP), 3-aminoquinoline (3-AQ), hydroxylphenylazo benzoic acid (HABA), and 4-hydroxy-α-alpha cyanocinnamic acid (HCCA) (Fluka), at 10 mg mL^{-1} in either aqueous solution and water/acetonitrile (50/50 v/v) trifluoroacetic acid (TFA) (0.1% v/v) mixture. A dried droplet sample preparation was adopted. Three replicated measurements were performed on each sample. External calibration was performed using the [M+H]+ ions of a peptide mixture (angiotensin I, angiotensin II, substance P, rennin, ACTH, insulin bovine, cytochrome c) (Sigma-Aldrich).

Samples

Different sources of fructans were analyzed both by HPAEC-PED and MALDI-TOF-MS: Raftiline ST, Raftilose P95 (Orafti, Tienen, Belgium), Actilight 950P (Beghin Mijie, Thumeries, France), Frutafit IQ, and Frutafit TEX (Sensus, Roosedaal, The Netherlands). All stock solutions were prepared at 1 mg mL^{-1} with HPLC-grade water and filtered on a 0.45 μm membrane filter. Cooked ham was kindly provided by I Fratelli Emiliani SpA (Langhirano, Parma, Italy). Symbiotic milk was purchased from the local market.

Sample Preparation

Symbiotic milk was analyzed for the separation of individual sugars and short-chain fructooligosaccharides (scFOS). Two milliliters of milk were transferred to a 25 mL volumetric flask. The sample was diluted in approximately 10 mL ethanol-water (1:1, v/v) and 300 μL Carrez I solution (stirred 1 minute) and 300 μL Carrez II solution (stirred 1 minute) were added at room temperature. Five milliliters of acetonitrile (HPLC-grade) were added. These reagents were used to precipitate the protein and noncarbohydrate fractions. The solution was made up to 25 mL with ethanol-water (1:1, v/v), then the solution was left for two hours until complete formation and precipitation of protein clot. The resulting solution diluted (1:1, v/v) with water, then was filtered through a filter paper and passed through a C18 Sep-Pak Plus cartridge Waters (Milford, Mass, USA) previously conditioned with 10 mL of methanol (HPLC-grade) and 10 mL of HPLC grade water. This filtered extract was forced through a 0.45 μm nylon filter and then injected into the HPLC system.

Cooked ham samples were prepared by blending two slices (2.0 cm in width) of each ham and homogenization. Ten grams of the homogenized sample were weighed and diluted with 50 mL of HPLC grade water and stirred with a magnetic stirrer. The beaker with the sample was placed in a shaking water-bath at 80°C for 60 minutes to denature proteins. The sample was centrifuged at 7000 xg for 45 minutes at 4°C. The clarified solution was removed and an aliquot (1 mL) was diluted with 12 mL of HPLC-grade water (final dilution 1:60). After filtration through a 0.45 μm membrane filter, sample was injected into HPLC.

Results and Discussion

In the first part of our work, the average degree of polymerization (DP) and distribution of oligo- and polysaccharides in commercial FOS and inulins having different molecular weight distribution were qualitatively evaluated by HPAEC-PED.

To develop an accurate, valid, and optimal chromatographic fingerprint for the quality evaluation of FOS and inulin at different degrees of polymerization (DP), the different HPAEC parameters including mobile phase composition, chromatographic column (CarboPac PA100 and CarboPac PA200), and flow rate of mobile phase were all examined and compared. The criterion used to evaluate the quality of a fingerprint was the number of peaks detected.

Various eluent combinations were tested using nitrate as pushing agent to enable the selective elution with high reproducible retention time of FOS and inulin with DP up to 80 or more. Gradient elution was advantageous for separating both oligo and polymers with different DP. Under gradient conditions, acetate ion is typically the preferred "pusher" ion used by HPAEC. Using nitrate instead of acetate as the pushing agent in gradient elution of carbohydrate by HPAEC, better resolution of polymers can be achieved [21]. In our work, the dual advantages of a nitrate gradient over elution using acetate ions were the simultaneous increase of the column peak capacity and the reduction of the analysis time. Employing a CarboPac PA200 column, under the chromatographic conditions described in method 1, Table 1, both low molecular weight and high molecular weight fructans were separated (see the chromatographic profiles reported in Figure 1). The assignment of the chromatographic peaks with DP higher than 3 was based on the generally accepted assumptions that the retention time of a homologous series of carbohydrates increased as the DP increased, and that each successive peak represented a glucofructan which had a fructose more than that of the previous peak. This is because retention time increases as the number of negatively charged functional groups concurrently increases [13]. Moreover, the individual peaks were sharp and well resolved, strongly suggesting that all the analyzed samples of FOS and fructans were, as expected, mainly linear.

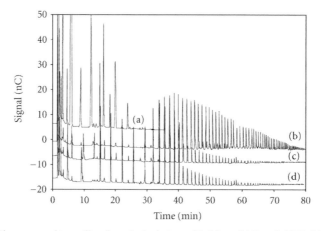

Figure 1. (a) Chromatographic profiles of standard solutions of Raftilose, (b) Frutafit TEX, (c) Frutafit IQ, (d) Raftiline. Chromatographic conditions as in the text.

Instrumental precision was checked from six consecutive injections of an inulin solution; the relative standard deviations (RSDs) obtained were better than 2.7%, as reported in Table 3.

Analyzing the chromatographic profiles depicted in Figure 1, the simpler chromatographic profile was observed in sample A, which corresponds to Raftilose, where only oligosaccharides from DP 3 to DP 9 were found.

This commercially FOS product consists of a powder composed of oligosaccharide fraction including the trisaccharide 1-kestose and short-chain FOS at higher degree of polymerization as well as the natural sugars glucose, fructose, and sucrose. In chromatographic profiles depicted in Figure 1, peaks with retention times longer than 30 minutes could be assigned to polysaccharides from DP 10 to DP 80, whereas the small peaks eluting among them could correspond to isomers composed only of fructose unit chains [20], as well as slightly branched fructans [21].

Table 3. Repeatability of retention time (sample Frutafit IQ).

N.Peaks	Retention time ±SD	cv(%)	N.Peaks	Retention time ±SD	cv(%)
1	1.90 ± 0.04	0.92	44	48.09 ± 0.32	0.40
2	1.97 ± 0.57	1.52	45	49.01 ± 0.41	0.33
3	2.10 ± 0.09	2.50	46	49.18 ± 0.38	0.48
4	2.60 ± 0.92	2.06	47	50.09 ± 0.40	0.46
5	3.12 ± 1.52	2.72	48	50.28 ± 0.33	0.47
6	4.70 ± 2.50	2.47	49	51.11 ± 0.48	0.45
7	5.73 ± 2.06	2.03	50	51.30 ± 0.46	0.35
8	9.11 ± 2.72	1.94	51	52.14 ± 0.47	0.33
9	12.38 ± 2.47	1.75	52	52.36 ± 0.45	0.30
10	15.15 ± 2.03	1.91	53	53.10 ± 0.35	0.30
11	16.38 ± 1.94	2.45	54	53.34 ± 0.33	0.32
12	18.82 ± 1.75	2.55	55	54.03 ± 0.30	0.32
13	20.36 ± 1.91	2.59	56	54.28 ± 0.30	0.25
14	22.64 ± 2.45	2.62	57	54.96 ± 0.32	0.23
15	24.52 ± 2.55	2.40	58	55.24 ± 0.32	0.21
16	26.40 ± 2.59	1.65	59	55.83 ± 0.25	0.10
17	27.00 ± 2.62	1.02	60	56.09 ± 0.23	0.14
18	28.50 ± 2.40	1.12	61	56.65 ± 0.21	0.04
19	29.92 ± 1.65	0.95	62	56.92 ± 0.10	0.41
20	31.26 ± 1.02	0.80	63	57.49 ± 0.14	0.52
21	31.55 ± 1.12	0.72	64	57.75 ± 0.04	0.62
22	32.37 ± 0.95	0.63	65	58.30 ± 0.41	0.51
23	33.79 ± 0.80	0.62	66	58.65 ± 0.52	0.45
24	34.31 ± 0.72	0.57	67	59.22 ± 0.62	0.39
25	35.64 ± 0.63	0.53	68	59.63 ± 0.51	0.35
26	35.99 ± 0.62	0.51	69	60.19 ± 0.45	0.33
27	37.33 ± 0.57	0.48	70	60.55 ± 0.39	0.29
28	37.57 ± 0.53	0.26	71	61.10 ± 0.35	0.04
29	38.89 ± 0.51	0.35	72	61.51 ± 0.33	0.24
30	39.05 ± 0.48	0.31	73	61.96 ± 0.29	0.28
31	40.26 ± 0.26	0.31	74	62.42 ± 0.04	0.20
32	40.43 ± 0.35	0.25	75	62.76 ± 0.24	0.15
33	41.62 ± 0.31	0.19	76	63.07 ± 0.28	0.29
34	41.76 ± 0.31	0.51	77	63.49 ± 0.20	0.27
35	43.04 ± 0.25	0.56	78	63.88 ± 0.15	0.36
36	43.21 ± 0.19	0.31	79	64.13 ± 0.29	0.43
37	44.21 ± 0.51	0.51	80	64.44 ± 0.27	0.47
38	44.37 ± 0.56	0.35	81	64.75 ± 0.36	0.48
39	45.52 ± 0.31	0.58	82	65.03 ± 0.40	0.45
40	45.78 ± 0.51	0.39	83	65.33 ± 0.47	0.43
41	46.72 ± 0.35	0.32	84	65.64 ± 0.48	0.35
42	47.01 ± 0.58	0.41	85	65.94 ± 0.45	0.37
43	47.87 ± 0.39	0.38	86	66.20 ± 0.43	0.35

Chromatographic profile of Frutafit TEX (Figure 1(b)) shows that most peaks were eluted with retention times between 32 to 80 minutes, revealing that it was mainly composed by polymeric fructans having DP higher than ten.

On the other hand, chromatographic profile depicted in Figure 1(c) was comparable to that of Figure 1(d), where a degree of polymerization number (DP) of about 80 was found for both products.

From the chromatograms generated by HPAEC, it was not possible to identify each observed component; however, by a qualitative comparison of the chromatographic profiles reported in Figure 1, it may be observed that the DP distribution of carbohydrates is very different. Frutafit IQ and Raftiline were found to be characterized by a polydisperse distribution of carbohydrates composed of both linear oligofructose (very similar to that of Raftilose, Figure 1(a)) and polifructose at higher DP, which correspond, regarding retention times, to that of Frutafit TEX.

On comparative grounds, it is evident from the separation profiles that HPAEC–PED allowed the study of a higher number of oligomers, and it is able to compare fructans with different degrees of polymerization. Furthermore, pulsed amperometric detection is highly selective and sensitive because only reactive compounds will give response and at very low concentrations.

To verify the chain length distribution of the analyzed FOS and inulin by HPAEC-PED, the same products were analyzed by MALDI-TOF MS.

As it is well known, the matrix plays a fundamental role in quality of the MALDI-TOF MS results both in terms signal-to-noise ratio and resolution. Before comparing the MALDI-TOF MS profile of the FOS investigated by LC, the effects of different matrices were tested. Various matrices, such as 2,5-DHB, 3-AQ, HCCA, and THAP, have been previously recommended for the analysis of carbohydrates. We started by testing these four different matrices.

After several experiments, DHB was found to be the best matrix for FOS analysis of Raftilose P95 (Figure 2), while 3-AQ seems to be the best matrix for inulin with a higher DP like Frutafit IQ (Figure 3) in which the spectrum was comparable with that obtained analyzing Raftiline inulin sample (data not shown).

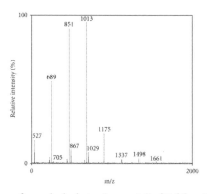

Figure 2. MALDI MS spectrum of a standard solution (x mg mL^{-1}) of Raftilose. DHB was used as matrix. "a.i." means arbitrary intensity and "m/z" means the mass-to-charge ratio. Other conditions as reported in Section 2.4.

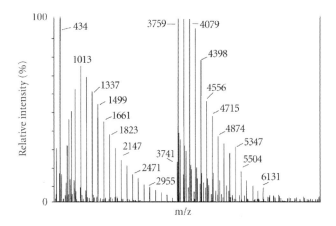

Figure 3. MALDI MS spectrum of a standard solution (1.0 mg mL^{-1}) of Frutafit IQ. 3-AQ was used as matrix. Other conditions as reported in Figure 2.

As reported in Figure 4, for a inulin mainly composed of fructans at high DP, such as Frutafit TEX, the best matrix resulted to be THAP with fast evaporation technique.

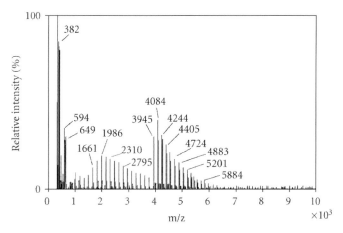

Figure 4. MALDI MS spectrum of a standard solution (1.0 mg mL^{-1}) of Frutafit TEX. THAP was used as matrix. Other conditions as reported in Figure 2.

The MALDI-TOF mass spectra exhibited the sodium and potassium adducts and ascribed the degree of polymerization of these fructans (Table 4). The oligomers showed the mass difference of 162 Da, which corresponds to hexose residues and, as expected, all spectra exhibited the monomodal mass distributions without any fragmentation [18].

Table 4. Example of sodium adduct and degree of polymerization of Raftiline standard.

Degree of polymerization (DP)	$[M + Na]^+$	Degree of polymerization (DP)	$[M + Na]^+$
DP3	527	DP17	2799.3
DP4	689	DP18	2961.6
DP5	852.1	DP19	3124
DP6	1014.6	DP20	3286.1
DP7	1176.9	DP21	3448.5
DP8	1339.4	DP22	3610.1
DP9	1501.7	DP23	3772.7
DP10	1664	DP24	3934.4
DP11	1826.3	DP25	4096.4
DP12	1988.5	DP26	4258.4
DP13	2150.9	DP27	4420.9
DP14	2313.1	DP28	4582.7
DP15	2475.3	DP29	4743.7
DP16	2637.4	DP30	4905.3

MALDI-TOF spectra elicited the same differences of DP and geometrical profiles between fructans, which were seen with HPAEC-PED.

In MALDI-TOF spectra, there are not only visibles Raftiline and Frutafit TEX Gaussian profiles like in chromatogram but also differences between Raftilose maximum relative intensity (in the ranges between DP 4–6) and Frutafit TEX maximum relative intensity (in the ranges between DP 12–14).

Many papers in literature cited the use of MALDI-TOF to analyze fructans present in natural sources as Jerusalem artichoke, onion, shallots, elephant garlic, and so forth [17–20]. MALDI-TOF MS gives better results for this type of food rather than HPAEC-PAD because fructans profiles are unknown and so MALDI-TOF MS identifications need less time to optimize analysis and assure correct molecular assignment.

Furthermore, MALDI-TOF MS is far less prone to contaminant influence and does not require a tedious purification of the analytes (which may cause selective losses of some compounds of the mixture to be analyzed) [8].

The limit of MALDI-TOF MS is the fact that similar mass branched and linear isomers cannot be distinguished. On the other hand, although HPAEC-PED does not allow for structure elucidation, it permits identification of unknown carbohydrates relative to standards whose retention behavior versus structure has been already established. A representative elution pattern of Actilight

950P, obtained by HPAEC-PED under the chromatographic conditions reported in the experimental part, is depicted in Figure 5. Actilight, which is industrially produced through fructosyl-transfer from sucrose using a fungal enzyme, is a commercial available food ingredient, having the following composition of dry substance: 0.3% fructose, 0.4% glucose, 3.0% sucrose, 36% 1-kestose (GF2), 49% nystose (GF3), and 12% fructosyl-nystose (GF4) [22]. From the chromatographic profile A depicted in Figure 5, it is possible to clearly identify the free monosaccharide glucose and fructose, the unreacted disaccharide sucrose, and the trisaccharide 1-kestose, which were identified eluting the corresponding standards. Other oligosaccharides were selectively eluted, but the identification of the individual oligosaccharides was a challenging task due to lack of suitable standards. However, the retention times of carbohydrates from the HPAEC column depend both on DP and structural differences (e.g., branching) and evaluating the chromatographic profiles of Figure 5, we can concluded that peak no. 6 and peak no. 7 can be assigned as the tetrasaccharide nystose (GF3) and the pentasaccharide 1F-fructosyfuranosyl-nystose (GF4), respectively.

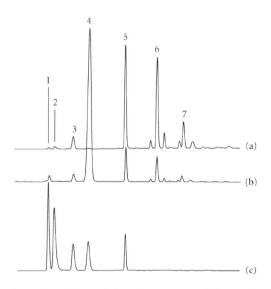

Figure 5. Chromatographic profiles of (a) symbiotic milk containing Actilight as prebiotic scFOS; (b) synthetic mixture of glucose, fructose, sucrose, lactose, and 1-kestose (GF3); (c) Actilight. Chromatographic conditions as reported in the text. Peak identification: (1) glucose, (2) fructose, (3) sucrose, (4) lactose, (5) 1-kestose, (6) GF4, (7) GF5.

Moreover, the chromatographic profile of the commercial scFOS preparation shows at least other four unknown oligomers that should be assumed as inulooligosaccharides ranging from F2 to F5 [23].

The unit chain length distribution of scFOS in the Actilight product was analyzed by MALDI-TOF MS, and the obtained spectrum confirmed that the analyzed product was a mixture of short-chain oligosaccharides ranging from DP 3 to DP 5, having a mass difference of 162 Da (results not shown). Although, a distinction of oligomers having similar masses branched and linear isomers is not possible by MALDI—TOF—MS, the use of this mass spectrometric technique was a very useful tool for the molecular weight measurement, whereas the heterogeneity of scFOS distribution in Actilight was confirmed through the chromatographic profile obtained by HPAEC-PED.

Food Applications

Symbiotic Milk

According to a widely accepted definition, a functional food is any modified food that has special effects on the human organism beyond the nutrients it contains [24].

Milk with added functional ingredients, but not fermented, is proposed in the market of functional foods. In these functional dairy, products are included unfermented milk with added probiotic and prebiotics, called symbiotic, due to their symbiotic functional action.

A representative pattern of a carbohydrate elution profile of commercial symbiotic functional milk is depicted by the chromatographic profile B of Figure 5. The main peaks in the chromatogram obtained on the CarboPac PA 100 column, using the elution program reported in method 2, Table 1, consist of glucose, sucrose, lactose, and scFOS. Besides the identified compounds, as reported in the chromatographic profile A, the chromatographic profile B shows the presence of minor peaks which can be identified as previously reported. Chromatographic profile C is referred to a standard solution of glucose, fructose, sucrose, lactose, and 1-kestose. The validation process of the optimized HPLC-PED method was carried out following the EURACHEM guidelines [25].

Table 5 summarizes the precision of retention times observed upon injecting the same sample of milk and using the optimized experimental conditions. The instrumental precision was evaluated by repeating the analysis of the same milk sample six times. As can be seen, relative standard deviations (RSDs) of retention times were lower than 2.10% (n=6) for the same day, while these values increase up to 2.55% when the same experiment was repeated in six different days (n=24). Furthermore, method repeatability was evaluated using the same data obtained for the accuracy study, where the RSDs of peak areas were in all experiments better than 2.85%.

Table 5. Reproducibility of retention times for intra- and interday analyses for the symbiotic milk sample by HPAEC-PED (separation conditions as in Figure 5).

Compound	Intraday ($n = 6$)		Interday (6 days, $n = 24$)	
	Retention time (min)	RSD (%)	Retention time (min)	RSD (%)
Glucose	5.75	1.88	5.93	2.24
Fructose	6.87	2.08	7.05	2.75
Saccharose	8.90	1.91	10.10	2.60
Lactose	12.41	1.82	12.62	2.44
1-kestose	15.90	1.77	16.10	2.10

Quantification was based on external standard method. The assay linearity was determined by the analysis of six different concentrations of the standard solutions. Each level of concentration was prepared in triplicate. The linearity of response for the analyzed sugars was demonstrated at six different concentrations from 50 to 300 µg/mL for glucose, fructose, sucrose, and 1-kestose and from 50 to 450 µg/mL for lactose, respectively. The standard curves were obtained by plotting peak area (y) versus nominal concentration x (µgmL–1) of each compound and were fitted to the linear regression. The standard deviation (SD) of slope and intercept was estimated at the 95% confidence level.

The limit of detection (LOD) was defined as three times the standard deviation of the blank values (Sb) divided by the slope of the calibration curves, whereas the limit of quantification (LOQ) was defined as 10 Sb divided by the slope of the calibration curve. LOD for all analyzed samples was ranging from 10.0 to 12.5 µg/mL and LOQ from 25.0 to 32.0 µg/mL, respectively.

Quantification of scFOS was determined from peak area using 1-kestose as the external standard. As commercial GF4 and GF5 standards were not available, they were not quantified and Actilight was quantified considering that 1-kestose represents the 36% of the whole product. The recoveries, measured at three concentration levels, varied from 97.0 to 104.3%. The validated method was successfully applied to the simultaneously determination of glucose, fructose, sucrose, lactose, and scFOS in symbiotic milk, obtaining the following results: glucose 0.58 (±0.08) mg/mL, sucrose 0.62 (±0.07) mg/mL, lactose 44.76 (±0.52) mg/mL, 1-kestose 7.28 (±0.11) mg/mL. Nystose (GF3) 78.7 mg/mL, fructofuranosylnistose (GF4) 56.8 mg/mL, and GF5 12.5 mg/mL. From the above results and data reported in literature, the content of scFOS in the examined symbiotic milk can be evaluated to be within 2.0 (±0.2)% (w/v). Furthermore, no variation of peak area ratio of 1-kestose-GF3 and 1-kestose-GF4 was noticed over ten days, demonstrating that scFOS composition did not change over the whole shelf life of the product.

Cooked Ham

Inulin was added to the brine used in the industrial process for preparation of commercial cooked ham, prior the cooking step, as a replacer of sugars with the aim of reducing caloric content. The inulin profile of ham was analyzed by the HPAEC PED method described in method 1, Table 1. Comparing chromatographic profiles reported in Figure 6, there were no observable differences in the inulin profiles between those extracted from (b) the cooked ham and (a) the control, corresponding to inulin which was added to the brine solution. Furthermore, the same chromatographic profiles were also observed during the whole shelf life of the product (data not shown).

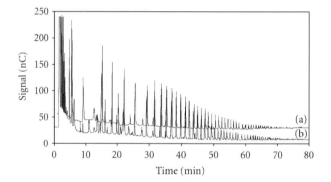

Figure 6. Chromatographic profiles of a standard solution (0.5 m gmL⁻¹) of (a) Frutafit IQ compared with (b) the chromatographic profile of an extract of inulin from a cooked ham sample. Chromatographic conditions as in the text.

To perform quantitative evaluation of inulin present in cooked ham, we selected six unidentified peaks of the oligosaccharide fraction, which were selectively eluted within 20 minutes using the gradient elution program reported in Table 2 (see Figure 7). To verify the quality and usefulness of the method, the analytical parameters linearity, sensitivity, precision, and percentage of recovery were determined. The linearity of response for the selected unidentified peaks was demonstrated at six different concentrations of inulin, ranging from 50 to 300 µg/mL. Higher concentrations were not assayed because we considered that the range was wide enough for the proposed applications. The linearity of the present method for all unidentified oligosaccharides selected as reference peaks was good, with correlation coefficients higher than 0.997. The limit of detection was evaluated for each of the eluted peaks, and results are summarized in Table 6. The recoveries, measured at three concentration levels of inulin added to the ham sample after homogenization, varied from 91 to 106%.

Table 6. Limits of detection and quantitation.

Peaks	1	2	3	4	5	6
Limit of detection (yD)[a] $\mu g/g$	12.21	13.65	12.45	13.22	14.27	16.03
Limit of quantification (yQ)[b] $\mu g/g$	43.32	48.62	49.89	40.67	50.77	51.21

[a] Concentration corresponding to signal and $yD = yb + 2t(95\%, n-1)sb$.
[b] Concentration corresponding to signal and $yQ = yb + 10sb$.

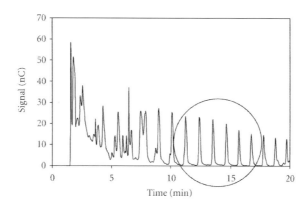

Figure 7. Chromatographic profile of an extract of inulin from a cooked ham sample, showing inside the circle the oligosaccharide fraction selected to perform quantitative evaluation of inulin present in the analyzed sample. Chromatographic conditions as in the text.

Result regarding quantitative determination of inulin in cooked ham gave results ranging from 3.20±0.20 to 3.00±0.42, dry matter and on the average 1.05 (±0.02)g on 100 g of cooked ham.

Conclusions

We have demonstrated the usefulness of MALDI-TOF MS, and that high-pH anion exchange chromatography with pulsed amperometric detection (HPAEC-PAD) provides powerful tools for the analysis of FOS and inulins with a high sensitivity and no need for derivatization.

HPAEC-PED methods were developed to determine chain length distribution of FOS and inulins, which were compared with the corresponding distributions from MALDI-TOF MS analyses. Although some differences were observed, the developed HPAEC-PED methods can be considered as secondary or orthogonal methods complementary to MALDI-TOF MS to evaluate oligo- and polysaccharides distribution in the analyzed samples. MALDI-MS gives better assurance of correct molecular assignment since the isotopic mass of each peak is available, although similar masses branched and linear isomers cannot be distinguished.

By using HPAEC and MALDI-TOF MS analysis, the presence of FOS and inulin at different degrees of polymerization could neither be demonstrated in ingredient preparations nor from functionalized foods.

Nomenclature

HPAEC-PED: High performance anion exchange chromatography with pulsed electrochemical detection

MALDI-TOF-MS: Matrix-assisted laser desorption/ionization time-of-flight mass spectrometry

MALDI-LR: Linear mode

DP: Degrees of polymerization

FOS: Fructooligosaccharides

DHB: Dihydroxybenzoic acid

THAP: Trihydroxyacetophenone

AQ: Aminoquinoline

HCCA: Hydroxy-α-alpha cyanocinnamic acid

TFA: Trifluoroacetic acid

ACTH: Adrenocorticotrophic hormone.

Acknowledgements

The project was funded by the Italian Ministry for the University and Research (MUR) with a PNR 2005-2007 Project no. RBIP06SXMR "Sviluppo di metodologie innovative per l'analisi di prodotti agroalimentari." The authors wish to thank Dr. Silvio Cardellini (I Fratelli Emiliani SpA, Langhirano, Italy) for providing cooked ham samples.

References

1. M. K. Ernst., N. J. Chatterto, P. A. Harrison, and G. Matitschka, "Characterization of fructan oligomers from species of the genus Allium L," Journal of Plant Physiology, vol. 153, no. 1-2, pp. 53–60, 1998.

2. I. Vijn and S. Smeekens, "Fructan: more than a reserve carbohydrate?," Plant Physiology, vol. 120, no. 2, pp. 351–359, 1999.

3. R. G. Crittenden and M. J. Playne, "Production, properties and applications of food-grade oligosaccharides," Trends in Food Science & Technology, vol. 7, no. 11, pp. 353–361, 1996.

4. M. B. Roberfroid, J. A. E. Van Loo, and G. R. Gibson, "The bifidogenic nature of chicory inulin and its hydrolysis products," The Journal of Nutrition, vol. 128, no. 1, pp. 11–19, 1998.

5. J. W. Yun, "Fructooligosaccharides—occurrence, preparation, and application," Enzyme and Microbial Technology, vol. 19, no. 2, pp. 107–117, 1996.

6. H. Hidaka, T. Eida, T. Takizawa, T. Tokunaga, and Y. Tashiro, "Effects of fructooligosaccharides on intestinal flora and human health," Bifidobacteria Microflora, vol. 5, no. 1, pp. 37–50, 1986.

7. H. Teeuwen, M. Thonè, and J. Vandorpe, "Inulin: a versatile fibre ingredient," International Food Ingredients, vol. 4, no. 5, pp. 10–14, 1992.

8. E. Chiavaro, E. Vittadini, and C. Corradini, "Physicochemical characterization and stability of inulin gels," European Food Research and Technology, vol. 225, no. 1, pp. 85–94, 2007.

9. G. R. Gibson and M. B. Roberfroid, "Dietary modulation of the human colonic microbiota: introducing the concept of prebiotics," The Journal of Nutrition, vol. 125, no. 6, pp. 1401–1412, 1995.

10. J. Huebner, R. L. Wehling, and R. W. Hutkins, "Functional activity of commercial prebiotics," International Dairy Journal, vol. 17, no. 7, pp. 770–775, 2007.

11. C. Corradini, F. Bianchi, D. Matteuzzi, A. Amaretti, M. Rossi, and S. Zanoni, "High-performance anion-exchange chromatography coupled with pulsed amperometric detection and capillary zone electrophoresis with indirect ultra violet detection as powerful tools to evaluate prebiotic properties of fructooligosaccharides and inulin," Journal of Chromatography A, vol. 1054, no. 1-2, pp. 165–173, 2004.

12. M. Rossi, C. Corradini, A. Amaretti, et al., "Fermentation of fructooligosaccharides and inulin by bifidobacteria: a comparative study of pure and fecal cultures," Applied and Environmental Microbiology, vol. 71, no. 10, pp. 6150–6158, 2005.

13. Y. C. Lee, "Carbohydrate analyses with high-performance anion-exchange chromatography," Journal of Chromatography A, vol. 720, no. 1-2, pp. 137–149, 1996.

14. K. K. Mock, M. Daevy, and J. S. Cottrell, "The analysis of underivatized oligosaccharides by matrix-assisted laser desorption mass spectrometry," Biochemical and Biophysical Research Communications, vol. 177, no. 2, pp. 644–651, 1991.

15. G. Jiang and T. Vasanthan, "MALDI-MS and HPLC quantification of oligosaccharides of lichenase-hydrolyzed water-soluble β-glucan from ten barley varieties," Journal of Agricultural and Food Chemistry, vol. 48, no. 8, pp. 3305–3310, 2000.

16. D. J. Harvey, "Matrix-assisted laser desorption/ionization mass spectrometry of carbohydrates and glycoconjugates," International Journal of Mass Spectrometry, vol. 226, no. 1, pp. 1–35, 2003.

17. B. Stahl, A. Linos, M. Karas, F. Hillenkamp, and M. Steup, "Analysis of fructans from higher plants by matrix-assisted laser desorption/ionization mass spectrometry," Analytical Biochemistry, vol. 246, no. 2, pp. 195–204, 1997.

18. M. Štikarovská and J. Chmelík, "Determination of neutral oligosaccharides in vegetables by matrix-assisted laser desorption/ionization mass spectrometry," Analytica Chimica Acta, vol. 520, no. 1-2, pp. 47–55, 2004.

19. J. Wang, P. Sporns, and N. H. Low, "Analysis of food oligosaccharides using MALDI-MS: quantification of fructooligosaccharides," Journal of Agricultural and Food Chemistry, vol. 47, no. 4, pp. 1549–1557, 1999.

20. M. Štikarovská and J. Chmelík, "Determination of neutral oligosaccharides in vegetables by matrix-assisted laser desorption/ionization mass spectrometry," Analytica Chimica Acta, vol. 520, no. 1-2, pp. 47–55, 2004.

21. Y. Zhang, Y. Inoue, S. Inoue, and Y. C. Lee, "Separation of oligo/polymers of 5-N-acetylneuraminic acid, 5-N-glycolylneuraminic acid, and 2-keto-3-deoxy-D-glycero-D-galacto-nononic acid by high-performance anion-exchange chromatography with pulsed amperometric detector," Analytical Biochemistry, vol. 250, no. 2, pp. 245–251, 1997.

22. B. Król and K. Grzelak, "Qualitative and quantitative composition of fructooligosaccharides in bread," European Food Research and Technology, vol. 223, no. 6, pp. 755–758, 2006.

23. S. N. Ronkart, C. S. Blecker, H. Fourmanoir, et al., "Isolation and identification of inulooligosaccharides resulting from inulin hydrolysis," Analytica Chimica Acta, vol. 604, no. 1, pp. 81–87, 2007.

24. A. T. Diplock, P. J. Aggett, M. Ashwell, F. Bornet, E. B. Fern, and M. B. Roberfroid, "Scientific concepts of functional foods in Europe: consensus document," British Journal of Nutrition, vol. 81, pp. SI–S27, 1999.

25. "The Fitness for Purpose of Analytical Methods: A Laboratory Guide to Method Validation and Related," Eurachem Guide, 1998, http://www.eurachem.ul.pt.

CITATION

Originally published under the Creative Commons Attribution License or equivalent. Borromei C, Careri M, Cavazza A, Corradini C, Elviri L, Mangia A, Merusi C. Evaluation of Fructooligosaccharides and Inulins as Potentially Health Benefiting Food Ingredients by HPAEC-PED and MALDI-TOF MS. Int J Anal Chem. 2009;2009:530639. doi: 10.1155/2009/530639.

Preparation, Characterization, and Analytical Application of Ramipril Membrane-Based Ion-Selective Electrode

Hassan Arida, Mona Ahmed and Abdallah Ali

ABSTRACT

The fabrication and electrochemical evaluation of two PVC membrane-based Ion-Selective electrodes responsive for ramipril drug have been proposed. The sensitive membranes were prepared using ramipril-phosphomolibdate and ramipril-tetraphenylborate ion-pair complexes as electroactive sensing materials in plasticized PVC support. The electrodes based on these materials provide near-Nernestian response (sensitivity of 53 ± 0.5–54 ± 0.5 mV/concentration decade) covering the concentration range of 1.0×10^{-2}–1.0×10^{-5} mol L–1 with a detection limit of 3.0×10^{-6}–4.0×10^{-6} mol L^{-1}. The suggested electrodes have been successfully used in the determination of ramipril drug in some pharmaceutical formulations using direct potentiometry with average recovery of >96% and mean standard deviation of <3% (n=5).

Introduction

Ramipril (see Scheme 1) contains not less than 98.0% of $C_{23}H_{32}N_2O_5$; (2S,3aS,6aS)-1-[(2S)-2-[[(2S)-1-ethoxy-1-oxo-4-phenylbutan-2-yl]amino] propanoyl]-3,3a,4,5,6,6a-hexahydro-2H-cyclopenta[d]pyrrole-2 carboxylic acid. It belongs in a class of drugs called angiotensin converting enzyme (ACE) inhibitors which are used for treating high blood pressure and heart failure and for preventing kidney failure due to high blood pressure and diabetes. ACE is important because it produces the protein, angiotensin II. Angiotensin II contracts the muscles of most arteries in the body, including the heart, thereby narrowing the arteries and elevating the blood pressure. In the kidney, the narrowing caused by angiotensin II also increases blood pressure and decreases the flow of blood. ACE inhibitors such as ramipril lower blood pressure by reducing the production of angiotensin II, thereby relaxing the arterial muscles and enlarging the arteries. The enlargement of the arteries throughout the body reduces the blood pressure against which the heart must pump blood, and it becomes easier for the heart to pump blood. The arteries supplying the heart with blood also enlarge. This increases the flow of blood and oxygen to the heart, and this improves further the ability of the heart to pump blood. The effects of ACE inhibitors are particularly beneficial to people with congestive heart failure. In the kidneys, the enlargement of the arteries also reduces blood pressure and increases blood flow.

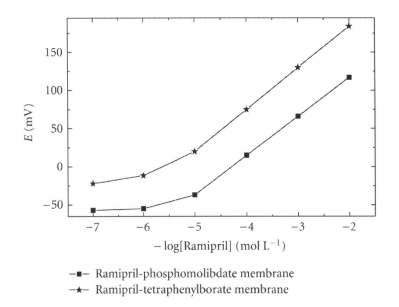

Figure 1. Potentiometric calibration response of ramibril-based selective electrodes.

Methods in current use for the assay of ramipril, in pharmaceutical prepara-tion, are based on spectrophotometry [1–6], liquid chromatography [7–12], gas chromatography [13, 14], enatioselective biosensors [15], enatioselective mem-brane [16], voltammetry [17], amperometric biosensor [18], and mass spectrom-etry [12]. However, most of these methods are sophisticated, tedious, and required many manipulation steps. On the other hand, although, ion-selective electrodes and potentiometric sensors are much simpler, fast, and inexpensive, only one ion-selective electrode has been reported for the determination of ramipril drug as anionic species [19].

In this paper, the preparation, characterization, and analytical application of new two ramipril cationic ion-selective electrodes based on phosphomolibdate and tetraphenylborate-ramipril ion-pair complexes have been reported. The mer-its offered by the proposed electrodes include the high cationic sensitivity of the drug in the acid media with fast response time (<30 seconds) and long life span (2 months).

Experimental

Apparatus

The potentiometric measurements were made at 25 ± 1C°, using an Orin (Model A720) digital pH/mV meter and Orion Ross Combination pH electrode (Model 81-02) for all pH measurements. The suggested ramipril PVC membrane-based electrode was used for all potentiometric measurements in conjunction with a double junction reference electrode (Orion Model 90-02) containing KNO$_3$ (10% w/v) in the outer compartment. Perkin-Elmer (Norwalk, Conn, USA) 1430 ratio recording IR spectrophotometer was used for structure elucidation of the ion-pair complexes.

Scheme 1: Structure of ramipril.

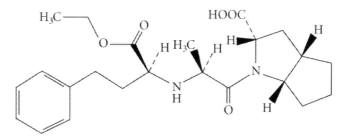

Reagents and Materials

All chemicals were of analytical reagent grade unless otherwise stated and doubly distilled water was used throughout. Poly (vinyl chloride) powder (PVC) of high molecular weight (10 000) and dioctyl phthalate plasticizer of purity 98% were obtained from Aldrich chemical company, Inc. (Milwaukee, Wis, USA). Tetrahydrofuran (THF) with a purity of 99% containing 0.025% butylatedhydroxytoluene inhibitor was used as solvent.

Ramipril stock solution (10–2 mol L–1) was prepared by dissolving 0.4165 g in a minimal volume of acetic acid with continuous stirring and then diluted to 100 mL with distilled water. The resulting clear solution of pH 3 was obtained by the addition of small aliquots of 10–2 mol L–1 HCl. Standard phosphomolibdate and tetraphenylborate solutions (0.2 mol L–1) were individually prepared by dissolving appropriate weight of sodium phosphomolibdate and tetraphenylborate, respectively, in a minimum volume of distilled water followed by filtration then completed to 100 mL with distilled water.

The Ramipril Membrane-Based Ion-Selective Electrodes

Ramipril-phosphomolibdate and ramipril-tetraphenylborate ion-pair sensing materials were individually prepared by mixing 30 mL aliquot of 10^{-2} mol L^{-1} ramipril solution with a 30 mL aliquot of 10^{-2} mol L^{-1} aqueous phosphomolibdate and tetraphenylborate solutions, respectively in l00 mL beaker. The obtained precipitates were filtered using a G4 sintered glass crucible, washed thoroughly with distilled water, and dried at room temperature.

Two PVC master membranes based on the suggested sensing materials containing 0.01 g ion-pair complex, 0.350 g dioctylphthalate (DOP), and 0.190 g of poly (vinyl chloride) were individually prepared. Each membrane contents were thoroughly mixed, dissolved in 6 mL aliquot of THF, and transferred to a glass Petri dish (3 cm diameter). The Petri dish was covered with a filter paper and left to stand overnight to allow slow evaporation of the solvent at room temperature. The membranes were sectioned with a cork borer (10 mm diameter) and attached to a length of polyethylene tubing (3 cm length, 8 mm i.d) by using THF.

A homemade electrode body was used, which consists of a glass tube, to one end of which the poly ethylene tubing was attached and filled with an equimolar mixture of 10–2 mol L–1 of potassium chloride and ramipril as the internal reference solution. An Ag/AgC1 internal reference wire electrode (1.0 mm diameter) was immersed in the internal solution. This assembly was used in the potentiometric characterization of the electrodes and the subsequent determination of the drug.

Electrochemical Evaluation of the Electrodes

In order to calibrate the suggested electrodes, aliquots (10 mL) of aqueous ramipril solutions (1.0×10^{-2}–1.0×10^{-7} mol L^{-1}) were transferred into 50 mL beakers, and the PVC membrane electrode, in conjunction with a double junction Ag/AgCl reference electrode, was immersed in the solution. Alternatively, the drug PVC membrane electrode in conjunction with a double junction Ag/AgC1 reference electrode was immersed in a 50 mL beaker containing a l0 mL aliquot of water. Aliquots (1 mL) of 10^{-2}–10^{-6} mol L^{-1} pure drug solution were successively added. The solutions were gently stirred during the measurements and the potential recorded after stabilization to ±0.2 mV and the e.m.f. plotted on semilogarithmic paper as a function of the drug concentration. The calibration graphs were used for subsequent determination of unknown concentration of the drug.

The potentiometric selectivity coefficient indicates the extent to which a foreign ion B interferes with the response of the electrode to its primary drug ion A. The potentiometric selectivity coefficients $K_{A,B}^{pot}$ for the suggested electrodes were measured by separate solutions method SSM [20]. In this method, the potential responses of the electrode in 10–2 mol L–1 solution of the interferents were measured and recorded. The potential response of the electrode for the drug was separately, obtained in a similar manner at the same concentration level. The selectivity coefficient values were calculated using the simplified form of Eisemnan Nicolsky equation:

$$K_{A,B}^{pot} = \frac{E_1 - E_2}{S} + \left(1 - \frac{1}{Z_2}\right)\log a_1, \tag{1}$$

where E_1 and E_2 are the potential readings of the electrode in separate solutions of the same concentration of the drug and interferants, respectively, S is the slope of the drug calibration graph (mV/concentration decade), a_1 is the activity of the drug, and Z_2 is the charge number of the interfering ion.

The effect of pH of the test solution on the potential reading of the suggested drug electrodes was studied by immersing a Ross combination glass electrode (Orion model 81-02), PVC membrane electrode, and a double junction Ag/AgC1 reference electrode in 50 mL beakers containing 30 mL aliquots of 10–3 and 10–4 mol L–1 drug solutions. The pH of each solution was gradually increased and decreased by adding small aliquots of dilute sodium hydroxide and hydrochloric acid, respectively. The potential at each pH value was recorded. The mV-pH profile at each drug concentration was plotted for the two electrode systems.

Analytical Application of the Ramipril Electrodes

In order to investigate the reliability of the proposed electrodes, they have been applied in the determination of the drug using direct potentiometry and potentiometric titration. In the direct potentiometry study, a PVC membrane-based ramipril electrode in conjunction with a double junction reference electrode was immersed in a 10 mL of the appropriate drug solutions of unknown concentration. The potential readings were recorded after stabilization to ±0.2 mV and compared with the calibration graph. The solutions were stirred during measurements, and the electrodes were thoroughly washed with distilled water between measurements.

In the potentiometric titration of ramipril, aliquots (2–6 mL) of 10–3 mol L–1 of ramipril solution were transferred to 50 mL beakers and diluted to 10 mL with distilled water. The solution was stirred and titrated with a standard 10–3 mol L–1 sodium tetraphenylborate solution using the suggested ramipril membrane electrode in conjunction with a double junction Ag/AgC1 reference electrode. The electrode potential (E) was recorded as a function of the titrant volume (v) added, (E versus v) curves were plotted. The end point was calculated from the maximum slope $\Delta E/\Delta v$ versus v.

Moreover, different forms and dosages of pharmaceutical preparations were assayed to determine ramipril in different formulations. In this study, aliquot of 5 mL of the drug was transferred to 25 mL beaker containing 15 mL of deionised water, and the pH of the solution was adjusted to pH 3 with a drop of dilute HC1 solution. The solution was then transferred to 25 mL volumetric flask, completed to the mark, shaken well, and transferred to a 100 mL beaker for ISE measurement. For the assay of tablet formulations, 10 tablets were finely powdered, mixed, and an accurate weight equivalent to 0.1% of the tablet was transferred into 50 mL beakers containing 20 mL of deionized water, the pH of the solution was adjusted to pH 3 with a drop of dilute HC1 solution, and transferred to 25 mL volumetric flask, completed with water to the mark, shaken well. The suggested ramipril-based membrane electrode was immersed in conjunction with a double junction Ag/AgC1 reference electrode into the solution. The potential was measured after a stable reading was obtained and compared with that on a calibration graph previously constructed for standard solutions.

Results and Discussion

The Nature and Composition of Ramipril-Based Membranes

Ramipril—phosphomolibdate and ramipril—tetraphenylborate ion-pair complexes have been prepared and examined as new electroactive ionophores in PVC matrix membranes responsive to ramipril cation. In acid media, ramipril cations readily react with phosphomolibdate and tetraphenylborate anions to form stable water insoluble ion association complexes. The IR studies confirm that the formation of 1:1 ramipril:anion ion-pair association. The potentiometric response characteristics of the proposed electrodes were evaluated, according to IUPAC recommendations [20] using the following electrochemical cell:

$$\text{Ag/AgCl} \left| \begin{array}{c} 10^{-2}\ \text{mol L}^{-1}\ \text{KCl} \\ 10^{-2}\ \text{mol L}^{-1}\ \text{ramipril} \end{array} \right| \begin{array}{c} \text{PVC} \\ \text{membrane} \end{array} \left\| \begin{array}{c} \text{Ramipril} \\ \text{test solution} \end{array} \right| \begin{array}{c} 10\% \\ \text{KCl} \end{array} \left| \begin{array}{c} \text{AgCl/} \\ \text{Ag} \end{array} \right.$$

Electrochemical Evaluation of Ramipril Electrodes

The potentiometric response characteristics of the investigated ramipril PVC membrane-based electrodes were evaluated from the data collected from four assemblies for each membrane electrode. The results are summarized in (Table 1). The responses of ramipril electrode systems are linear over the concentration range $1 \times 10^{-2} – 1 \times 10^{-5}$ mol L^{-1}. The slopes of the calibration plots (Figure 1) are typically 53 ± 0.5 mV and 54 ± 0.5 per concentration decade for ramipril—phosphomolibdate and ramipril—tetraphenyle borate membrane electrode, respectively. Deviation from the ideal Nernstian slope (59.2 mV/concentration decade) stems from the fact that most potentiometric drug sensors respond to the activity of the drug cations rather than the concentration.

Table 1. Response characteristics of ramipril membrane-based selective electrodes.

Parameter	Ramipril phosphomolibdate	Ramipril tetraphenylborate
Slope, mV/decade	53 ± 0.5	54 ± 0.5
Linear range, mol L^{-1}	$1 \times 10^{-2} – 1 \times 10^{-5}$	$1 \times 10^{-2} – 1 \times 10^{-5}$
Lower limit of detection, mol L^{-1}	4.0×10^{-6}	3.0×10^{-6}
Response time, (s)	<20	<15
Life time, (d)	60	50
Working pH range	1–3.8	1–4

The response time and stability of the membranes have been investigated. In these studies, the time required for ramipril poly (vinyl chloride) membrane electrodes to reach a value of ±1 mV from the final equilibrium potential in the same day after successive immersion in different ramipril solutions was measured. The results obtained (Figure 2) show that the electrode attain stable potential values within 20 seconds. In addition, the potentials displayed by the electrode in the linear concentration range of ramipril in the same day do not vary by more than

±0.5 mV. The stability of the potential reading for the ramipril-based electrodes is within ±2 mV during the lifetime (2 months) of the electrodes.

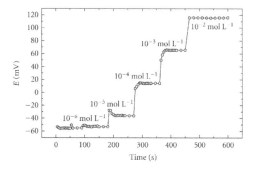

Figure 2. Potentiometric dynamic response of ramipril-phosphomolibdate membrane electrode.

The effect of pH on the potential readings of the proposed electrode was also examined by recording the e.m.f. of ramipril test solutions (10–3 and 10–4 mol L–1) at various pH values, which were obtained by the addition of very small volumes of hydrochloric acid and/or sodium hydroxide solutions 10–1 mol/L of each. The e.m.f-pH plots presented in Figures 3 and 4 revealed that the potential readings are insensitive to pH changes in the ranges 1.0–3.8 and 1.0–4.0 for the ramipril-phosphomolibdate and ramipeil-tetraphenylborate, respectively. The shape of mV-pH profile depends on the stability of the ion-pair in the membrane as a function of pH, the nature of the drug (protonation or complexation equilibrium) in the test solution, and the sensitivity of the membrane to either [H+] or [OH−] at low or high pH values, respectively. It was observed that ramipril ion-pair complexes deteriorate in alkaline media causing severe potential shift. At relatively higher pH values, the e.m.f decreased this is probably due to deprotonated species in test solution.

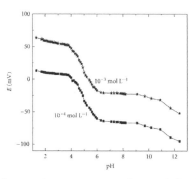

Figure 3. Effect of pH on the potentiometric response of ramipril-phosphomolibdate PVC membrane electrode.

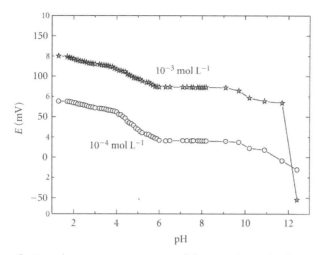

Figure 4. Effect of pH on the potentiometric response of the ramipril-tetraphenylborate PVC membrane electrode.

The selectivity coefficients of the investigated electrodes were evaluated using separate solution method (SSM). In this study, the performance of the ramipril electrodes in the presence of some tested organic and inorganic cations was studied. The selectivity coefficient values ($K_{A,B}^{pot}$) are calculated and presented in Table 2. As can be seen, the electrode offers a reasonable selectivity for the ramipril cation over most of the tested species.

Table 2. Potentiometric selectivity coefficients ($K_{A,B}^{pot}$) for the ramipril phosphomolibdate and ramipril-tetraphenylborate selective electrodes.

Interfering species, B	Ramipril-phosphomolibdate electrode	Ramipril-tetraphenylborate electrode
Ramipril	1	1
NH_4^+	8.8×10^{-3}	9.1×10^{-3}
Co^{2+}	8.7×10^{-3}	8.8×10^{-3}
Cd^{2+}	8.8×10^{-3}	8.7×10^{-3}
Ni^{2+}	8.8×10^{-3}	8.7×10^{-3}
Cr^{3-}	4.2×10^{-2}	4.3×10^{-2}
Cu^{2+}	6.6×10^{-3}	9.0×10^{-3}
Ca^{2+}	4.6×10^{-3}	4.8×10^{-3}
Mg^{2+}	9.8×10^{-3}	9.8×10^{-3}
Sn^{2-}	5.1×10^{-2}	5.3×10^{-2}
K^+	8.8×10^{-3}	8.6×10^{-3}
Na^+	8.1×10^{-3}	8.5×10^{-3}
Fe^{3+}	8.8×10^{-3}	8.8×10^{-3}
Glycine	7.8×10^{-3}	8.1×10^{-3}
Phenyl hydrazine	3.2×10^{-2}	3.2×10^{-2}
Hydroxylamine	2.2×10^{-2}	2.7×10^{-2}

Analytical Application of Ramipril Electrodes

In order to investigate the analytical usefulness of the proposed electrodes, they have been successfully applied in the potentiometer determination of ramipril in aqueous samples as well as in some local pharmaceutical formulations. In the former study, solutions of concentration in the liner range of the tested electrode are prepared from pharmaceutical grade of ramipril and determined by direct potentiometry and potentiometric titration using the proposed ramipril electrodes. The results obtained are summarized in Table 3. In the later study, some local pharmaceutical formulations have been prepared and analyzed by direct potentiometry using the investigated ramipril electrodes. The results obtained are summarized in Table 4. As can be seen, the two electrodes provide a good accuracy (average recovery >96%) and high precision (RSD <3%, n = 5) in both studies.

Table 3. The accuracy and reliability of result obtained with ramipril PVC matrix membrane electrode.

Sample	Concentration, mol L^{-1}	Ramipril-phosphomolibdate		Ramipril-teraphenylborate	
		Found	Recovery %	Found	Recovery %
1	1×10^{-3}	0.95×10^{-3}	95.0	0.95×10^{-3}	95.0
2	2×10^{-3}	1.87×10^{-3}	93.5	1.90×10^{-3}	95.0
3	3×10^{-3}	2.91×10^{-3}	97.0	2.92×10^{-3}	97.3
4	4×10^{-3}	3.94×10^{-3}	98.5	3.94×10^{-3}	98.5
5	5×10^{-3}	4.98×10^{-3}	99.6	4.93×10^{-3}	98.6
	Averagere covery		96.7		96.9

Table 4. Determination of ramipril in some pharmaceutical preparations.

Sample[a]	Nominal content mg/tablet	Ramipril-phosphomolibdate		Ramipril-teraphenylborate	
		Found	Recovery %	Found	Recovery %
1	1.25	1.23	98.4	1.24	99.2
2	2.50	2.43	97.2	2.46	98.4
3	5.0	4.95	99.0	4.95	99.0
	Average recovery		98.2		98.8

[a] Different pharmaceutical formulations (tritace; aventis).

Conclusion

Two ramipril selective electrodes have been prepared and electrochemicaly evaluated. They provide analytical, useful sensitivity to the drug with fast response time, reliable, and reproducible response. The proposed electrodes have been successfully applied in the determination of ramipril in some pharmaceutical formulations with good accuracy and high precision.

References

1. S. M. Blaih, H. H. Abdine, F. A. El-Yazbi, and R. A. Shaalan, "Spectrophotometric determination of enalapril maleate and ramipril in dosage forms," Spectroscopy Letters, vol. 33, no. 1, pp. 91–102, 2000.

2. H. H. Abdine, F. A. El-Yazbi, R. A. Shaalan, and S. M. Blaih, "Direct, differential solubility and compensatory-derivative spectrophotometric methods for resolving and subsequently determining binary mixtures of some antihypertensive drugs," S.T.P. Pharma Sciences, vol. 9, no. 6, pp. 587–591, 1999.

3. D. Bonazzi, R. Gotti, V. Andrisano, and V. Cavrini, "Analysis of ACE inhibitors in pharmaceutical dosage forms by derivative UV spectroscopy and liquid chromatography (HPLC)," Journal of Pharmaceutical and Biomedical Analysis, vol. 16, no. 3, pp. 431–438, 1997.

4. H. E. Abdellatef, M. M. Ayad, and E. A. Taha, "Spectrophotometric and atomic absorption spectrometric determination of ramipril and perindopril through ternary complex formation with eosin and Cu(II)," Journal of Pharmaceutical and Biomedical Analysis, vol. 18, no. 6, pp. 1021–1027, 1999.

5. M. M. Ayad, A. A. Shalaby, H. E. Abdellatef, and M. M. Hosny, "Spectrophotometric and AAS determination of ramipril and enalapril through ternary complex formation," Journal of Pharmaceutical and Biomedical Analysis, vol. 28, no. 2, pp. 311–321, 2002.

6. N. Rahman, Y. Ahmad, and S. N. H. Azmi, "Kinetic spectrophotometric method for the determination of ramipril in pharmaceutical formulations," AAPS PharmSciTech, vol. 6, no. 3, pp. E543–E551, 2005.

7. H. Y Aboul-Enein and C. Thiffault, "Determination of ramipril and its precursors by reverse phase high performance liquid chromatography," Analytical Letters, vol. 24, no. 12, pp. 2217–2224, 1991.

8. M. Ito, T. Kuriki, J. Goto, and T. Nambara, "Separation of ramipril optical isomers by high-performance liquid chromatography," Journal of Liquid Chromatography, vol. 13, no. 5, pp. 991–1000, 1990.

9. U. J. Dhorda and N. B. Shetkar, "Reverse phase HPLC determination of ramipril and amlodipine in tablets," Indian Drugs, vol. 36, pp. 638–641, 1999.

10. R. Bhushan, D. Gupta, and S. K. Singh, "Liquid chromatographic separation and UV determination of certain antihypertensive agents," Biomedical Chromatography, vol. 20, no. 2, pp. 217–224, 2006.

11. F. Belal, I. A. Al-Zaagi, E. A. Gadkariem, and M. A. Abounassif, "A stability-indicating LC method for the simultaneous determination of ramipril and hy-

drochlorothiazide in dosage forms," Journal of Pharmaceutical and Biomedical Analysis, vol. 24, no. 3, pp. 335–342, 2001.

12. Z. Zhu, A. Vachareau, and L. Neirinck, "Liquid chromatography-mass spectrometry method for determination of ramipril and its active metabolite ramiprilat in human plasma," Journal of Chromatography B, vol. 779, no. 2, pp. 297–306, 2002.

13. K. M. Sereda, T. C. Hardman, M. R. Dilloway, and A. F. Lant, "Development of a method for the detection of angiotensin-converting enzyme," Analytical Proceedings, vol. 30, no. 9, p. 371, 1993.

14. D. Schmidt, K.-U. Weithmann, V. Schlotte, and D. Riecke, "Hochempfindliche GC/MS-Methode zur Bestimmung des stabilen Prostacyclin Metaboliten 6-oxo-PGF1α in Humanurin," Fresenius' Journal of Analytical Chemistry, vol. 320, no. 7, p. 732, 1985.

15. R.-I. Stefan, J. F. van Staden, and H. Y. Aboul-Enein, "Analysis of chiral drugs with enantioselective biosensors. An overview," Electroanalysis, vol. 11, no. 16, pp. 1233–1235, 1999.

16. P. Shahgaldian and U. Pieles, "Cyclodextrin derivatives as chiral supramolecular receptors for enantioselective sensing," Sensors, vol. 6, no. 6, pp. 593–615, 2006.

17. A. A. Al-Majed, F. Belal, A. Abadi, and A. M. Al-Obaid, "The voltammetric study and determination of ramipril in dosage forms and biological fluids," Farmaco, vol. 55, no. 3, pp. 233–238, 2000.

18. R.-I. Stefan, J. F. van Staden, C. Bala, and H. Y. Aboul-Enein, "On-line assay of the S-enantiomers of enalapril, ramipril and pentopril using a sequential injection analysis/amperometric biosensor system," Journal of Pharmaceutical and Biomedical Analysis, vol. 36, no. 4, pp. 889–892, 2004.

19. H. Y. Aboul-Enein, A. A. Bunaciu, C. Bala, and S. Fleschin, "Enalapril and ramipril selective membranes," Analytical Letters, vol. 30, no. 11, pp. 1999–2008, 1997.

20. Y. Umezawa, K. Umezawa, and H. Sato, "Selectivity coefficients for ion-selective electrodes: recommended methods for reporting $K_{A,B}^{pot}$ values," Pure and Applied Chemistry, vol. 67, no. 3, pp. 507–518, 1995.

CITATION

Originally published under the Creative Commons Attribution License or equivalent. Arida H, Ahmed M, Ali A. Preparation, Characterization, and Analytical Application of Ramipril Membrane-Based Ion-Selective Electrode. International Journal of Analytical Chemistry, Volume 2009 (2009), Article ID 954083, 7, http://dx.doi.org/10.1155/2009/954083.

GC-MS Studies of the Chemical Composition of Two Inedible Mushrooms of the Genus Agaricus

Assya Petrova, Kalina Alipieva, Emanuela Kostadinova,
Daniela Antonova, Maria Lacheva, Melania Gjosheva,
Simeon Popov and Vassya Bankova

ABSTRACT

Background

Mushrooms in the genus Agaricus have worldwide distribution and include the economically important species A. bisporus. Some Agaricus species are inedible, including A. placomyces and A. pseudopratensis, which are similar in appearance to certain edible species, yet are known to possess unpleasant odours and induce gastrointestinal problems if consumed. We have studied the chemical composition of these mushrooms using GC-MS.

Results

Our GC-MS studies on the volatile fractions and butanol extracts resulted in the identification of 44 and 34 compounds for A. placomyces and A. pseudopratensis, respectively, including fatty acids and their esters, amino acids, and sugar alcohols. The most abundant constituent in the volatiles and butanol were phenol and urea respectively. We also identified the presence of ergosterol and two Δ7-sterols. In addition, 5α,8α-Epidioxi-24(ξ)-methylcholesta-6,22-diene-3β-ol was isolated for the first time from both mushrooms. Our study is therefore the first report on the chemical composition of these two species.

Conclusion

The results obtained contribute to the knowledge of the chemical composition of mushrooms belonging to the Agaricus genus, and provide some explanation for the reported mild toxicity of A. placomyces and A. pseudopratensis, a phenonomenon that can be explained by a high phenol content, similar to that found in other Xanthodermatei species.

Background

Mushrooms in the genus Agaricus have a worldwide distribution, with up to 90 species recorded in Europe. The genus includes the most economically important and commercially cultivated mushroom in the world, A. bisporus (button mushroom) as well as many other edible species [1]. Some Agaricus species are inedible, including A. placomyces and A. pseudopratensis Bohus, which are similar in appearance to certain edible species, yet are known to possess unpleasant odours and result in gastrointestinal problems if consumed [2,3]. To the best of our knowledge, there is no information available in the literature concerning the chemical composition of these two species.

In this article, we report the results of the GC-MS analyses of volatile and polar compounds, in addition to the sterol fraction obtained from the fruiting bodies of A. placomyces and A. pseudopratensis. The results can help characterise the species investigated, indicate the presence of some biologically active compounds and shed light upon their reported mild toxicity.

Results and Discussion

The volatile fractions were obtained from the fresh mushrooms' fruiting bodies and analysed as described in the experimental section. The results obtained are outlined in Table 1. The most abundant constituent of the volatiles in both species was phenol: over 40% in A. placomyces and over 80% in A. pseudopratensis. Besides phenol, the main groups of compounds in the volatile fractions were fatty acids and their esters. Because the extraction was performed using ethanol, it is possible that the ethyl esters are artefacts. Our findings offer the first analytical proof of the presence of phenol in these two species.

Table 1. Volatile constituents of A. placomyces and A. pseudopratensis (% of the total ion current, GC-MS).

Compound	A. placomyses (%)	A. pseudopratensis (%)
Alcohols and phenols		
Phenol	41.5	81
Benzyl alcohol	-	tr
Aldehydes		
Hexanal	0.9	-
2-decenal (90)	0.2	-
2,4-decadienal	<0.1	-
Undecenal (95)	0.5	-
Ketones		
2-undecanone	0.2	-
Acids		
Acetic acid	1.7	-
Hydroxy acetic acid	0.3	-
Pentadecanoic acid (93)	0.3	-
Hexadecanoic acid	4.5	5.5
Hexadecanoic acid – isomer	1.1	
9,12-Octadecadienoic acid (linoleic acid)	2.2	9.0
9-Octadecanoic acid (oleic acid)	3.7	-
Ethers		
Hydroquinone monopropyl ether (95)	3.3	-
Estres		
Acetic acid phenyl ester	0.5	-
Tetradecanoic acid ethyl ester	0.2	-
Hexadecanoic acid ethyl ester	5.1	8.4
Octadecanoic acid ethyl ester 18:0 (98)	0.5	1.2
Octadecenoic acid ethyl ester 18:1	1.1	-
9,12-Octadecadienoic acid ethyl ester (89)	10.1	-
Terpenoids		
Squalene	0.4	-

*The ion current generated depends on the characteristics of the compound concerned and is not a true quantification.
** In parenthesis: extent of matching (as a percentage) of MS data with the literature values, where the matching is not 100%

The phenol levels detected are not hugely surprising, given the numerous reports of the mushrooms' phenol-like odour, in addition to their belonging to the Agaricus section, Xanthodermatei, which is characterised by this typical unpleasant odour. It has been suggested that the production of phenol originates from an evolutionary ancestral biochemical shift, also demonstrated by the Agaricus species of the section Xanthodermatei in defence mechanisms [4]. According to Del Signore et al. [5] phenolic substances in mushrooms may be involved in a chemical defence mechanism against insects and micro-organisms, very similar to those described for certain plants (e.g. Salicaceae). Recently, it has been suggested that species with higher evolutionary positions in the section Xanthodermatei demonstrate higher phenol contents [4]. For this reason it could therefore be surmised that A. pseudopratensis is a more recently differentiated species than

A. placomyces. However, this is only a preliminary hypothesis that will require further more detailed study.

The polar fractions obtained from the fungal species investigated were analysed by GC-MS after silylation. The derivatisation was necessary in order to increase the volatility of the polar fraction, thereby enabling its analysis by gas chromatography. The results obtained are outlined in Table 2. The presence of phenolics is clearly evident, as is the higher amount of phenol observed in A. pseudopratensis. This finding supports the suggestion of the defensive role of phenolics in section Xanthodermatei [6]. Another characteristic was the presence of the aromatic amino acids tryptophan and tyrosin, but only in A. pseudopratensis.

Table 2. Constituents of the polar fraction of A. placomyces and A. pseudopratensis (% of the total ion current, GC-MS).

Compound**	Agaricus placomyses (%)	Agariricus pseudopratensis (%)
Alcohols and phenols		
Ethylene diol	0.1	0.1
1,3-butane diol (96)	1.1	1.6
Phenol	0.6	1.0
Catechol	0.3	1.0
Resorcinol (98)	0.2	-
Hydroquinone	-	1.2
Aminoacids		
Alanine	1.5	1.3
Glycine (98)	0.2	0.2
Valine (96)	2.8	3.0
Leucine	1.5	2.7
Isoleucine(95)	0.5	1.8
Proline	0.3	0.9
Proline-5-oxo (90)	3.0	4.2
Threonine (95)	0.8	0.6
Phenylalanine	2.5	2.3
Tyrosine	-	1.4
Triptophan	-	1.4
N – containing compounds		
Urea	29.5	29.0
9-H-purine-6-OH (hypoxanthine)	0.4	-
6-amino-9-β-D-ribofuranosyl-9H-purine (91)	<0.1	-
Uracyl	-	0.1
Xanthine	-	0.1
Acids		
Butanedioic acid	0.5	1.1
Hydroxy butanedioic acid	1.6	2.6
2-butenedioic acid	0.7	0.6
Glutamic acid	0.4	-
Panthotenic acid	-	0.3
9,12-Octadecadienoic acid	-	0.1
Hexadecanoic acid	-	0.2
H_3PO_4	1.6	0.9
Sugars and sugar alcohols		
Glucitol (98)	35.0	10.0
Manitol (98)	-	9.8
Disaccharide	-	1.2

*The ion current generated depends on the characteristics of the compound concerned and is not a true quantification.
** In parenthesis: extent of matching (as a percentage) of MS data with the literature values, where the matching is not 100%

Sugar alcohols are amongst the major soluble carbohydrates found in fungi [7]. In the butanol fractions of both species we detected a sugar content of around 30%: glucitol in A. placomyces and a mixture of glucitol and mannitol in A. pseudopratensis. The most abundant constituent of the polar extracts was urea.

This finding once again confirms that higher fungi of the family Agaricaceae accumulate substantial amounts of urea in their fruiting bodies [8]. It has been suggested that urea acts as an osmotically favourable nitrogen reserve for fungi [9].

We also investigated the sterol composition of the two Agaricus species. As expected the predominant sterol in A. placomyces, was ergosterol, although two others were detected, each possessing a Δ7 double bond. In A. pseudopratensis only the latter two sterols were present in substantial quantities, with ergosterol detected in trace amounts.

Using column and preparative thin-layer chromatography, a crystalline compound was isolated from the extracts of both species and identified as 5α,8α-epidioxi-24(ξ)-methylcholesta-6,22-diene-3β-ol through comparison of its spectral data (MS, 1H- and 13C-NMR) with those reported in the literature [10]. The compound was first isolated from these species and might be regarded as an artefact produced from the oxidation of the corresponding Δ5,7 sterol. It is noteworthy that this type of compound has recently been shown to express inhibitory behaviour against the HTLV-1 virus, in addition to demonstrating cytotoxic activity against human breast cancer cell line (MCF7WT) [10].

Conclusion

The results obtained add to the knowledge on the chemical composition of mushrooms belonging to the genus Agaricus, and help provide further explanation for their reported mild toxicity, which we attribute to their high phenol content similar to that of other Xanthodermatei species [11].

Experimental

Collection of the Samples

The macromycetes (age 3 – 5 days), were collected near Plovdiv, Bulgaria, in October 2005.

Extraction

The fresh fruiting bodies of macromycetes (A. pseudopratensis – 105 g; A. placomyces – 245 g) were cut into small pieces and consecutively extracted with ethanol, ethanol-chloroform (1:1) and chloroform. The extracts were combined, water was added and the chloroform layers were removed and evaporated in vacuo to yield 1.88 g of dry residue for A. placomyces and 0.6 g for A. pseudopratensis.

Isolation and Analysis of Volatile Compounds

A portion of the chloroform extract (about 300 mg) was subjected to a four-hour distillation-extraction on Lickens-Nickerson apparatus [12]. The volatile compounds were extracted from the distillate with diethyl ether (yields of volatiles in Table 1), and analysed by GC/MS on a GC Hewlett Packard 6890 + MS 5973 (Hewlett Packard, Palo Alto, California, USA), with a HP5-MS capillary column (30 m × 0.25 mm, 0.25 μm film thickness, Agilent Technologies, Wilmington, Delaware, USA). The ion source was set at 250°C and the ionisation voltage at 70 eV. The temperature was programmed from 40 – 280°C at a rate of 6°C min-1, and a helium carrier gas was used used.

Analysis of Polar Compounds

A portion of the n-butanol extract (5 mg) was dissolved in 50 μL of dry pyridine, before the addition of 75 μL of bis-(trimethylsilyl)-trifluoroacetamide (BSTFA). The mixture was heated at 80°C for 30 min and analyzed by GC/MS. The silylated extract was investigated by GC/MS using the same instrument described above, with a capillary column HP-5 (23 m × 0.2 mm, 0.5 μL film thickness, Agilent Technologies, Wilmington, Delaware, USA). A helium carrier gas was used. The temperature programmed at 100 – 315°C at a rate of 5°C min-1, with a 10 min hold at 315°C.

Identification of Compounds

Identification was carried out by searching commercial library databases. Some components remained unidentified, however, owing to both a lack of authentic samples and library spectra of the corresponding compounds.

Isolation and Identification of Sterols

The chloroform extracts was subjected to column chromatography on silica gel with an n-hexane – acetone gradient (30:1 – 5:1) to produce several fractions. The third set of fractions, after further purification by preparative TLC (silica gel G, n-hexane – acetone 10:1), yielded sterol mixtures: 23 mg from A. placomyes and 7.9 mg from A. pseudopratensis, which were analysed by GC-MS. A gas chromatograph (Hewlett Packard 5890) linked to a mass spectrometer (Hewlett Packard 5972) with a capillary column SPB-50 (30 m × 0.32 mm, 0.25 μm film thickness) was employed. A helium carrier gas was used with a temperature programme set at 270°C – 290°C at a rate of 4°C min-1 with a 20 min hold. The ion source was set at 250°C with the ionisation voltage at 70 eV.

From the fourth set of preparative TLC fractions (silica gel, n-hexane – methyl ethyl ketone 10:1) obtained from both species 5α,8α-epidioxi-24(ξ)-methylcholesta-6,22-diene-3β-ol was isolated (10.4 mg from A. pseudopratensis and 81 mg from A. placomyces) and characterised through comparison of its EIMS, 1H- and 13C-NMR spectra with literature data [10].

Authors' Contributions

AP performed the extractions, obtained the volatiles and isolated sterol mixtures, as well as participated in the data analysis and interpretation. KA participated in the data analysis and interpretation. EK performed the isolation and structural identification of the epidioxysterols. DA performed the GC-MS analyses. ML and MG participated in the collection and identification of mushroom material. SP conceived the study and helped draft the manuscript. VB participated in the design and coordination of the study, worked on the data analysis and interpretation and helped draft the manuscript. All authors read and approved the final manuscript.

Acknowledgements

The authors gratefully acknowledge the partial support offered by the National Science Fund (Bulgaria), Contract X-1415.

References

1. Calvo-Bado LC, Noble R, Challen M, Dobrovin-Pennington A, Elliott T: Sexuality and Genetic Identity in the Agaricus Section Arvenses. Appl Environl Microbiology 2000, 66:728–734.

2. Kuo M: Agaricus placomyces. www.mushroomexpert.com/agaricus_placomyces.html (accessed on 19.10.2007)

3. Agaricus pseudopratensis var. pseudopratensis www.deltadelpo.it/leggi.asp?articolo=463&posizione=426 (accessed on 19.10.2007).

4. Pallaci Callaci, Guinbertau J, Rapior S: New Hypotheses from Integration of Morphological Traits Biochemical Data and Molecular Phylogeny in Agaricus spp. The 5th ICMBMP April 2005. www.shroomtalk.com/forum/lofiversion/index.php/t7455.html (accessed on 21.10.2007).

5. Del Signore A, Romeo F, Giaccio M: Content of phenolic substance in basidiomycetes. Mycol Res 1997, 101:552–556.

6. Stoop JMH, Mooibroek H: Advances in genetic analysis and biotechnology of the cultivated button mushroom, Agaricus bisporus. Appl Microbiol Biotechnol 1999, 52:474–483.

7. Lewis DH, Smith DC: Sugar Alcohols (Polyols) in Fungi and Green Plants. I. Distribution, Physiology and Metabolism. New Phytologist 1967, 66:143–184.

8. Wagemaker MJM, Welboren W, van der Drift C, Jetten MSM, Van Griensven LJLD, Op den Camp HJM: The ornithine cycle enzyme arginase from Agaricus bisporus and its role in urea accumulation in fruiting bodies. Biochim Biophys Acta (BBA) – Gene Structure and Expression 2005, 1681:107–115.

9. Wagemaker MJM, Eastwood DC, van der Drift C, Jetten MSM, Burton K, Van Griensven LJLD, Op den Camp HJM: Expression of the urease gene of Agaricus bisporus: a tool for studying fruiting body formation and post-harvest development. Appl Microbiol Biotechnol 2006, 71:486–492.

10. Gauvin A, Smadja J, Aknin M, Faure R, Gaydou E-M: Isolation of bioactive 5a,8a-epidioxy sterols from the marine sponge Luffariella cf. variabilis. Can J Chem 2000, 78:986–992.

11. Gill M, Strauch RJ: Constituents of Agaricus xanthodermus Genevier: the first naturally endogenous azo compound and toxic phenolic metabolites. Z Naturforsch [C] 1984, 39:1027–1029.

12. Hendriks H, Geerts HJ, Malingre M: The occurrence of valeranone and crypto-fauronol in the essential oil of Valeriana officinalis L. s. l. collected in the northern part of The Netherlands. Pharmac Wekblad Scient Edition 1981, 116:1316–1320.

CITATION

Originally published under the Creative Commons Attribution License or equivalent. Petrova A, Alipieva K, Kostadinova E, Antonova D, Lacheva M, Gjosheva M, Popov S, Bankova V. GC-MS studies of the chemical composition of two inedible mushrooms of the genus Agaricus. Chem Cent J. 2007; 1: 33. doi:10.1186/1752-153X-1-33.

Structural Analysis of Three Novel Trisaccharides Isolated from the Fermented Beverage of Plant Extracts

Hideki Okada, Eri Fukushi, Akira Yamamori, Naoki Kawazoe, Shuichi Onodera, Jun Kawabata and Norio Shiomi

ABSTRACT

Background

A fermented beverage of plant extracts was prepared from about fifty kinds of vegetables and fruits. Natural fermentation was carried out mainly by lactic acid bacteria (Leuconostoc spp.) and yeast (Zygosaccharomyces spp. and Pichia spp.). We have previously examined the preparation of novel four trisaccharides from the beverage: O-β-D-fructopyranosyl-(2->6)-O-β-D-glucopyranosyl-(1->3)-D-glucopyranose, O-β-D-fructopyranosyl-(2->6)-O-[β-D-glucopyranosyl-(1->3)]-

D-glucopyranose,   O-β-D-glucopyranosyl-(1->1)-O-β-D-fructofuranosyl-
(2<->1)-α-D-glucopyranoside and O-β-D-galactopyranosyl-(1->1)-O-β-
D-fructofuranosyl-(2<->1)- α-D-glucopyranoside.

Results

Three further novel oligosaccharides have been found from this beverage and
isolated from the beverage using carbon-Celite column chromatography and
preparative high performance liquid chromatography. Structural confirma-
tion of the saccharides was provided by methylation analysis, MALDI-TOF-
MS and NMR measurements.

Conclusion

The following novel trisaccharides were identified: O-β-D-fructofuranosyl-
(2->1)-O-[β-D-glucopyranosyl-(1->3)]-β-D-glucopyranoside (named "3G-
β-D-glucopyranosyl β, β-isosucrose"), O-β-D-glucopyranosyl-(1->2)-O-
[β-D-glucopyranosyl-(1->4)]-D-glucopyranose   (4¹-β-D-glucopyranosyl
sophorose) and O-β-D-fructofuranosyl-(2->6)-O-β-D-glucopyranosyl-(1-
>3)-D-glucopyranose (6²-β-D-fructofuranosyl laminaribiose).

Background

A beverage was produced by fermentation of an extract from 50 kinds of fruits
and vegetables [1,2]. The extract was obtained using sucrose-osmotic pressure in
a cedar barrel for seven days and was fermented by lactic acid bacteria (Leuconostoc spp.) and yeast (Zygosaccharomyces spp. and Pichia spp.) for 180 days. The
fermented beverage showed scavenging activity against 1,1'-diphenyl-2-picrylhy-
drazyl (DPPH) radicals, and significantly reduced the ethanol-induced damage
of gastric mucosa in rats [1]. Analysis by high performance anion exchange chromatography (HPAEC) showed that this beverage contained high levels of saccharides, estimated between 550 and 590 g/L; mainly glucose and fructose, and
a small amount of undetermined oligosaccharides. Recently, it was reported that
different positions of glycosidic linkage of oligosaccharide isomers affected physiological properties as well as physical properties [3-5]. Development of HPLC
analysis with high sensitivity and separation ability enables the detection and isolation of oligosaccharides in the fermented beverage.

We have previously examined the preparation of saccharides of the
fructopyranoside series from the fermented beverage of plant extracts,
such as O-β-D-fructopyranosyl-(2->6)-D-glucopyranose [2], O-β-D-fr-
uctopyranosyl-(2->6)-O-β-D-glucopyranosyl-(1->3)-D-glucopyranose and O-
β-D-fructopyranosyl-(2->6)-O-[β-D-glucopyranosyl-(1->3)]-D-glucopyranose

[6]. The characteristics of O-β-D-fructopyranosyl-(2->6)-D-glucopyranose were non-cariogenicity and low digestibility, and the unfavorable bacteria that produce mutagenic substances did not use the saccharide [7,8]. Recently, we have studied isolation and identification of novel non-reducing trisaccharides, such as O-β-D-glucopyranosyl-(1->1)-O-β-D-fructofuranosyl-(2<->1)-α-D-glucopyranoside and O-β-D-galactopyranosyl-(1->1)-O-β-D-fructofuranosyl-(2<->1)- α-D-glucopyranoside from the beverage [9], and those saccharides were confirmed to be produced by fermentation.

In this paper, we have confirmed structures of the novel trisaccharides (Fig. 1): O-β-D-fructofuranosyl-(2->1)-O-[β-D-glucopyranosyl-(1->3)]-β-D-glucopyranoside (named "3G-β-D-glucopyranosyl β, β-isosucrose"), O-β-D-glucopyranosyl-(1->2)-O-[β-D-glucopyranosyl-(1->4)]-D-glucopyranose (41-β-D-glucopyranosyl sophorose) and O-β-D-fructofuranosyl-(2->6)-O-β-D-glucopyranosyl-(1->3)-D-glucopyranose (62-β-D-fructofuranosyl laminaribiose), isolated from the fermented beverage using methylation analysis, MALDI-TOF-MS and NMR measurements.

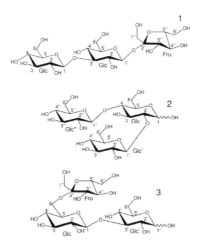

Figure 1. Structures of O-β-D-fructofuranosyl-(2->1)-O-[β-D-glucopyranosyl-(1->3)]-β-D-glucopyranoside (1), O-β-D-glucopyranosyl-(1->2)- O-[β-D-glucopyranosyl-(1->4)]-D-glucopyranose (2) and O-β-D-fructofuranosyl-(2->6)- O-β-D-glucopyranosyl-(1->3)-D-glucopyranose (3).

Results and Discussion

Saccharides 1, 2 and 3 were isolated from the fermented beverage of plant extracts using carbon-Celite column chromatography, and were shown to be homogeneous using anion exchange HPLC [tR, sucrose (relative retention time; retention time of sucrose = 1.0): 1.89, 2.23 and 2.40 respectively]. The retention time of saccharides 1, 2 and 3 did not correspond to that of any authentic saccharides [glucose (0.62), fructose (0.68), sucrose (1.00), maltose (1.43), trehalose (0.58), laminaribiose (1.33), raffinose (1.23), 1-kestose (1.47), 6-kestose (1.75), neokestose (1.90), maltotriose (2.59), panose (1.87), nystose (2.06), fructosylnystose (3.81), O-β-D-fructopyranosyl-(2->6)-D-glucopyranose (0.83) [2], O-β-D-fructopyranosyl-(2->6)-O-β-D-glucopyranosyl-(1->3)-D-glucopyranose (1.74) [6], O-β-D-fructopyranosyl-(2->6)-O-[β-D-glucopyranosyl-(1->3)]-D-glucopyranose (1.72) [6], O-β-D-glucopyranosyl-(1->1)-O-β-D-fructofuranosyl-(2<->1)- α-D-glucopyranoside (1.24) [9], O-β-D-galactopyranosyl-(1->1)-O-β-D-fructofuranosyl-(2<->1)-α-D-glucopyranoside (0.84) [9], 2(2-α-D-glucopyranosyl)isokestose (1.57) [10], 2(2-α-D-glucopyranosyl)2isokestose (1.79) [10], 2(2-α-D-glucopyranosyl)3isokestose (2.09) [10], 2(2-α-D-glucopyranosyl)nystose (2.17) [10], 2(2-α-D-glucopyranosyl) 2nystose (2.63) [10], O-α-D-glucopyranosyl-(1->2)-O-α-D-xylopyranosyl-(1->2)-β-D-fructofuranoside (1.51) [11], O-α-D-glucopyranosyl-(1->2)-O-α-D-glucopyranosyl-(1->2)-O-α-D-xylopyranosyl-(1->2)-β-D-fructofuranoside (1.80) [11].

The degree of polymerization of saccharides 1, 2 and 3 was established as 3 by measurements of [M+Na] ions (m/z: 527) using TOF-MS (see Fig. 2), and analysis of the molar ratios of D-glucose to D-fructose in the acid hydrolysates. Acid hydrolysates of saccharides 1 and 3 were liberated to glucose and fructose, and saccharide 2 was liberated to glucose. From the GC analysis, relative retention times of the methanolysate of the permethylated saccharides were investigated [tR (relative retention time; retention time of methyl 2, 3, 4, 6-tetra-O-methyl-β-D-glucoside = 1.0; retention time, 9.60 min)]. The methanolysate of permethylated saccharide 1 exhibited six peaks corresponding to methyl 2,3,4,6-tetra-O-methyl-D-glucoside (tR, 0.94 and 1.48), methyl 2,4,6-tri-O-methyl-D-glucoside (tR, 3.27 and 4.81) and methyl 1,3,4,6- tetra-O-methyl-D-fructoside (tR, 1.06 and 1.32). The methanolysate of permethylated saccharide 3 also exhibited six peaks corresponding to methyl 2,3,4-tri-O-methyl-D-glucoside (tR, 2.58 and 3.59), methyl 2,4,6-tri-O-methyl-D-glucoside (tR, 3.22 and 4.73), and methyl 1,3,4,6-tetra-O-methyl-D-fructoside (tR, 1.07 and 1.29). On the other hand, the methanolysate of permethylated saccharide 2 exhibited two peaks corresponding to methyl 2,3,4,6-tetra-O-methyl-D-glucoside (tR, 0.97 and 1.47). GC-MS analysis on the retention times and fragmentation patterns of the methyl glucosides [12] showed the two peaks (10.08 min and 10.21 min) from the methanolysate of permethylated saccharide 2 to be methyl 3,6-di-O-methyl-D-glucoside. From these findings above, saccharides 1, 2 and 3 were proved to be, O-D-fructofuranosyl-(2->1)-O-[D-glucopyranosyl-(1->3)]-D-glucopyranoside, O-D-glucopyranosyl-(1->2)-O-[D-glucopyranosyl-(1->4)]-D-glucose and O-D-fructofuranosyl-(2->6)-O-D-glucopyranosyl-(1->3)-D-glucose, respectively.

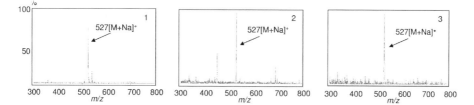

Figure 2. MALDI-TOF-MS spectra of saccharides 1, 2, and 3. 1: saccharide 1, 2: saccharide 2, 3: saccharide 3.

The structural confirmations of saccharides 1, 2 and 3 according to 1H and 13C NMR analyses and the subsequent complete assignment of 1H and 13C NMR signals of the three saccharides were carried out using 2D-NMR techniques.

First, the NMR spectra of saccharide 1 were analyzed. The HSQC-TOCSY spectrum revealed the 1H and 13C signals of each Glc, Glc' and Fru. The isolated methylene was assigned as H-1 and C-1 in Fru. The other three methylene carbons were assigned as C-6 in these residues. The COSY spectrum assigned the spin systems of these residues; from H-1 to H-3 and H-1' to H-3' (Fig. 3(a)), and from H-3" to H-6". The corresponding 13C signals were assigned by HSQC spectrum (Fig. 3(b)). These results clarified the assignment of 1H and 13C NMR signals of each residue. The position of the glucosidic linkage and fructosidic linkage was analyzed as follows. The C-3' showed the HMBC [13,14] correlations between H-1 (Fig. 3(c)). The J (H-1/H-2) value was 7.9 Hz. These results indicated the Glc 1β ->3' Glc linkage, namely the laminaribiose moiety. The C-2" showed the HMBC correlations to H-1'. The J (H-1'/H-2') value was 7.4 Hz. These results indicated the Glc' 1β ->2 β Fru linkage.

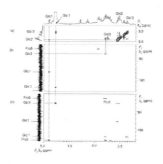

Figure 3. Part of COSY (a), HSQC (b) and HMBC (c) spectra of saccharide 1.

The coupling patterns of overlapped 1H were analyzed by the SPT method [15,16]. Due to strong coupling between H-4' and H-5', these couplings could not be analyzed in first order.

The NMR spectra of saccharide 2 showed that it was an anomeric mixture at the Glc. The α anomer was predominant. The COSY spectrum was assigned from H-1 to H-6. The C-4 showed the HMBC correlations between H-1" (Fig. 4(a) and 4(b)). The J (H-1"/H-2") value was 7.6–7.8 Hz. These results indicated the Glc" 1β ->4 Glc linkage, namely the cellobiose moiety. The C-2 showed the HMBC correlations to H-1'. The J (H-1'/H-2') value was 7.6 Hz. These results indicated the Glc' 1β ->2 Glc linkage.

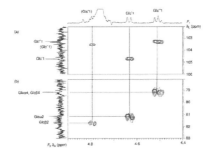

Figure 4. Part of HSQC (a) and HMBC (b) spectra of saccharide 2. ( ) = minor anomer.

The NMR spectra of saccharide 3 were analyzed in the same manner as those of saccharide 2. Saccharide 3 was also an anomeric mixture at the Glc'. The β anomer was predominant. The HSQC-TOCSY spectrum revealed the 1H and 13C signals of each Glc, Glc' and Fru. The isolated methylene was assigned as H-1" and C-1". The other three methylene carbons were assigned as C-6 in these residues (Fig. 5(a)). The position of the glucosidic linkage and fructosidic linkage was analyzed as follows. The C-3' showed the HMBC correlations between H-1 (Fig. 5(b)). The J (H-1/H-2) value was 7.9 Hz. These results indicated the Glc 1β ->3 Glc' linkage, namely the laminaribiose moiety. The C-2 showed the HMBC correlations to H-6. These results indicated the Glc 6 <-2 β Fru linkage.

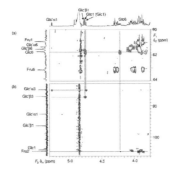

Figure 5. Part of HSQC-TOCSY (a) and HMBC (b) spectra of saccharide 3. ( ) = minor anomer.

From all of these findings, saccharides 1, 2, and 3 from the fermented beverage of plant extracts were confirmed to be new oligosaccharides (Fig. 1):O-β-D-fructofuranosyl-(2->1)-O-[β-D-glucopyranosyl-(1->3)]-β-D-glucopyranoside (named "3G-β-D-glucopyranosyl β, β-isosucrose"), O-β-D-glucopyranosyl-(1->2)-O-[β-D-glucopyranosyl-(1->4)]-D-glucopyranose (41-β-D-glucopyranosyl sophorose) and O-β-D-fructofuranosyl-(2->6)-O-β-D-glucopyranosyl-(1->3)-D-glucopyranose (62-β-D-fructofuranosyl laminaribiose).

Synthesis of the saccharides by fermentation of plant extracts was investigated using HPAEC. Almost all of the monosaccharides were removed from the fermented and unfermented beverages of plant extracts by the batch method with Charcoal. The saccharides 1, 2, and 3 were observed in the fermented beverage, but were not present in the unfermented one. Therefore, saccharides 1, 2, and 3 were confirmed to have been produced during fermentation of the beverage of plant extracts (Fig. 6).

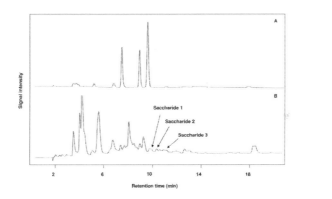

Figure 6. High performance liquid chromatogram of fermentation products. A: The beverage of plant extract was fermented for 0 days. B: The beverage of plant extract was fermented for 180 days. The beverage (100 mL) fermented for 0 or 180 days was mixed with charcoal (10 g), stirred for 3 h. and filtered. The charcoal was extracted with 30% ethanol (500 mL) three times. The ethanol extracts were combined, concentrated to dryness and solubilized with one mL of distilled water. The sugar solution was analyzed by HPAEC.

Conclusion

We have previously found that the fermented beverage contained the novel saccharide, O-β-D-fructopyranosyl-(2->6)-D-glucopyranose, which is produced by fermentation. The saccharide showed low digestibility. The saccharide was selectively used by beneficial bacteria, Bifidobacterium adolescentis and B. longum, but was not used by unfavorable bacteria, Clostridium perfringens, Escherichia coli and Enterococcus faecalis that produce mutagenic substances [8]. It is interesting to study the biological functions of other oligosaccharides existing in the beverage. In this report, three novel oligosaccharides have been found from this beverage, and isolated from the beverage using carbon-Celite column chromatography and preparative high performance liquid chromatography. Structural confirmation of the saccharides was provided by methylation analysis, MALDI-TOF-MS and NMR measurements. These saccharides were identified as new trisaccharides:O-β-D-fructofuranosyl-(2->1)-O-[β-D-glucopyranosyl-(1->3)]-β-D-glucopyranoside (named "3G-β-D-glucopyranosyl β, β-isosucrose"), O-β-D-glucopyranosyl-(1->2)-O-[β-D-glucopyranosyl-(1->4)]-D-glucopyranose (41-β-D-glucopyranosyl sophorose) and O-β-D-fructofuranosyl-(2->6)-O-β-D-glucopyranosyl-(1->3)-D-glucopyranose (62-β-D-fructofuranosyl laminaribiose). These saccharides were confirmed to be produced during fermentation.

Experimental

Preparation of Fermented Beverage of Plant Extract

For preparation of the initial juice, 50 kinds of fruits and vegetables were used to produce the final extract as shown in a previous paper [1,2]. The 50 fruits and vegetables were cut, sliced or diced into small pieces, mixed and put in cedar barrels. Afterwards, an equivalent weight of sucrose was added to the samples, mixed well to allow high contact between the samples and sucrose, and then the barrels were left for one week at room temperature. The juice exudate was then separated without compression from solids and used for fermentation. The fermented beverage was obtained by incubation of the juice at 37°C in the dark by natural fermentation using yeast (Zygosaccharomyces spp. and Pichia spp.) and lactic acid bacteria (Leuconostoc spp.). After 7 days, the fermented beverage was kept in a closed enameled tank at 37°C for 180 days for additional maturation and ageing, finally obtaining a brown and slightly sticky liquid.

High Performance Anion-Exchange Chromatography (HPAEC)

The oligosaccharides were analyzed using a Dionex Bio LC Series apparatus equipped with an HPLC carbohydrate column (Carbo Pack PA1, inert styrene divinyl benzene polymer) and pulsed amperometric detection (PAD) [17,18]. The mobile phase consisted of eluent A (150 mM NaOH) and eluent B (500 mM sodium acetate in 150 mM NaOH) with a sodium acetate gradient as follows: 0–1 min, 25 mM; 1–2 min, 25–50 mM; 2–20 min, 50–200 mM; 20–22 min, 500 mM; 22–30 min, 25 mM; using a flow rate through the column of 1.0 mL/min. The applied PAD potentials for E1 (500 ms), E2 (100 ms), and E3 (50 ms) were 0.1, 0.6, and -0.6 V respectively, and the output range was 1 μC.

Isolation of Saccharides

The fermented beverage of plant extracts (1000 g) was loaded onto a carbon-Celite [1:1; charcoal (Wako Pure Chemical Industries, Ltd; Osaka, Japan) and Celite-535 (Nacalai Tesque Inc, Osaka, Japan)] column (4.5 × 35 cm), and was successively eluted with water (14 L), 5% ethanol (30 L) and 30% ethanol (10 L). Almost all of the glucose and fructose were eluted with water (4 L), and then saccharides 1, 2 and 3 were eluted with 30% ethanol (1–2 L). The 30% ethanol fraction containing saccharides 1, 2 and 3 was concentrated in vacuo and freeze-dried to give 894 mg of sample. Subsequently, the 30% ethanol fraction was successfully repeatedly purified using an HPLC system (Tosoh, Tokyo, Japan) equipped with an Amide-80 column (7.8 mm × 30 cm, Tosoh, Tokyo, Japan) at 80°C, and eluted with 80% acetonitrile at 2.0 mL/min, and using refractive index detection. Furthermore, the saccharides were purified by HPLC with the ODS-100 V column (4.6 mm × 25 cm, Tosoh, Tokyo, Japan) at room temperature, and eluted with water at 0.5 mL/min. Purified saccharides 1 (2.5 mg), 2 (2.2 mg) and 3 (2.0 mg) were obtained as white powders.

Methylation and Methanolysis

Methylation of the oligosaccharides was carried out by the method of Hakomori [19]. The permethylated saccharides were methanolyzed by heating with 1.5% methanolic hydrochloric acid at 96°C for 10 or 180 min. The reaction mixture was treated with Amberlite IRA-410 (OH-) to remove hydrochloric acid, and evaporated in vacuo to dryness. The resulting methanolysate was dissolved in a small volume of methanol and analyzed using gas liquid chromatography.

Gas Liquid Chromatography (GLC)

For the analysis of the methanolysate, GLC was carried out using a Shimadzu GC-8A gas chromatograph equipped with a glass column (2.6 mm × 2 m) packed with 15% butane 1,4-diol succinate polyester on acid-washed Celite at 175°C. Flow rate of the nitrogen gas carrier was 40 mL/min.

GC-MS Analysis

GC-MS analysis was performed using a JMS-AX500 mass spectrometer (JEOL, Japan) using a DB-17HT capillary column (30 m × 0.25 mm I.D., J & W Scientific, USA). Injection temperature was 200°C. The column temperature was kept at 50°C for 2 min after sample injection, increased to 150°C at 50°C/min, kept at

150°C for 1 min, and then increased to 250°C at 4°C/min. The mass spectra were recorded in the electron ionization (EI) mode.

MALDI-TOF-MS

MALDI-TOF-MS spectra were measured using a Shimadzu-Kratos mass spectrometer (KOMPACT Probe) in positive ion mode with 2.5%-dihydroxybenzoic acid as a matrix. Ions were formed by a pulsed UV laser beam (nitrogen laser, 337 nm). Calibration was done using 1-kestose as an external standard.

NMR Measurement

The saccharide (ca. 2 mg) was dissolved in 0.06 mL (saccharide 1) and 0.4 mL (saccharide 2 and 3) D2O. NMR spectra were recorded at 27°C with a Bruker AMX 500 spectrometer (^{1}H 500 MHz, ^{13}C 125 MHz) equipped with a 2.5 mm C/H dual probe (saccharide 1), a 5 mm diameter C/H dual probe (1D spectra of saccharide 2 and 3), and a 5 mm diameter TXI probe (2D spectra of saccharide 2 and 3). Chemical shifts of ^{1}H (δ_H) and 13C (δ_C) in ppm were determined relative to an external standard of sodium [2, 2, 3, 3-^{2}H$_4$]-3-(trimethylsilyl)-propanoate in D$_2$O (δ_H 0.00 ppm) and 1, 4-dioxane (δ_C 67.40 ppm) in D$_2$O, respectively. 1H-^{1}H COSY [20,21], HSQC [22], HSQC-TOCSY [22,23] CH$_2$-selected E-HSQC-TOCSY [24], HMBC [13,14] and CT-HMBC [13,14] spectra were obtained using gradient selected pulse sequences. The TOCSY mixing time (0.15 s) was composed of DIPSI-2 composite pulses.

Competing Interests

The authors declare that they have no competing interests.

Authors' Contributions

HO, AY and NK performed data analysis, and contributed to drafting the manuscript. EF and JK collected the NMR data. NS and SO conceived of the study, participated in its design and contributed to drafting the manuscript. All authors read and approved the final manuscript.

References

1. Okada H, Kudoh K, Fukushi E, Onodera S, Kawabata J, Shiomi N: Antioxidative activity and protective effect of fermented plant extract on ethanol-induced damage to rat gastric mucosa. J Jap Soc Nutr Food Sci 2005, 58:209–215.

2. Okada H, Fukushi E, Yamamori A, Kawazoe N, Onodera S, Kawabata J, Shiomi N: Structural analysis of a novel saccharide isolated from fermented beverage of plant extract. Carbohydr Res 2006, 341:925–929.

3. Kohmoto T, Fukui F, Takaku H, Machida Y, Arai M, Mitsuoka T: Effect of isomalto-oligosaccharides on human fecal flora. Bifidobacteria Microflora 1988, 7:61–69.

4. Murosaki S, Muroyama K, Yamamoto Y, Kusaka H, Liu T, Yoshikai Y: Immuno-potentiating activity of nigerooligosaccharides for the T helper 1-like immune response in mice. Biosci Biotechnol Biochem 1999, 63:373–378.

5. Murosaki S, Muroyama K, Yamamoto Y, Liu T, Yoshikai Y: Nigerooligosaccharides augments natural killer activity of hepatic mononuclear cells in mice. Int Immunopharmacol 2002, 2:151–159.

6. Kawazoe N, Okada H, Fukushi E, Yamamori A, Onodera S, Kawabata J, Shiomi N: Two novel oligosaccharides isolated from a beverage produced by fermentation of a plant extract. Carbohydr Res 2008, 343:549–554.

7. Okada H, Kawazoe N, Yamamori A, Onodera S, Shiomi N: Structural analysis and synthesis of oligosaccharides isolated from fermented beverage of plant extract. J Appl Glycosci 2008, 55:143–148.

8. Okada H, Kawazoe N, Yamamori A, Onodera S, Kikuchi M, Shiomi N: Characteristics of O-β-D-fructopyranosyl-(2->6)-D-glucopyranose isolated from fermented beverage of plant extract. J Appl Glycosci 2008, 55:179–182.

9. Kawazoe N, Okada H, Fukushi E, Yamamori A, Onodera S, Kawabata J, Shiomi N: Structural analysis of two trisaccharides isolated from fermented beverage of plant extract. Open Glycosci 2008, 1:25–30.

10. Okada H, Fukushi E, Onodera S, Nishimato T, Kawabata J, Kikuchi M, Shiomi N: Synthesis and structural analysis of five novel oligosaccharides prepared by glucosyltransfer from β-D-glucose 1-phosphate to isokestose and nystose using Thermoanaerobacter brockii kojibiose phosphorylase. Carbohydr Res 2003, 338:879–885.

11. Takahashi N, Fukushi E, Onodera S, Benkeblia N, Nishimato T, Kawabata J, Shiomi N: Three novel oligosaccharides synthesized using Thermoanaerobacter brockii kojibiose phosphorylase. Chem Cent J 2007, 1:18.

12. Funakoshi I: Mass spectrum. In Seikagaku data book. Edited by: Yamashina I. Tokyo: Tokyokagakudojin; 1979:606–668.

13. Bax A, Summers MF: 1H and 13C assignments from sensitivity-enhanced detection of heteronuclear multiple-bond connectivity by 2D multiple quantum NMR. J Am Chem Soc 1986, 108:2093–2094.

14. Hurd RE, John BK: Gradient-enhanced proton-detected heteronuclear multiple-quantum coherence spectroscopy. J Magn Reson 1991, 91:648–653.

15. Pachler KGR, Wessels PL: Selective Population Inversion (SPI). A pulsed double resonance method in FT NMR spectroscopy equivalent to INDOR. J Magn Reson 1973, 12:337–339.

16. Uzawa J, Yoshida S: A new selective population transfer experiment using a double pulsed field gradient spin-echo. Magn Reson Chem 2004, 42:1046–1048.

17. Rocklin RD, Pohl CA: Determination of carbohydrate by anion exchange chromatography with pulse amperometric detection. J Liq Chromatogr 1983, 6:1577–1590.

18. Johnson DC: Carbohydrate detection gains potential. Nature 1986, 321:451–452.

19. Hakomori S: A rapid permethylation of glycolipid and polysaccharide catalyzed by methylsulfinyl carbanion in dimethylsulfoxide. J Biochem 1964, 55:205–208.

20. Aue WP, Batholdi E, Ernst RR: Two-dimensional spectroscopy. Application to nuclear magnetic resonance. J Chem Phys 1976, 64:2229–2246.

21. von Kienlin M, Moonen CTW, Toorn A, van Zijl PCM: Rapid recording of solvent-suppressed 2D COSY spectra with inherent quadrupole detection using pulsed field gradients. J Magn Reson 1991, 93:423–429.

22. Willker W, Leibfritz D, Kerssebaum R, Bermel W: Gradient selection in inverse heteronuclear correlation spectroscopy. Magn Reson Chem 1993, 31:287–292.

23. Domke T: A new method to distinguish between direct and remote signals in proton-relayed X, H correlations. J Magn Reson 1991, 95:174–177.

24. Yamamori A, Fukushi E, Onodera S, Kawabata J, Shiomi N: NMR analysis of mono- and difructosyllactosucrose synthesized by 1F-fructosyltransferase purified from roots of asparagus (Asparagus officinalis L.). Magn Reson Chem 2002, 40:541–544.

CITATION

Originally published under the Creative Commons Attribution License or equivalent. Okada H, Kawabata J, Shiomi N. Structural analysis of three novel trisaccharides isolated from the fermented beverage of plant extracts. Chem Cent J. 2009; 3: 8. doi:10.1186/1752-153X-3-8.

Copyrights

19. © 2007 Petrova et al; This is an Open Access article distributed under the terms of the Creative Commons Attribution License (http://creativecommons.org/licenses/by/2.0), which permits unrestricted use, distribution, and reproduction in any medium, provided the original work is properly cited.

20. © 2009 Okada et al

Index

X

Z